Genetic Alchemy

Genetic Alchemy

The Social History of the Recombinant DNA Controversy

Sheldon Krimsky

The MIT Press
Cambridge, Massachusetts
London, England

Second Printing, 1983

This book was set in Meridien by Graphic Composition, Inc., and printed and bound by Halliday Lithograph in the United States of America.

Library of Congress Cataloging in Publication Data

Krimsky, Sheldon
Genetic alchemy.

Bibliography: p.
Includes index.
1. Recombinant DNA—Social aspects. I. Title
QH442.K74 174′.25 82–207
ISBN 0–262–11083–0 AACR2

For my mother, Rose Krimsky

Contents

III
The Science and Policy Debates

List of Figures and Tables

Figures

Tables

Acknowledgments

This book grew out of a two-and-a-half-year study funded by the Program in Ethics, Values in Science and Technology (EVIST) of the National Science Foundation. I am grateful to EVIST Directors William Blanpied and Arthur Norberg for their support of the study titled "Value Issues in the Controversy over Recombinant DNA Research."

My research was helped considerably by the groundbreaking work of Charles Weiner of MIT who set up an archival collection on the recombinant DNA controversy at his host institution. The primary source material in that collection proved to be an invaluable resource for this contemporary social history of science.

MIT's Institute Archivist, Helen Slotkin, and her staff provided excellent support services to me and my assistants during the long hours we spent poring through manuscripts. I thank them heartily for encouraging the use of their collections.

David Ozonoff of Boston University's School of Public Health was the principal author of chapters 1–3 and 15–17. I am indebted to him for his impeccable scholarship and exceptional insights and for serving as my principal collaborator on the EVIST/NSF project.

I want to thank Edward T. McCarthy for the laborious work he did as a research assistant in the preparation of earlier manuscripts.

Critical input for selected parts of this book was provided by Daniel F. Liberman, Kostia Bergman, and William Petri, who also served on the technical advisory committee of the EVIST/NSF project.

I want to acknowledge the contributions of Philip Bereano and his two assistants Constantine Stergides and Anita Mires of the Program in the Social Management of Technology, University of Washington. This team collected information about the principal actors and institutions involved in the development of recombinant DNA technology and its controls.

I also want to acknowledge the help of Frances Tanaka, who typed drafts of this manuscript, Donna Gold, who typed, edited and proofread the four-volume EVIST/NSF study.

Finally, I am forever grateful for the support and understanding of my wife Carolyn, who helped me through the turbulent periods of this task.

Genetic Alchemy

Introduction

Early in the 1970s scientists discovered a technique for cutting and splicing the basic substance of all life forms—DNA (the abbreviation for deoxyribonucleic acid). They also learned how to transport functional segments of DNA between cells from different species of living things. The hybrid molecules that resulted from this process are called recombinant DNA molecules. A controversy over the use of this technique dates as far back as 1971 and has continued in some form or other to the present time. In the decade that has elapsed, there has been a voluntary moratorium on some classes of experiments, an assessment of the technique's risks and benefits by various scientific bodies and study panels, state and local enactments, over a dozen bills introduced into Congress to regulate all use of gene splicing, and guidelines issued by the National Institutes of Health.

Each year we have been accustomed to hearing about major controversies that have far-reaching consequences for the public's health or well-being. To cite a few examples, let us recall the siting of the Seabrook nuclear power plant, the drilling for offshore oil in Georges Bank, the landing of the SST, banning the sale of saccharin, the setting of generic carcinogen standards, and the association of fluorocarbons with the depletion of the ozone layer. All of these issues involved an assessment of risks and benefits under conditions of uncertain knowledge. While scientific testimony played an important role in each decision-making process, these debates were ultimately resolved in the public arena, in the courts, through congressional action, or by a decision of a public-agency head.

In contrast to these public controversies, the recombinant DNA issue exhibited some unique features. First, at least initially, the subject of concern in the latter issue was a scientific research tool. And the scientists who stood to gain the most from the research were the ones to alert the public about its potential adverse consequences. Second, the potential hazards discussed and debated for a decade have not been confirmed. Third, the public debate incorporated issues involving both environmental health and ethics. For some, there was only one issue—Will the research be hazardous to the public? Others believed that the primary hazards were social and political in that the research technique bridges a significant gap in our knowledge of human genetic engineering.

The goal of this study is to examine the many areas of controversy over the scientific use, development, and application of gene-transplantation research. The study will be looking at how the risks and benefits were conceptualized through time. It will trace the locus of the controversy and examine the social and political events that influenced its course.

The analysis offers the reader a perspective on why the experts disagreed. It highlights the special interests and reconstructs the key arguments in the debate, bringing out the scientific and value assumptions in which they were grounded. It is my hope that this study will provide a look into the world of science under stress. It should be evident to the reader who follows the analysis that the technical discussions of the debate cannot be fully understood or appreciated apart from the social and political context. It should also be clear that risk assessment transcends the sphere of pure reason. Decision makers, whether they be scientists or government bureaucrats, have interests that cannot easily be sorted out from their reasoning to yield a neutral scientific rationality. Those with self-interests may find themselves in a compromising role.

There have been several notable attempts throughout history on the part of our intellectual forebears to solve the self-interest problem in public decision making. The Greek philosopher Plato advanced the idea of a philosopher-king.[1] According to Plato, such men must exhibit the highest intellectual abilities, while divesting themselves of worldly possessions. Thus, a small group of enlightened patriarchs would care for society as a herdsman cares for his flock.

Karl Mannheim, the German sociologist, promoted the idea of a detached intelligentsia who could avoid the kind of thinking he identified as obstructionist, namely, ideological or utopian thinking.[2] According to his view, a civil service made up of such intelligentsia would be called upon to make policy in the public interest.

Modern political theorists have given up on the idea of the disinterested decision maker who follows the dictates of reason in selecting a course of action in the face of uncertainty. The term "pluralistic decision making" has gained some currency. In essence it calls upon diverse interest groups to have a role in the decision-making process.[3] Rationality makes way for fairness and the distribution of power.

In these times of strategic environmental decisions, however there is renewed interests in the extent to which social choices can be rational and objective. It must be recognized that a considerable schism exists between the normative models of decision analysis and the means by which social choices are actually made. Theories are invaluable as standards or guides for improving our problem-solving capacities. Nevertheless, there is much to be learned from looking at an actual process. The recombinant DNA story can be viewed as a sequence of decisions. Some were made by individuals responding to a feeling of personal responsibility or fear that science was getting out of control. From that point other decisions followed: by small informal groups of scientists; by larger, organized collectives of scientists at professional meetings; by scientific panels; and by officials of government at the federal, state, and local levels. The aggregate of these decisions culminated in a public policy addressed to the regulation of research involving the manipulation of genes.

The fundamental methodological assumption of this investigation is that scientific decisions, whether involving assessments of theory or of risk, cannot be fully understood without taking account of the factors exogenous to scientific inquiry. These include the social institutions of science, the education of scientists, the political economy of the funding mechanism, and the interactions of scientists and the public. There are occasions when I shall show how these factors affected the way that scientists perceived the risks or responded to governmental regulations. Similarly, scientific arguments also affect public policy, and there too I shall examine the types and sources of information that played an influential role during the time that policy was in the making.

When science is under stress it becomes more self-conscious, causing its defenders to sharpen their intellectual sabers and refine their concepts. It is a time when foundational questions receive more attention than during periods of calm, unanimity, and optimism. Two major periods of disorder or crisis can be distinguished in the development of science. The first has its roots in antiquity, while the second is characteristic of modern science.

Disorders of the first kind are associated with the emergence of anomalies in a theory. We owe a debt of gratitude to Thomas Kuhn for articulating the problems that result when these anomalies give rise to a new theory which then competes with the old theory for support. In his masterful work *The Structure of Scientific Revolutions*, we find case examples from the history of science that illustrate such critical events, which are referred to as paradigm replacements.[4]

The crisis is over which paradigm (a general term incorporating, among other concepts, theory and world view) prevails. Frequently, younger scientists promoting the change in scientific paradigm square off with the venerable masters. As the new paradigm takes hold, the intellectual forces of science coalesce, internal criticism diminishes, and normal science once again proceeds.

Another type of crisis I shall call disorders of the second kind. This refers to those controversies that develop within the scientific community over risk—in particular, the risk associated with a new research technology or research methodology. It is my view that both types of disorders or crises force science to higher stages of self-criticism, by which I mean its practitioners must justify themselves in ways that were never before required.

The recombinant DNA debate is not the first example in modern times when scientists were divided over the potential hazards of the work they were doing. One of the most frequently cited precedents is the development of the atomic bomb. When science was divided over the hazards for civilization of the research in nuclear physics, there was a reasonably clear distinction drawn between the methods of inquiry and the methods of destruction. One did not need an atomic bomb to learn about the structure of the nucleus. Moreover, the hazards associated with the use of radioactive materials in scientific inquiry were overshadowed by the magnitude of the destruction that atomic energy was capable of unleashing. And by the time Congress was contemplating whether to establish controls on scientific research in 1945, the awesome power of the energy released from the fission reaction had been verified as a tool of mass destruction.

In contrast to the atomic-energy episode, there has been no evidence that recombinant DNA research can be used to produce a catastrophic event.[5] The recombinant DNA technology has two potential risks associated with it. The first involves the use of the technique in research and the potential hazards linked to the process of inquiry. The second concerns how the technique might be developed for its use in human genetic manipulation and industrial genetic engineering. A situation in atomic-energy research analogous to the first potential risk would be bombarding heavy metals with elementary particles, which sometimes results in the creation of hazardous substances. In addition, to supply the needs of the research community with its radioactive sources, for example, highly purified uranium, an entire industry developed with its attendant dangers of mining, processing, and transporting radioactive materials.

While the overriding national debate in the atomic-energy episode was over the nation's willingness and the ability to control the new power of the atom, there were also discussions on the kinds of controls necessary for the research activities of scientists who would be using radioactive materials to investigate atomic processes. The recombinant DNA controversy has focused mainly on the investigative techniques and far less on the use to which the techniques may be put.

In both the atomic-energy and molecular-genetics debates we see what philosophers call "the egocentric predicament": The knower cannot understand the world without interacting with it, and thereby affecting it. Another way of saying this is that the knowledge-acquisition process is partially constitutive. One cannot know reality as a passive agent.

As scientists investigate the world, they change it. If the system they work in is not closed, these changes will be released into the larger environment. It is certainly not new to science that the tools of investigation alter a portion of the reality being investigated. But when the objects of inquiry become unleashed and introduce hazards to the investigator or the environment outside the laboratory, social risks have been generated that are inextricably connected to the investigative methods. I have termed disorders of the second kind the crises that develop when science becomes self-conscious of its own perilous or potentially perilous investigative processes.

Certain cataclysmic or rare events provide unique opportunities for the scientific investigator. For example, for an astronomer, the eclipse of the sun is an opportune time to study the properties of the stellar universe. Similarly, for the philosopher, sociologist, or historian of science, certain unique periods in science can aid in one's investigation. During times of crisis, when debates take place over competing theories or hypotheses, philosophical concepts such as "explanatory simplicity" or "coherence" or "inductive support" are more likely to be brought to the foreground of discussion. As a general rule, stress stimulates self-conscious activity.

This study will be examining how the scientific community reacted to stress that originated from within and was subsequently imposed from without. It began over the risks of a new investigative technique that involved the creation of hybrid organisms. As the issue developed, factors external to science began to exert additional stress on one sector of the scientific community and its funding agencies.

On first examination, the disorders I have termed of the first kind and of the second kind appear distinctive. When a conflict arises within a scientific discipline over two competing theories or hypotheses, it is generally agreed that those experts in the discipline are best suited—indeed, the only ones suited—to participate in the debate. The problems, by and large, are internal to science. Therefore, where there is disagreement over which of two paradigms is more tenable or more fruitful, observers wait with anticipation for a resolution. Should there be a paradigm shift, there will be questions whether the shift was warranted in light of the evidence presented, and debates will persist, primarily among scholars in the nonscientific community, whether the new paradigm is more successful.

While it may be true that the active participants in scientific disorders of the first kind are almost exclusively members of the discipline, it is not the case that the decisions made by individual scientists on the acceptance or rejection of a new paradigm are exclusively of a scientific nature. Other considerations entering into the decision might include sociological factors, such as the extent to which a senior scientist built a reputation on the old paradigm or, alternatively, a junior scientist sees an opportunity to establish a reputation based on the new paradigm; aesthetic factors, such as one's view of simplicity or coherence; psychological factors, such as one's predilection for change.

The analysis of acceptable risk applied to scientific investigations is a relatively new one. The issues themselves are not new. During the period when Galileo was piercing the medieval veil of ignorance about the nature of celestial bodies and proposing heliocentric theories of the universe, the risks to the social and religious orders were clearly perceived. Similarly, Darwin's evolutionary universe was threatening to the creationist theories of his time.

I previously referred to the differences between disorders of the first and second kind. Where they are similar, however, is in the fact that the outcome of a particular controversy in both cases involves factors external to science or principles that are exogenous to the boundaries of a single discipline. The recombinant DNA debate is an exemplification of these ideas. In this instance a crisis developed within the scientific community. It was only after the persistence of scientific controversy that the public became involved. Once the issues were opened up to public scrutiny, there were permanent changes that developed in the relations between this area of science and society.

This study examines the recombinant DNA controversy with two kinds of questions of paramount concern. The first question asks, What were the arguments and counterarguments, and how did they change through time? The second question asks, How did the events external to the scientific developments affect the form and outcome of the debate?

Questions of the first type lend themselves to an internal examination of the arguments. A valuable tool for this type of analysis is logical reconstruction. This method involves selecting certain seminal arguments, identifying the conclusions and the supporting statements, and then supplying the missing or hidden assumptions, making them an explicit part of the discussion. In this manner, the arguments are reconstructed so as to have strict logical validity. One can then study the epistemological criteria applied or the supporting evidence.

The positions that scientists take on controversial issues, whether they be associated with competing theories or a risk assessment of some program of research, are influenced by factors external to the science. These factors include one's position in the scientific community and proximity to the research in question, societal changes such as social unrest, and one's associates and institution of affiliation. Sometimes these factors are acknowledged by actors in a debate; more frequently, however, their influence on the scientific arguments are difficult to assess.

The external context has played an important role in the recombinant DNA debate. Therefore, to look at the arguments in isolation would be one dimensional. On the other hand, we recognize the difficulties in attempting to assign a causal nexus between external events and the attitudes and behavior of scientists on technical issues. This difficulty is met in part through an examination of the reflections of the scientists themselves. Faculty and staff of the Oral History Program at the Massachusetts Institute of Technology (MIT) interviewed over one hundred key actors in the recombinant DNA debate. In many instances scientists were asked to identify external events or factors that influenced their views of the risks or their ideas on social responsibility. These oral histories are housed at MIT's Institute Archives and Special Collections.

In terms of the rationality of science, it may be important to know whether evidence and arguments make a difference. Scientists may have decided quite intuitively that the risks are great (or small), whereupon they proceeded to justify their positions. The same issue can be raised about the controversies over paradigm shifts: Are the positions set before the arguments are made, before all the data are in, or do the arguments help determine the positions? If positions are taken a priori, independent of the empirical factors that emerge in a debate, it suggests rather ominously that risk analysis is fundamentally a rationalization. In Kuhn's analysis of paradigm replacement, he makes the point that some scientists become blind to new data or cannot see the anomalies. They become fixed to a paradigm for reasons that are sociological or psychological in nature.

Ultimately, it may not be possible to determine how an individual arrived at a position on the risks of recombinant DNA research and what factors were responsible for shifts in that position. While political forces may be at work, such as the possibility of legislating basic research, one can always maintain that a changeover in position resulted from an awarensss of new evidence. Nevertheless, the recombinant DNA controversy can teach us something about the changes that occur in science during a crisis period. What explanations are advanced? What form does criticism take? How deep does the criticism within the field penetrate into the methodological norms of the science? To what extent do disagreements touch upon the dominant paradigm of biological processes? One may be aware that external factors shape the nature and distribution of viewpoints in a scientific debate while also finding value in a structural analysis of the key arguments that enter into the controversy.

One of the objectives of this investigation is to examine certain archetypal arguments as they emerge in different phases of the controversy. According to a process that I have referred to as logical reconstruction, the arguments are fully realized through a construction and elaboration of their premises. By such an approach we can illuminate tacit assumptions and make the standards of validation for various claims explicit. Here is a simple example of how the process works: Suppose we reach a critical discussion on the exchange of genetic material between prokaryotes (lower-order organisms) and eukaryotes (higher-order organisms). It is argued that there is a low probability of exchange of genetic information between the higher- and lower-order organisms. The discussion is pertinent to the issue of risk assessment. A reconstructed argument might take the following form:

1. Similar species have a greater degree of similarity (homology) in their DNA sequences than nonsimilar species.
2. The greater degree of similarity that exists in the DNA sequences of two species, the greater the likelihood of natural genetic exchange.

Recombinations occur more frequently between organisms of similar species.

The next step in the construction is to justify a low frequency of genetic exchange occurring between eukaryotes and prokaryotes:

1. Organisms classified as eukaryotes are vastly different in their species characteristics than organisms classified as prokaryotes.
2. The probability of gene exchange between two organisms of vastly different species type is exceedingly low.

The probability of genetic exchange between eukaryotes and prokaryotes is exceedingly low.

For illustrative purposes this example was taken out of context. When the various periods in the controversy are analyzed, the appropriate logical reconstructions will be executed in the context of the debate. Sources of arguments will also be identified. Supporting evidence for critical arguments will be cited when pertinent, and the link between the evidence and the claims will be examined.

It should be emphasized to the reader who has had no experience with this analytical technique that in normal discourse and in most written works people do not present arguments in a rigorous form. Conclusions are advanced with supporting evidence. But in strictly logical terms where the conclusion follows *necessarily* from the premises, most arguments that we come across are not valid. What we usually find are argument sketches in which some of the key elements are present. The method of logical reconstruction uses the datum of the argument sketch and adds the necessary components to provide a valid argument form. The dissection and refinement of arguments in this way gives the analyst an opportunity to draw out the implicit assumptions and the sufficient conditions for a particular claim. The final outcome should provide greater clarity and precision to the position advanced.

I

The Genesis of Concern

1

The Social and Political Climate

Scientific and social controversies, like living organisms, have not only birth dates but also gestation periods. The birth date of the debate over what is now known as recombinant DNA technology would seem to be the announcement at the 1973 Gordon Conference on Nucleic Acids[1] that segments of hereditary material from disparate organisms could be hooked together in the test tube at will and then reinserted into a host organism. The news provoked an immediate and dramatic response from the assembled scientists. Through the medium of a widely read professional journal, an urgent call was issued for wider discussion of this new-found ability to create unnaturally hybridized or deliberately engineered genetic systems.[2] But just as a birth is not only the beginning of a new process but also the fruition of a previous one, the apparently spontaneous concern did not arise de novo. Indeed, the very vigor of the response at the Gordon Conference and the rapidity with which it set in motion a chain of events that were to have local, national, and international repercussions suggests that the ground had already been prepared.

For several years prior to the advent of the specific technology later designated as recombinant DNA (abbreviated rDNA) techniques, there had been concerns about related research problems within the scientific community.[3] These concerns, which were precursors to many or most of the elements in the later debate, reflected more general problems related to science's ethical, social, and political ramifications. Thus, while the previously inchoate state of these anxieties came to a focus in the gene-splicing technology reported at the Gordon Conference, they certainly existed prior to that meeting. In many respects these precursor concerns provided the opportunity for a dress rehearsal for the debate that was to follow. As such, they offer important clues as to why that debate took the form that it did.

The 1960s and early 1970s have almost taken their place as a discrete historical epoch in this country, much like the Great Depression

I gratefully acknowledge the contribution of David Ozonoff to this chapter and also chapters 2, 3, 15, 16, and 17.

or the Roaring Twenties. Whether one believes the era is behind us or not, it most certainly marked a watershed in the political consciousness of many who were later to be embroiled in the rDNA controversy. That consciousness was formed under the twin impacts of the abrupt tearing of the American social fabric as civil disorder and violent protest hit city after city and the consequences of American military adventure abroad. The impacts of the war in Southeast Asia reached deeper and deeper into American college campuses. Science, technology, and the talents of researchers inevitably became involved in the turmoil. The Massachusetts Institute of Technology in Cambridge was in many ways typical. The institute seemed to be torn apart as students and faculty violently opposed to the war confronted their classmates and colleagues who, if not exactly in favor of the war, at least saw nothing wrong with working on research projects whose sole aim was to further US capabilities to wage it. At the same time that some faculty and students organized to resist the policy of conscription that provided the human resources for the war effort, others were helping prepare the technological means that seemed to offer nothing but an endless prolongation of the agony.

The events that took place on the campus during these tumultuous years were the major influence on those who were later to be actors in the rDNA debate. Even when participation widened to include members of the lay public, the drama was played out in academic communities like Cambridge where town-gown tensions were an important element. Yet because the campus conflicts were part of a debate that incorporated issues of domestic turmoil in addition to foreign policy, there was a tendency for Viet Nam War-related issues to take on a much larger and deeper significance.

During this period, technology took on a schizophrenic nature to those working closely with it. On one hand, it offered the means for reducing US casualties. But on the other side, for a war of this nature, it meant prolonging the engagement and increasing the effectiveness of weapons intended for the destruction of the environment and human life. Some believed technology could make war more humane. At MIT, for example, there was an attempt to design and build a zone of electronic surveillance across the DMZ between North and South Viet Nam. This idea was first suggested in 1966 by, among others, MIT's Jerome Wiesner, who in 1971 was installed as president of the institute. It raised forcefully the question of the proper use of technical skills and the responsibility of scientists during struggles between states. The fact that the suggestion had first been made to provide a "humane" alternative[4] to aerial bombing only served to emphasize the extent to which good instincts and technical abilities had become twisted by the presuppositions of what was proper, permissible, and effective for scientists to do. Joseph Weizenbaum, a computer scientist at MIT, expressed the reaction of many of his colleagues:

> The intention of most of these men was not to invent or recommend a new technology that would make warfare more terrible and, by the way, less costly to highly industrialized nations at the expense of "underdeveloped" ones. Their intention was to stop the bombing. In this they were wholly on the side of the peace groups and of well-meaning citizens generally . . . . However, these enormously visible and influential people could have instead simply announced that they believed the bombing, indeed the whole American Viet Nam adventure, to be wrong, and that they would no longer "help." I know that at least some of the participants believed that the war was wrong; perhaps all of them did. But, as some of them explained to me later, they felt that if they made such an announcement, they would not be listened to, then or ever again. Yet, who can tell what effect it would have had if forty of America's leading scientists had, in the summer of 1966, joined the peace groups in coming out flatly against the war on moral grounds? Apart from the positive effect such a move might have had on world events, what negative effect did their compromise have on themselves and on their colleagues and students for whom they served as examples?[5]

Moral imperatives are surer guides than the technological fix. It is a position that will be taken up again in the rDNA debate.

In addition to the electronic battlefield, of course, there were a host of other war-related projects that inspired the opposition of many on the campus, from the technology required for "smart" (laser-guided) bombs to counterinsurgency studies. Nor were physical scientists and engineers the only ones whose work was being drawn into the maw of the war. Biologists had to confront the wide use of chemical and biological warfare in Viet Nam, for example, and doctors, the use of Green Beret medical teams for political purposes ("winning the hearts and minds of the people").[6]

The border between antiwar and domestic concerns was a labile one, however. As explicitly radical groups like the Black Panther party, the Young Lords, and the Socialist Workers party became more visible, so too did FBI and police "red-squad" harassments. The police killing of Black Panther leader Fred Hampton in Chicago was announced to the MIT community as antiwar students and faculty gathered to protest university involvement in the war effort. Rumors, since confirmed, of police *agents-provocateurs* within the ranks of the antiwar movement abounded. In such an atmosphere, the technology developed for counterinsurgency abroad had immediate domestic implications as well. Nor was the impact of antiwar politics on the home front confined to the radical movement. The more moderate segment of the university was already questioning the government's priorities, taking up an old theme that held that domestic strife was a result of too much attention to guns and not enough to butter. In this heady atmosphere there functioned an array of scientist-oriented political-interest groups whose modes of operation were also to be recreated by others in the rDNA debate.

The oldest, largest, and strongest of these groups was the politically moderate Federation of American Scientists (FAS).[7] Stemming from the Federation of Atomic Scientists and the postwar struggle to establish international rather than national controls over atomic energy, the FAS had stuck to the arms-control question fairly narrowly for most of its history. It was especially active in the years from 1967 to 1970 in the fight over deployment of an antiballistic-missile (ABM) system. The political assumption that guided FAS legislative lobbying, its main activity, was that scientists had a special responsibility to warn government and the general public of the perils of a technology about which they were better informed than others. For the FAS, ignorance was an obstacle for the well functioning of democracy in the United States. One of the early participants in the scientists' lobby expressed the FAS analysis of the nuclear weapons danger in this way: "We were troubled . . . by the fact that except for the scientists and the Army there was no foreknowledge of the tremendous implications of the atom bomb."[8] To call attention to these implications was the duty of the responsible scientist. But deeper political intentions were disavowed. As the executive director of the FAS, Jeremy Stone, put it in a 1970 policy statement:

> [Many] of the ills which we do attack might be better resolved without attempting changes in structure or law. There is nothing wrong with the political and economic structure of this country that could not be fixed if its institutions stopped operating in such a mindless way. Corporations can be run for the public interest rather than in the narrow economic interest; they *would* be if the individuals running them felt that this was expected of them. Why not begin to expect it

of them? Such expectations are far more effective than law and they may be easier to generate.[9]

Some participants in the rDNA debate have drawn explicit parallels between the gene-splicing technology and the early years of the application of atomic energy. The secrecy shrouding the development of a nuclear arsenal was a great source of guilt and embarrassment to some members of the scientific establishment. Henceforth, that period became an effective reference point for looking at policy issues involving the use of rDNA technology.

Several other groups were similar to the FAS in their conception of the "special responsibility" of scientists to inform and, if need be, instruct the public on matters of technoscientific importance. One was the Society for Social Responsibility in Science (SSRS). Conceived of in 1949 as an organization that would try to push for "constructive alternatives to militarism," SSRS also stressed the need to create an informed public opinion.[10] In 1969 it called for complete US withdrawal from Viet Nam and sent a team to investigate the effects of defoliants on that country. SSRS became concerned with environmental issues through its principled opposition to the war. The Scientists Institute for Public Information (SIPI) arrived at the same position via the nuclear issue.[11] A descendant of local antinuclear fallout committees of the 1950s, SIPI had, by the late sixties and early seventies, turned its attention to broader environmental issues. Riding the crest of "ecology awareness," local SIPI scientists drew attention to the frequency with which unanticipated consequences followed irresponsible tampering with the biosphere. Those cited as irresponsible often turned out to be well-meaning scientists who held a more limited perspective on the impact of their work.

More issue oriented, and a model for the later Coalition for a Responsible Genetic Research, was the Citizens' League against the Sonic Boom (CLASB).[12] A single-issue group devoted to combating plans to build an American SST, CLASB banded together in 1970–1971 with a number of others, including the Sierra Club, the Friends of the Earth, Common Cause, and the Environmental Defense Fund, to form the Coalition against the SST. In its successful fight to defeat congressional funding for the SST, CLASB functioned with its coalition partners primarily as a lobbying force.

All of these moderate-scientist political-interest groups had certain common features. David Nichols's analysis of their membership indicates that they were primarily composed of professionally established scientists from recognized elite institutions like Harvard and MIT.[13] They concentrated on public education, emphasizing action in the legislative arena using traditional lobbying methods. Their critiques of government policy generally excluded critiques of the economic and political system itself. This is not true of the radical groups that were also active in this period. Two of these groups deserve special mention: The Medical Committee for Human Rights (MCHR) and Scientists and Engineers for Social and Political Action (SESPA).

Both MCHR and SESPA were products of the sixties, but as radical scientific groups they were not without precedent.[14] In 1938 the American Association of Scientific Workers (AASW) was formed to allow scientists, through collective action, to assert greater control over the social use of their own labor. Although there is little or no continuity between the AASW and either MCHR or SESPA, its left-wing reputation and the resultant relations with the FAS in the latter's early years is important for what it reveals about the character of the moderate political movement among scientists. It is probable that many members of the AASW were also members of the Communist Party of the USA, which often worked with such groups as part of its Popular Front strategy of the thirties. On the other hand, Nichols's analysis of the politics of scientists in the sixties and seventies recognizes that:

> The important point about the A.A.S.W. is not that of its exact relationship to the Communist Party. The point is, rather, that the group represented an incipient radical politics which the postwar scientists' movement rejected. . . . F.A.S. was anticommunist from its organizational inception. Its first secretary was dismissed because she was a member of the Communist Party. The chairman of the F.A.S. chapter at Fort Monmouth, New Jersey, resigned because he questioned the loyalties and sympathies of the members. A radical group in the New York City F.A.S. chapter was barely defeated for leadership of that important body. All suggestions that the F.A.S. ally with . . . the A.A.S.W. . . . were rejected . . . .
>
> As for the moderate political movement, it can now be seen that it was not born in a political vacuum; rather a specific rejection of a radical policy outlook took place at its inception. Such radicalism as existed among American scientists was not organized through a national political interest group again until the mid-1960's.[15]

One of these national organizations was MCHR. Formed in 1964 by doctors and nurses to provide medical presence during the southern civil-rights marches and demonstrations, MCHR had by the late sixties expanded to include a wide range of health-care professionals: doctors, nurses, dietitians, medical social workers, lab technicians, and so forth. After losing its early male-dominated and doctor-dominated character, MCHR had by 1972 twenty-four local chapters, mostly based in medical-school constituencies around the country. Thus, even these campuses, often isolated from the liberal-arts turmoil by virtue of professional and geographical isolation, received their share of political activism. Three features characterized MCHR activity: (1) local chapter autonomy; (2) an emphasis on the use of medical skills in the service of progressive causes;[16] and (3) a radical critique of American society and its profit-based health-care system.[17]

MCHR brought to the medical faculty the same questions that were put to their scientist and engineering colleagues by radical activists on the liberal-arts campus: Who will profit from your work? Who might be harmed by it? Whose interest does it serve? What alternatives are there? The pharmaceutical industry and its complicity with the medical establishment was a special target. MCHR was successful in raising many troubling questions about the relation of the medical schools and profession to the drug companies and sensitized many to the implicit conflict of interest inherent in it.

Another radical, university-based group was the New University Conference (NUC). Attracting faculty from all disciplines, this nationwide campus organization developed position papers on issues ranging from campus politics, to domestic policies, to the Viet Nam war. As with the other more issue-oriented radical organizations, political economy formed the keystone of their analysis. Campus issues of tenure, ROTC, or grading were examined in the broader context of the university's meaning and function in contemporary society. The NUC operated around faculty affinity groups that could mobilize support when critical issues of university policy surfaced.

Faculty who were bent toward more activist roles found the NUC too cerebral. Struggling with their sense of helplessness in the face of the Viet Nam war, they put their efforts toward well-focused strategies that were designed either to draw attention to the university's complicity with the war effort or to undertake some small act of support for the new socialist societies. At MIT, the Science Action Coordinating Committee had organized with the exclusive task of orchestrating an all-university strike by scientists, researchers, and other supporting personnel. The one-day work stoppage, held on 4 March 1969, which drew wide support from institute personnel, was a means by which scientists who opposed US foreign policy in East Asia could issue a vote of nonconfidence in government policy.

An illustration of the less publicized actions of scientists can be seen in the organization of the Red Crate Collective. Created by a small group of activist scientists and centered in the Boston area, Red Crate was established to undertake small humanitarian actions for developing socialist countries. In one such action, they sent used textbooks packed in red crates to Cuba and Viet Nam.

The other major radical organization of the period will figure more directly in the rDNA story. SESPA, now more generally known by the slogan they adopted in 1970 as Science for the People, was formed by physicists in 1969.[18] A decentralized, loosely organized network of local chapters without officers or membership criteria, SESPA/Science for the People rapidly progressed from questioning the scientist's moral and social responsibilities to a full-blown critique of American capitalism. Confronting the scientific establishment at professional meetings like the American Association for the Advancement of Science (AAAS) became a favorite tactic. Just before National Academy of Sciences President Philip Handler was to address the 1970 AAAS meeting in Chicago, SESPA/Science for the People made the following (unauthorized) statement to the audience: "Philip Handler is going to talk to you . . . about . . . how the scientific community can help prop up the ruling class' corporate profit . . . . We're here in the interest of those people who are not interested in rationalizing their rule, but in destroying it . . . . What is needed now is not liberal reform . . . but a radical attack . . . . Scientific workers must develop ways to put their skills at the service of the people . . . ."[19]

At the same meeting, the Dr. Strangelove Award was presented to (but not accepted by) Dr. Edward Teller. Besides these well-publicized media events, Science for the People members also confronted selected seminar sessions at the meeting that seemed to them to represent typical examples of how science was used to prop up and rationalize the US ruling class and the form of repression it required to stay in power. At the session on "Crime, violence and social control," for example, "SESPA/Science for the People members succeeded in changing the structure of the panel, making comments critical of the fundamental assumptions of some panelists and eliciting extensive discussion from the audience."[20]

The tactic of sharp confrontation based on a radical critique and bolstered by impressive technical depth has characterized much of

Science for the People's style to this day. Several of its supporters who were later involved in the rDNA debate, like Jonathan Beckwith and Jonathan King, had earlier experience with such issues as the relation of XYY genotype to criminal behavior (Science for the People denied that criminality was a genetic anomaly), or the sociobiology debate (Science for the People believed sociobiology to be just another attempt to use science to ratify sexism, racism, and economic exploitation as biological inevitabilities). Similar issues were raised in their activities with respect to the "IQ controversy." Not incidentally, their focus on the tendencies in science to justify the status quo and the legitimate social and economic meritocracies made genetics and genetic engineering one of the natural issues to attract Science for the People members; and the later rDNA debate proved to be no exception to this general pattern.

A rough comparison of the memberships of the moderate and radical groups shows the radical members to be substantially less well established professionally.[21] Even more to the point was the general radical stance of distrust toward the methods and good will of the established profession, which the radicals saw as part of the problem rather than the means to its solution. Perhaps this, more than anything else, separates the basic assumptions about risks and benefits that were to become apparent when scientists with attitudes more akin to the moderate style confronted others whose style was more radical. Throughout the years of turmoil, both groups had become accustomed to thinking of science as having a social impact, and both recognized that a scientist had a "responsibility to society." But they interpreted these propositions very differently.

For the moderates, the most feared impacts of science on society were the unanticipated or unintended ones, and the responsibility of the scientist consisted in calling the attention of the public to these dangers and designing technical safeguards to prevent them. Indeed, in the rDNA debate, the concerns of this group centered on the problem of laboratory accidents or the ecological effect of some hybrid organism that might escape from the laboratory. The proposed solutions took for granted both the ability and the good faith of the scientific community to neutralize these possibilities.

The radicals, on the other hand, looked at the impact of science on society principally in terms of how science served to strengthen existing economic and social relations and saw the responsibility of the scientist to be active opposition to those conditions. The aspect of gene-splicing technology on which they focused initially was the potential it held for genetic engineering, and this was not something that could be dealt with by a simple technological fix.

In addition to these political forces, there were those whose preoccupation was oriented more to ethical than social questions. They represented a heterogeneous collection of individuals who saw the problem of science and its power as posing a fundamental dilemma. Paying less attention to the question of how and whether something *could* be done, they addressed the question of *ought* it be done. By the time of the 1973 Gordon Conference, there was a considerable body of literature on bioethics in general and on genetic engineering in particular. Most scientists active in the area had been exposed to at least some speculation about the moral dilemmas inherent in the development of the field. While there was a good deal of private skepticism on the part of practicing scientists about the usefulness of this kind of speculation, ethicists were granted, at least in public, an attitude of respect and tolerance. What the ethics issue did was to raise again in abstract fashion the general question whether the governance of science was too important to be left to scientists.

One of the leading centers organized around the issue of biomedical ethics was started in 1969 by the philosopher Daniel Callahan. Its full title is the Institute of Society, Ethics and the Life Sciences; for brevity it is referred to as the Hastings Center (its location is Hastings-on-the-Hudson, New York). The Center publishes teaching materials, sponsors workshops, promotes interdisciplinary research groups, and issues its Hastings Center report, a bimonthly journal that contains essays on legal and ethical issues in science and medicine. When the first international risk-assessment conference was held on rDNA research in 1975, members of Hastings played a key role in providing input from the legal community. In March of 1977, the National Academy of Sciences devoted an Academy Forum to rDNA research. Once again the resources of Hastings were tapped with a lecture by director Daniel Callahan on the involvement of the public in the rDNA debate. Throughout the period during which discussions were taking place both inside and outside the scientific community about the social impact of the new advances in molecular biology, the Hastings Center again had a role in getting scientists, philosophers, and legal experts communicating with one another. There were also important informal connections between scientists involved in the debate and members of the Hastings board.

A second example of a setting within which ethicists study the value issues emanating from the biomedical sciences is the Center for Bioethics of the Kennedy Institute, located in Georgetown University. Hastings and the Kennedy Institute share similar goals. They both support research fellows who write on areas where values and science interface. Unlike the analyses of the radical groups previously mentioned, these writings generally assume that the basic institutions of government are sound and that social decisions can best be made incrementally and with informed public participation. At a time when the director of the National Institutes of Health (NIH) was seeking to expand the membership on the federal advisory board on rDNA research to include representatives of the bioethicists, LeRoy Walters, director of the Center for Bioethics, was selected. It was not until January 1979 that participation on the Recombinant DNA Advisory Board was broadened by Secretary of Health, Education, and Welfare Joseph Califano to include members from more radical organizations and environmental groups.

This was the general social and political situation within science and the university in the years just before the Gordon Conference announcement in the summer of 1973 brought the DNA issues to the public's attention. Two research controversies conceptually related to the rDNA issue arose in this early period and provide a useful index to the extent to which this context was already influencing events in this area. In each of these cases, the tendency was to push the issue out of the scientific realm; but in each case the matter was settled by the scientists themselves. But just barely. Much of the post-1973 political development is clearly visible in these precursor cases, emphasizing once again that much of what took place later was a natural outcome of prior conditions and not by any means a unique and idiosyncratic affair.

2

A Troublesome Experiment

In 1970 the Stanford biochemist Paul Berg was embarking on a major line of research involving animal tumor viruses. Forty-four years old and already at the top of his field, Berg had made his name in the area of protein synthesis, using bacterial systems for his experimental material. In 1965 he began to wonder whether the mechanisms of protein synthesis and gene expression differed in higher organisms. At the time a great deal was being learned about lower organisms by studying the genetics of the bacterial viruses called phages. These organisms, after infecting suitable bacteria, often picked up pieces of their victim's genetic apparatus and carried them off to another host bacterium, where the effects of these pieces could be isolated and studied.

It had been suggested to Berg that perhaps the animal tumor viruses could be used in mammalian cells as experimental probes in the same way as the phages in the bacterial systems. In 1967 Berg took a sabbatical leave from Stanford and went to Renato Dulbecco's Salk Institute Laboratory to learn the techniques of animal tissue culture, the growing of animal cells in the test tube using artificial media for nourishment. Back again at Stanford, Berg began to phase out his bacterial work, gradually adding people interested in animal tumor viruses. By 1970, three of these new additions, David Jackson, John Morrow, and Janet Mertz, were hard at work on the animal virus problem.[1]

The relation of viruses to human cancer was (and is) unclear, but it had been known for a long time that they could cause tumors in animals. Many viruses could also change the properties of cells in tissue culture so that they began to behave like "transformed" or malignant cells. Such cells, when injected into living animals, could then produce tumors. The Berg group had chosen to work with a monkey virus called SV40 (so designated because it was the fortieth simian, or monkey, virus to be discovered), and it was known that this virus was a particularly efficient transformer of human cells in tissue culture.

Despite these "theoretical" dangers, researchers had not been accustomed to taking these viruses, or tissue cultures possibly contaminated with them, very seriously. Janet Mertz recalled that when she was an undergraduate at MIT in the late 1960s and taking a course with future Nobelist David Baltimore in animal-cell virology, they "did everything out on the bench, along with mouth pipetting and everything else."[2] But by 1969 or 1970, there began to be a new interest and sensitivity to possible dangers of working with animal cells and viruses. Worry was "in the air." Baltimore's recollection of that period is similar to others: "I guess you couldn't be a newcomer to the field of tumor viruses, especially at MIT, and not feel the hot breath of worry . . . . I mean, when I first started working in tumor viruses I had people down on my back almost immediately about whether it was safe to work with them, and wouldn't I give cancer to everybody who walked down the hall? So the atmosphere here sensitized me to that kind of question."[3]

The Rockefeller University microbiologist Norton Zinder concurred:

[A] large number of people were suddenly working with so-called oncogenic viruses and other viruses of that ilk and there was a potential biohazard. And a large number of these people were trained in biochemistry or molecular biology and had very little microbiological training, and therefore were not familiar with the procedures you'd generally use to keep the investigator himself from getting infected, no less the general population. And so in study sections [peer review committees that looked at grant proposals for the NIH] in places like that, these particular grants were flagged, and there was a biohazards committee at the NIH set up. We were concerned that people be made aware of what they were handling.[4]

The gradual change over to safer procedures in the 1969–1971 period was not something that happened in isolation from the political climate, however. Baltimore is especially explicit about this:

I guess I had kind of undergone a certain transformation over the period from probably '68 to '70, which a lot of people did, from being involved or trying to be involved in larger political issues—as a speaker I'd . . . been involved in the Left Wing in San Diego [while at the Salk Institute], and here [MIT], to a certain extent, I'd been involved in the March 4th organization [a one-day work stoppage to protest university involvement in the war], that kind of thing—to the feeling that if I was going to do anything, it ought to be within the field I know best, because I'd been . . . ineffective outside of it. Like everybody else was, or almost everybody else. And so I was sensitized to issues that involved the biological community, and felt that if I was going to put in political time, it should be there rather than anywhere else. And this specific issue of biohazards became a concern to me because I was a newcomer to a field which presented the greatest potential hazard at that time, which was tumor viruses. And I saw people being incredibly sloppy about things, but also I found a kind of know-nothingism among a lot of people working with phages and bacteria and biochemistry who didn't have anything to do with animal virology who were just on me all the time.[5]

As Baltimore makes clear, the whole issue of biohazards was connected with a range of nonscientific issues by a set of nonlogical but very real links. Zinder reports what happened at a 1972 conference on restriction enzymes in Basel, Switzerland, when the question of biohazards arose: "[We] started off on talking about biohazards, and the next thing I knew, I was discussing the American position in Vietnam, which I had nothing to do with."[6]

Despite the widespread worry, the only controls on potentially hazardous research were informal and depended on the concern of the research community itself. Many labs, including Baltimore's and Berg's, had already installed safer equipment such as laminar-flow hoods and negative-pressure rooms. But other labs were not so conscientious. The biohazard advisory committee at NIH continued to flag certain grant proposals that the original reviewers viewed as potentially hazardous, but nothing formal was ever built into the review mechanism.[7] The problem was on many people's minds, but no one seemed to know what to do about it.

When Janet Mertz joined the lab as a graduate student in 1970, Berg was well on his way to changing over his facilities to work solely on SV40 tumor virus. Those who chose to work with Berg seemed to accept whatever hazard there might be, but Mertz felt others were not so sure. "There were other people in the departments who weren't in Berg's group who may have disliked the idea of SV40. I think that partly because of that there was pressure on Berg to go ahead and have a safer facility."[8] Mertz makes clear, however, that Berg himself was quite safety conscious and had one of the safest facilities around.

Mertz's first order of business was to choose a thesis topic. As an interim experiment to get her started, Berg had set her to work on developing an assay procedure for the virus, but even as she was setting it up she became interested in the work of two other scientists in Berg's lab, David Jackson and Robert Symons. Their work was part of the general idea Berg had had some years earlier—to use tumor viruses as probes to study cell function in animal cells, much as bacterial viruses had been used to study bacteria. The bacterial studies had enabled scientists to discover a great deal about the mechanisms of gene expression in those organisms by using the viruses to transfer genes from one bacterium to another. These viruses picked up bits of chromosome from one infected cell and carried them to the new cell, where the viruses and their accompanying genes reintegrated themselves into the new host cell chromosome. Berg knew that tumor viruses, and SV40 was one example, also integrated themselves into the chromosome of *animal* cells, and he wondered whether the same trick that had yielded so much information about bacteria might not also be made to work for the cells of higher organisms. In effect, he wanted to use these viruses as mediators, or vectors, to transfer genes from one animal cell into another. It did not take him long to convince himself that the frequency with which the SV40 virus picks up genes from a host cell naturally was too low to be experimentally useful. It then occurred to Berg that if SV40 was to be used as a vector, the addition of a new gene as a "hitchhiker" for later integration into an animal cell's chromosome would have to be done artificially, in the test tube.

[W]e began to think in terms of developing a purely unnatural, if you will, system for using the SV40 chromosome as a vector for bringing new genes into cells. It was known that the SV40 chromosome integrates into the cellular chromosome when it transforms the cell, converts it into a tumor cell. If it can do that, and if genes can be inserted into the chromosome of SV40 and use SV40's capability of integrating into the host chromosome, one would transport new genes as well. So we began to think about ways we could chemically, or enzymatically, introduce new DNA segments into the SV40 chromosome, *in vitro* [in a test tube].[9]

What Berg and Jackson planned to do was formally equivalent to what was later to come under the heading of recombinant DNA technique. But their method was considerably more cumbersome than the later refinements—so much so that it is doubtful whether many other labs could even attempt such a bold experiment.

The bacterial gene that Jackson, Symons, and Berg picked to hook up to the SV40 animal tumor virus chromosome was called the galactose operon; it had become attached to part of a bacterial virus called the lambda phage. This particular phage was defective because it had lost some of its functions and lived as a satellite outside the chromosome, replicating independently as a small circular segment of DNA called a plasmid.[10]

The technology to form the hybrid of SV40 and the defective lambda phage DNA with the galactose operon attached took time to develop. When the hybrid DNA molecule was completed, Jackson, Symons, and Berg planned to put it into an animal cell to see whether the bacterial galactose gene would be "expressed" (functional). This was one way of seeing whether genes from lower organisms could function in cells of higher organisms. While work on creating the hybrid DNA molecule was in progress, Berg suggested to Mertz that she might like to work on the reverse of Jackson's project for her thesis: instead of putting the bacterial-animal virus hybrid into an animal cell to see whether bacterial genes would be expressed, she could try putting it into bacteria to see whether the animal-virus genes would be expressed. If SV40 virus chromosomes could actually be reproduced in bacteria, it would also be a useful way to isolate mutants of the virus. Mutants are extremely useful or even necessary for many kinds of studies, but isolating them in cumbersome animal cell systems was tedious and difficult. If they could be selected for in the rapidly growing and easily handled bacterial cultures, the work on the virus could be made considerably easier.[11]

Mertz apparently hesitated at first. Fresh from the atmosphere of MIT of the late sixties, she had become somewhat sensitized to the social and political implications of science. Still vivid in her mind was a course on molecular genetics given by MIT's Ethan Signer in the spring of 1970, which Mertz attended not long before coming to Berg's lab. One session particularly remained with her. That winter researchers Jim Shapiro, Jon Beckwith, and Larry Eron from the nearby Harvard Medical School had isolated a bacterial gene as a segment of a pure double-stranded DNA. They called a news conference—not to discuss the science of the matter, but to announce that the first steps toward genetic engineering had been taken and to warn of government misuse of technology in general.[12] Signer had devoted an entire class to this topic, and the social and political implications of genetic engineering in general. Signer was very active in antiwar circles and he drew many other connections between the general political situation and the misuse of science.

In any event, Mertz thought of herself as an MIT-type "middle-of-the-road radical" and was similarly disposed to view genetic engineering with disfavor. When Berg suggested the lambda phage-SV40 experiment to her, she wondered to herself whether she were not now about to take another step toward that undesirable possibility: "I remember after talking with [Berg] . . . I went off and spent a long time thinking about this whole thing because . . . I wanted to work for [him] at that . . . time . . . . But there was the whole question that bothered me, in terms of thinking back to M.I.T., of whether I wanted to be involved in developing genetic engineering techniques."[13]

On balance, her interest in working with Berg prevailed: "[I] spent a lot of time thinking about it and, in the end, decided that for the time being I would go ahead and work on that problem anyway on the grounds that people would probably say, 'Being able to put some DNA into *E. coli* [in the bacterium that was to be used] is still a long way from being able to change people's genes.' I don't know if I was right or wrong, but it was sort of a compromise between my radical point of view and wanting to work on specific types of projects and have Paul Berg as my thesis advisor."[14]

Mertz's first six months in Berg's lab were spent devising a means of getting a hybrid back into the bacterium, should one be developed. By June of 1971, all that was necessary was the actual preparation of the lambda-SV40 molecule, and that seemed near fruition. At that point, Mertz went off to take a three-week course on cell culture techniques given by Robert Pollack at James Watson's Cold Spring Harbor Laboratory on Long Island, New York. Pollack was a cell biologist who had had considerable experience with SV40 and viewed it with great respect. At that time he was not working on the virus, but others at Cold Spring Harbor were. A senior research associate at the laboratory, Carel Mulder, had just obtained some newly discovered biological agents, called restriction enzymes, which had the nice property of chopping up long strands of DNA into well-defined fragments that could then be studied more closely. In order to do the work, large amounts of DNA were needed; so when it was suggested that SV40 DNA might be investigated by chopping it up with restriction enzymes, Pollack became concerned. What would be the effect of raising the infectious concentration of SV40 by a factor of a hundred million or so? Could normal defense mechanisms be overwhelmed by the huge exposure made possible by the massive concentration of virus needed for research purposes? That Pollack recognized he did not need the virus for his own work made it easier for him to criticize those who did:

[I] could safely, without threatening my own career, say I was worried about the consequence of raising the infectious concentration of a tumor virus gene by a factor of a hundred million . . . . That worried me, and I could afford to be worried about it . . . . I'm trying to make [it] clear, that I see in my own work, and I see in other people's work, that a shade comes over your eyes when the problems affect your own work. I'm just lucky that by being at Cold Spring Harbor I was exposed to all of this kind of technology without having to use it. I think if I were using it, I probably would have found my own rationalization for not worrying about it. But to be constantly right at the edge of where these things are going on, and to know exactly what was going on in a technical sense, and not to need it, gave me a chance to worry about it. I think that, more than any higher ethical development, is what got me into this business of making noises.[15]

The "noises" Pollack was making about SV40 at Cold Spring Harbor were heard also in the short course that Janet Mertz attended that June. One afternoon session was devoted to a discussion of laboratory safety. But her mind was still occupied with the genetic-engineering problem. Here you are worrying about hazards to the investigator, she said in effect, and you are not at all worried about the possible impact on society of the results and capabilities demonstrated in the laboratory.

I then proceeded to mention the fact that we were presently working out a technology for making recombinant DNA molecules and, within a few months, would have the technique available. I told him that all the technology, other than that needed for making the molecules, was already available for taking the DNA and replicating it in *E. coli*. In other words, we were now almost at the point where we should be able to take mammalian DNA's, replicate them in *E. coli*, and then have these DNA's as clones for use in a variety of experiments. So here we were, almost on the verge of genetic engineering.[16]

Pollack was apparently dumbfounded by news of the experiment. "Rather than seeing my statement in terms of the potential hazards versus advantages of genetic engineering, what struck Bob Pollack was the danger of having a tumor virus growing in *E. coli*, a hypothetical danger I hadn't seriously thought about before."[17]

In their oral-history accounts of the events, both Pollack and Mertz agree that it was he who picked up on the biohazard implicit in putting a tumor virus into a bacterium that was a normal inhabitant of the human gut. "She was bragging about how they're going to make Russian dolls: SV40 in lambda, the lambda in *coli*, *coli* in something else. I said, well it's *coli* in people. She didn't know that, as far as I could see. It never occurred to her that that would be a problem."[18]

Mertz's recollection is that there was about a week left in the course at that point, and in that time she was involved in many discussions, not just with Pollack, about the safety of her proposed experiment. No one seemed to care about the broader question of genetic engineering, but many were concerned with the immediate biohazard problem. She recalls particularly MIT's David Botstein spinning out scenario after scenario of how a human bacterium with a tumor virus inside could escape from the lab and create a public-health catastrophe. Mertz countered these challenges with ways that the danger could be reduced. What if she put mutations into the *E. coli* so that it could not survive outside the laboratory? Botstein's reply was that that would only *reduce* the danger, not eliminate it. If there really were a danger, you would only have to have one escape to cause a worldwide disaster. She could never be absolutely sure.

Therefore, I made the decision not to do the experiment, even though I was quite upset about the whole matter and was thinking to myself, "Well, here is a really good thesis project that I've gotten started on and these guys are telling me I can't do my project." On the other hand, coming from a radical-type background, I figured, "Well, even if it's only a one-in-10^{30} chance that there's actually something dangerous that could result, I just don't want to be responsible for that type of danger." I started thinking in terms of the atomic bomb and similar things. I didn't want to be the person who went ahead and created a monster that killed a million people. Therefore, pretty much by the end of that week, I had decided that I wasn't going to have anything further to do with this project, or for that matter, with *anything* concerned with recombinant DNA.[19]

Mertz called Berg and told him about the furor that had erupted over the experiment, but he told her not to worry about it, that they would discuss it further when she returned to Palo Alto. But shortly afterward, Pollack also called Berg. Berg remembers that Pollack raised the concern that introducing SV40 DNA into bacterium could result in escape of such bacteria and colonization of human beings. "[I] think I reacted negatively to his suggestion. I raised counterarguments, he would make other arguments, and the debate moved back and forth; I remember specifically Pollack saying that he thought the experiment shouldn't be done. I recall that I was quite annoyed."[20]

But Pollack pressed him on the matter, and Berg decided to consult with several colleagues; their reaction was generally negative. Even some of those he did not speak to personally, such as Cold Spring Harbor's Jim Watson of double-helix fame, were reported to have been critical. On one of Berg's trips to the east coast, he discussed it with Baltimore.

> [Paul] . . . was [being] attacked from many sides . . . and we talked about it at some length . . . . I remember one night [at Cold Spring Harbor] we went out for pizza and we spent the whole night talking . . . . [T]he discussion [by others] had at first shocked Paul, but later he realized the basic validity of the worries . . . . I believe that I advised him not to do the proposed experiment . . . . We felt that the probability of hazard was small, but could not argue that it was zero. That was the beginning of a line of reasoning that has led us to where we are now.[21]

Most of the concerns were about the biohazard aspects of the experiment. But there were those besides Janet Mertz who looked to the larger issues, and two of these were Daniel Singer, a real-estate lawyer and former general counsel of FAS, who is the husband of biochemist Maxine Singer—one of those from whom Berg had solicited advice, and a friend of the Singers, a physician and bioethicist Leon Kass. Maxine Singer was later to be cochairperson of the Gordon Conference session that started the rDNA issue on its public career. At this time she was an NIH researcher and personal friend of Berg. Kass and Daniel Singer were also friends and associates of the Hastings Institute of Society, Ethics and the Life Sciences, a bioethics think tank located outside New York City. Singer was especially interested in safe-guarding the rights of the human subjects of research, so that his ethical interests in the Berg experiment were still connected to the biohazard problem.

> To some extent I viewed the recombinant problems only partly as a biohazard problem, but partly also as a problem in subjecting human subjects to risk, the subject in this case being only marginally the principal investigator, but much more importantly the other laboratory workers, whether or not involved directly in laboratory work: custodians, secretaries, peripheral personnel, as somehow becoming subjects of a research project, albeit inadvertent and, as such, entitled to more protection than they were at the time being afforded, and certainly being afforded some opportunity to grant or withhold consent to participation.[22]

Kass's interest was much more wide ranging. He had received an MD degree from the University of Chicago and a PhD from Harvard University in biochemistry and molecular biology. Early in his career

Kass began directing his research to the ethical and social implications of advances in biochemical science and the possible implications of genetic manipulation. Berg's intention was to develop a technology that could eventually be used to move genes in and out of animal cells at will. But some people saw the technique's potential for use in "gene therapy," a speculative medical treatment of genetic disorders like sickle-cell anemia or Tay-Sachs disease. Knowing of these possibilities and of Kass's interest in them, the Singers had invited Berg to have dinner with them at their home in Washington in the fall of 1970; Leon Kass was also invited. At that time, Kass held the position of executive secretary of the Committee on the Life Sciences and Social Policy, National Research Council, National Academy of Sciences.

They talked at great length about the "ethical basis of science."[23] A few days later, Kass sent Berg a letter, with a copy to the Singers, recapping the discussion and raising some questions he felt merited further discussion "sooner rather than later."[24] The issues brought up by Kass were certainly among the most specific and wide ranging of any reaction to Berg's work up to and including controversy over the experiment. Because of the significance of Kass's remarks to Berg and the scope of issues he raises, the letter, dated 30 October 1970, is reproduced in its entirety with the omission only of the salutation and two concluding paragraphs.

Here is a sketchy outline and fragmentary discussion of some of the questions which occur to me and which I think merit serious consideration, sooner rather than later.

1. Ethical questions related to safety-and-efficacy. These are merely sophisticated versions of general problems related to clinical trials of a new therapy or experiments to discern whether a new agent is potentially therapeutic.
a. Is the procedure efficacious? Or more modestly, have we ruled out obvious reasons why it might be useless?
Here appropriate trials in tissue culture and animals should precede first trial in humans. Special attention needs to be given to the problem of delivery of the viral vector to the appropriate target organ or tissue.
b. Is the procedure safe? How safe? Have cellular and animal studies been performed to detect possible deleterious effects of introducing the carrier viral DNA? Has a prospective study been designed to accompany the administration of the viral therapy to the first human patients?
c. Difficult judgements concerning:

(1) comparative value of other available therapy—e.g., organ or tissue transplantation;
(2) weighing likely chance of success, natural history of the untreated condition, likely and suspected harmful "toxic" effects.

2. Medical-ethical questions dependent upon the stage of "life" treated.
a. In the case of a child or adult known to have a particular genetic defect, or even in the case of an embryo or fetus on whom a definite diagnosis is made, the difficult judgements (1.c. above) are ethically no different than for any other form of therapy. To be sure, unforeseen tragic consequences may ensue, but the physician would have acted rightly because he was seeking to treat a known serious disease in an existent human being or existent fetus.
b. Far greater certainty with respect to safety and efficacy would be required to perform the same manipulations on the germ cells prior to fertilization. Here, by no stretch of argument can it be said that one is engaged in therapy of existing persons with known disease. To manipulate germ cells is a form of experimentation, albeit well-meaning, on a not-yet-existent and not-yet-afflicted human being. *Ignorance* of untoward consequences that might result is here no excuse; considerable *knowledge* that *no* such consequences will follow gene manipulation would be a prerequisite for going ahead. Childlessness or adoption are to be preferred to subjection of the unconceived to potentially hazardous manipulation.

3. Therapeutic and other purposes. The ethical questions about genetic manipulation will be dependent in part upon the purpose served. Obviously, once a technique is introduced for one purpose, it can then be used for any purpose. Therapeutic use is one thing, eugenic, scientific, frivolous, or even military are quite another. Thus there are two questions to be considered:
a. What would be the range of ethically legitimate purposes?
b. How could one limit use to those purposes? I have my own views on the first for which I would argue ("therapeutic use only"), but more importantly, I would insist that we need to foster public deliberation about this question, since I don't think this is a matter to be left to private tastes or to scientists alone. I defer the question of control until later.

4. Possible undesirable consequences of ethical use for ethical purposes.

In the previous paragraph, I considered the problem of so-called "good" *vs.* "bad" ends. We have also to consider "bad" consequences of a technique used only for "good" purposes. This is a far more difficult problem, and unfortunately, I think, a more pervasive one in the whole biomedical area. The inevitable social costs of desired progress are probably higher than the costs of progress willfully perverted

by bad men. We must consider and weigh the following kinds of questions in deciding about the *first* use of a new technique:
a. What are the biological consequences in future generations of widespread use of gene therapy on afflicted individuals? Anything but treatment of gonads, gametes, and zygotes will work to increase the frequency of the given gene in the population.
(1) Are we wise enough to be tampering with the balance of the gene pool? That we do so inadvertently already is no argument for our wisdom to do so deliberately.
(2) What are our obligations to future generations? Do we want to commit them to the necessity of more and more genetic screening and genetic therapy to detect and correct the increasing numbers of defects which our efforts will have bequeathed to them?
b. Do we ourselves wish to embark upon a massive program of screening and intervention? With federal support? Under compulsion of law?
c. Are we not moving toward more and more laboratory control over procreation? What are the *human* costs of this development, especially for marriage and the family?

5. How can these developments be monitored and kept under control?

The more I think about this question, and the more I contemplate the possible widespread consequences of genetic manipulation, the more I believe that all decisions to employ new technologies and even to develop them for employment in human beings should be public decisions. How to do this is not obvious, although the question is now being actively explored by various groups. Regardless of who should decide and who should control, the problem of whether control is possible remains. Here, much depends upon the demand for the new technology, its expense, the scale on which it will be used, and the know-how needed to use it. The smaller the demand and scale or the greater the cost and know-how, the better the possibility for control.

6. What are the obligations of the basic scientist whose research brings closer a new biomedical technology? What would be the ingredients of an ethical warrant for him to go ahead? Let me suggest the following.
a. Consideration of the kinds of questions outlined above (they are meant to be suggestive and exemplary, not definitive) by himself, with his co-workers, and with appropriate colleagues, scientific and non-scientific.
b. To be responsible for helping to set forth in advance standards and procedures for testing safety and efficacy.
c. Calling the attention of responsible publics to the technological possibilities his research (and that of the field generally) makes more imminent. This would best be done, it seems to me, merely by writing

articles in a responsible journal (e.g., *Science*), outlining the technological possibilities and some possible social and ethical problems they present, and then inviting sober and responsive public deliberation concerning implementation and control of the technology.
d. To be willing and prepared to abide by a public judgement which may undermine his own research (especially true in applied research or in those areas of basic research where there may be little or no gap between knowledge and use—as in psychopharmacology).

Kass had opened his letter by complimenting Berg on the openness and sensitivity to ethical questions that he showed when they had talked. He may have overestimated Berg's receptivity, however. Recalling their conversation later, Berg remarked, "[I] was unpersuaded. I was pretty much convinced that many of the fears people were expressing were red herrings . . . ."[25] Maxine Singer, to whom a copy of the letter was also sent, may have taken it more to heart. The action she took at the Gordon Conference in the summer of 1973 in publicizing the potentially adverse consequences of a new scientific development was precisely one of the recommendations that Kass had made in his letter.

Berg, on the other hand, was sufficiently persuaded by the arguments over hazard, and by his own assessment of the possible payoff of the experiment, to postpone the planned studies. What if the viral gene, once introduced into *E. coli*, did not function (was not expressed)? In that case the reason for doing the experiment would be nullified. But the bacterium would still be a carrier of the tumor virus's genes, whether it expressed them or not, and from there it might spread to the population at large, possibly gaining access to a human cell where it might be expressed as a cancer.

So Berg and Jackson began to consider other riskless experiments that might tell them whether animal genes could be expressed in bacterial systems. "[O]ur rationale for not doing the experiment . . . was: we're not sure it will give us the answers we sought; it carries some risk which we can't assess; we can hold off doing the experiment and find out whether the benefit . . . is going to be high enough to match that kind of risk."[26] This seemed to leave the door open, should Berg become convinced that the benefits of the experiment would warrant the risk.

In fact, Berg called Pollack back several months after Pollack's original call and told him that the experiment was not going to be done. Mertz, who had already turned her attention to other matters (ironically, she was soon to make a discovery that would be of crucial importance for the further development of the rDNA technique), recalls that Berg's attitude at the time seemed to be that since the methodology for joining the bacterial and viral DNA segments was not yet in hand, the whole issue was moot. For the moment, Mertz remembers Berg could argue, "'No, we're not working on it. We don't have the technology to do the experiment yet.'"[27] However, Jackson and Symons were trying to make the hybrid DNA molecule. Their intention was not to put it into *E. coli*, but into an animal cell. Besides, *making* the hybrid molecule was not the dangerous thing; putting it into a gut bacterium was what concerned some people.[28]

Then, during the fall and winter of 1971–1972, Jackson and Symons completed the construction of the SV40-galactose operon rDNA, getting up to 20% yield. Shortly thereafter, Jackson went off to the University of Michigan to start his own lab. Therefore, no one in Berg's lab was working on these hybrid ("recombinant") DNA molecules. Mertz recalls,

> The whole thing [work with the SV40-galactose operon DNA] was just dropped. The technology that had been developed up to that point in time was very hard to use because it involved a whole series of enzymes. There were very few people anywhere in the world who were willing to attempt the types of experiments that had been postulated or that were being planned to be done. Therefore, Berg pretty much told everybody, "Well, we're not working on it anymore," which was perfectly true: There *was* nobody in his lab working on it. So the whole issue just dropped at that point . . . . Berg assured Pollack no one was going to work on it. So everything just stopped at that point.[29]

The affair seemed to be over. "[T]he whole discussion and the debate had gone back and forth, largely not in public, but just among scientists, myself, and a group of people who felt strongly about it . . .," Berg said. "I remember Pollack and others saying, 'Good, great, nobody else is going to do it if you're not . . . .' In fact there were people who said, nobody, no other lab *could* do it but ours. And so the debate just sort of died."[30]

But the news of its death was greatly exaggerated, as Mark Twain might have said. For starters, the controversy led directly to a conference on biohazards held in 1973, known as the first Asilomar Conference. It is discussed in Chapter 4. In addition, it prompted many to pursue a line of reasoning that, as Baltimore remarked, has led us to where we are today.

But it was not the only event in the years before the Gordon Conference to have had this impact. Another episode involving a hybrid virus also figures into the story.

3

The Adeno-SV40 Virus Episode

The furor over the Berg experiment had begun because of the possibility of producing an organism that could carry a tumor virus into the human population, perhaps becoming endemic there and causing cancer years or decades later. But at the same time that this debate was going on, there actually existed a candidate for such an organism, and its "keeper" was wondering what to do about it. The organism in question was a virus that was a hybrid between the monkey tumor virus SV40 and an agent of acute respiratory disease in young adults and children called an adenovirus. There are many different varieties, or serotypes, of adenovirus—at least thirty-one serotypes for humans and seventeen for animals.

Adenovirus infection occurs most commonly in children and next most commonly in military recruits, where it is believed to be responsible for 15 to 50 percent of acute respiratory disease.[1] Because of its importance for the military, an attempt was made in the 1950s to produce a vaccine against the principal adenovirus serotypes. Like all viruses, it could only be grown in living cells. In this case, the virus for the adenovirus vaccine, and also the one for the polio vaccine, were grown in rhesus monkey kidney cells, cells which turned out to be contaminated with SV40 virus. But SV40 virus itself was not discovered until 1960, after the vaccines had been in use for some four years. Several manufacturers immediately undertook to clear their vaccine of SV40 virus, and it was apparently eliminated in types 3 and 7.[2]

Meanwhile, two teams at the National Institutes of Allergy and Infectious Diseases (NIAID) laboratories in Bethesda, Maryland, were converging on the problem from different directions. M. D. Hoggan and D. H. Black were working on the immune response of hamsters carrying SV40-initiated tumors. They were especially interested in characterizing a substance, called SV40 T-antigen, that seemed to be associated with SV40 transforming activity. At the same time, R. H. Huebner was interested in the ability of the adenoviruses to cause tumors in hamsters on their own. To accomplish this, he put different strains of the virus into hamsters and observed the animals for tumor development. Much to his surprise, the strain from the supposedly SV40-free Adeno 7 vaccine produced tumors characteristic of SV40 virus infection, not adenoviruses. Moreover, it produced SV40 associated T-antigen. Huebner immediately guessed that he was dealing with a hybrid molecule composed of pieces of both SV40 and Adenovirus 7.[3]

This discovery was made between 1963 and 1964. At about this time, Andrew Lewis, Jr., a young physician fairly new in Huebner's lab, began to look at the other vaccine strains to see whether they too contained hybrids, or recombinants as they are also called. This was not Lewis's first research with NIAID, however. In 1964 he investigated an in-house epidemic of lymphocytic choriomeningitis (LCM) infection among Huebner's animal colony. Although the hazards of LCM infection in hamsters is now well recognized,[4] it was not known at the time. Apparently, a tumor obtained from a commercial source was contaminated with LCM virus and it spread to the colony through the tumor transplants that were made. When Lewis looked carefully at the circumstances of the outbreak, he discovered that a half-dozen of the animal keepers had also had an influenzalike illness, and later showed evidence, through blood tests, of having been infected with LCM virus. "Before it was all over we had . . . 10 or 12 people in the laboratory, both technical and professional staff, who had been infected with LCM virus and had been clinically ill. One of the secretaries was hospitalized for 3 weeks. We wound up having to kill many thousands of animals as a result of this."[5]

So Lewis had already had a significant experience with a biohazard when he began work on the adenovirus-SV40 recombinant. Moreover, as a physician, he had a markedly different background than many others later involved in the recombinant DNA debate. "Anybody with a medical background that goes into microbiology is certainly aware of the potential complication of handling microbes in the laboratory. We all know that people can get sick from this type of thing. Tuberculosis is an occupational hazard of pathologists, . . . and so I think that people with a medical background especially have a heightened awareness of infectious complications of microbiological settings."[6]

Not long after this episode, Lewis began to work on the other strains of adenovirus that had been stored in colleague Wallace Rowe's freezer since 1956. Lewis and his coworkers quickly discovered that types 1, 2, 4, and 5 all contained SV40 DNA along with adenovirus DNA, both within a wrapping or envelope characteristic of the adenovirus alone, but capable of producing infectious SV40 virus. They all turned out to contain the entire SV40 genome. (The genome is the totality of an organism's genetic information.) Types 3 and 7 of the vaccine strains contained only a portion of the SV40 genome and did not produce infectious SV40. The hybrid virus *itself* (with two kinds of DNA within the adenovirus envelope) did not reproduce unless the cells were simultaneously infected with normal adenovirus: in the language of virology, it was defective.[7]

Other NIAID labs worked on the hybrids that contained only part of the SV40 genome (designated E46 hybrids), while Lewis and his lab worked on the Adeno 2 hybrids which apparently had all of the SV40 genome. The idea behind these studies was to probe the genetic makeup of tumor viruses. The hybrids seemed to provide a good opportunity to compare viruses that had different parts of the SV40 genome. Soon Lewis learned that his hybrids, too, had different segments of the SV40 genome. At first the work did not seem to be very hazardous:

> When the idea was initially conceived that we should do genetic studies on these hybrids, I very frankly admit that we weren't really thinking in terms of potential biohazards. In the first place, we had no evidence that such recombinants occurred [i.e., they were not sure how the two DNAs were hooked together.] But I think, more specifically, at that point . . . all seven of the hybrids that we had been studying, were defective. In other words, they would not replicate independently of nonhybrid [Adeno 2] helper. . . . So there seemed to be no health hazard associated with these materials, at least as near as we could determine, because if an individual got infected, it's very unlikely that the hybrid [virus] would even grow in a single individual, let alone be spread around the community.[8]

The first intimations that something might be amiss came slowly in the period from 1967 to early 1969. Lewis isolated a virus descended from his original stock that produced SV40 T-antigen, was infective for monkey cells, *and* grew on human cells without any help from simultaneous infection with nonhybrid Adeno 2 virus: in other words, Lewis had a nondefective Adeno 2-SV40 hybrid (designated Adeno 2 + ND_1) that could replicate in human cells on its own. Before they could become seriously concerned, however, the whole phenomenon began to change. "We no longer got viruses that induced SV40 T-antigens . . . [b]ut they did induce some kind of antigen that seemed specific for SV40."[9] So most of 1967 and 1968 were spent trying to characterize this new antigen, and to show that it did really come from SV40.

In early 1969 they collected together their material for a paper, which was submitted that April. When they stopped to look at what they had, they realized it was an unusual agent:

> [T]his agent was inducing . . . an early SV40 antigen. This antigen is . . . something that [was] an SV40-specific function. Once we were aware that this was now a human adenovirus, which had all the properties of the human adenoviruses, but in addition contained this sequence of the early SV40 genome which we subsequently demonstrated to be in cells that had been rendered malignant by infection with SV40 virus, we had all the information that was needed to begin to worry about . . . what this was and the consequences of it.[10]

The 1969 paper on this nondefective hybrid, in fact, contained a concluding statement that indicated Lewis's concern: "The pathogenicity of non-defective viruses which are hybrids between human pathogens and other viral agents such as SV40 must be considered. Such viruses could be maintained in human and sub-human populations, and as pathogens would represent unknown hazards."[11] Lewis believed that this was the first statement in the virus literature to suggest "that agents which people were creating in the laboratory could be considered as pathogens."[12] It was not, of course, to be the last.

Worried or not about the hazards of such recombinants, no one in Lewis's lab seemed to know what to do about it. The unusual antigen that the ND_1 hybrid produced, dubbed by Lewis the U-antigen, was not as impressive as the T-antigen. Like the U-antigen, the T-antigen was controlled by a gene in the "early region" of SV40 (i.e., the region of SV40 that begins to function first, ahead of other gene functions), but it was the T-antigen that was most clearly associated with cell transformation (and hence tumor production). No one knew what the significance of the U-antigen might be. So Lewis and his coworkers just worried.[13]

It soon became clear that ND_1 was not the only nondefective hybrid that Lewis had. In sum there turned out to be five, all descended from the same virus pool that yielded the ND_1 agent. Moreover, they all were discovered to have different segments of the part of the SV40 genome—the early region—that most people were interested in because of its association with cell transformation. But by the same token, nondefective viruses containing this region were a matter of concern. A principal element in this concern was the fact that the adenovirus is responsible for chronic infections in the tonsils and adenoids of young children. Adenovirus infection was endemic in the pediatric population. What would be the result of a chronic respiratory infection that also carried animal tumor virus? Lewis did not know. There was more discussion about the problem in his laboratory, but again nothing was done about it except to handle the agents with care and respect.[14]

The paper on the ND_1 hybrid was published in 1969, but until 1971 no one outside Lewis's lab expressed any interest in working on it. It was the custom, however, to supply an experimental agent to whosoever wished to work on it. In the spring of 1971 such a request was made by Carel Mulder at Cold Spring Harbor, the same scientist who wanted to work with restriction enzymes and large amounts of SV40 virus. At first Lewis hesitated, but then agreed to supply Mulder with ND_1 by September of that year. The reasons for his initial hesitation, and the subsequent delay in supplying Mulder with the virus, seemed to have less to do with biohazard than with the sociology of science. The fact that no one had inquired about the virus for two years after publication of the original paper meant, on the one hand, that Lewis and his group had a two-year head start toward characterizing it; but it meant also that Lewis and the six young colleagues in his lab had *invested* several years of their new careers in the project. As a group, they were much less experienced in the techniques of molecular biology than were Mulder and his people at Cold Spring Harbor. "By September, I felt like we would have done as much with the ND_1 hybrid as we possibly could, given the circumstances under which we were operating. . . ."[15] So Lewis had put Mulder off until then.

The delay apparently did not sit well at Cold Spring Harbor, however, at least as far as its director, James Watson, was concerned. That became clear in August, just before the 1 September date that Lewis and Mulder had agreed as the time he would supply him with the virus. That month found Lewis presenting unpublished data on all the hybrids to a tumor virus workshop at Cold Spring Harbor. The meeting was actually the climax of the program that had begun with the series of summer short courses at the laboratory—one of which, in June, had been the scene of controversy over Janet Mertz's thesis experiment for Paul Berg. Lewis remembers that Mulder prefaced the session, where he presented his data, with the disclaimer that it was a very informal conference and that nobody would be held responsible for what they said, which is "as workshops are supposed to be."[16]

But after Lewis's presentation, a remarkable scene took place. Watson, whom Lewis did not at first recognize, confronted him during the coffee break with the claim that Lewis had no right to talk about the hybrids if he was not also willing to send them to Cold Spring Harbor. Watson then accused Lewis of conspiring with Lewis's former boss, Huebner, to keep the agents at NIH. (Huebner had actually left NIAID some years before and was then working on something completely different.) Watson then threatened to force release of the virus by one of three mechanisms: (1) He would personally write to the director of the NIH to say that there was a conspiracy to keep valuable agents from the scientific community. (2) He would write a letter to *Science* telling them, Lewis later reported, "what kind of guy I was, and that I was sitting on these agents and had no right to be doing this when other people were interested in working with them." (3) He would write to Congress saying that "public money was being spent to develop reagents not being made available to other people."[17]

Lewis was stunned by this outburst, the more so when he realized who it was making the threats. The result seemed to be to precipitate the hazard question. Lewis and his coworkers had had additional discussions about their responsibilities for supplying the virus after Mulder called that spring; and by way of reply to Watson's onslaught, Lewis asked him bluntly whether he considered the agents to represent a hazard. He then recapped the discussions that had been taking place in his lab on the subject. Watson's answer was that if one worried about things like that, science would quickly grind to a halt. "And that was the end of that. We didn't pursue it any further. He walked away and I went on back into the meeting. . . ."[18] In the space of less than thirty minutes, what had started out as a question of scientific etiquette had been transformed into a major issue over biohazards.

Others soon became aware of the confrontation. According to Lewis, "The most important result of my conversation with Watson was . . . the immediate furor it provoked over the biohazard problem at the meeting."[19]

[T]here was a tremendous diversity of opinion as to the . . . problem of hazard raised by these nondefective hybrid viruses. I was confronted then by other people whose opinions ranged from [those] . . . essentially like Watson's: there's no problem here that we can't handle, or we have nothing to worry about, because SV40 doesn't seem to do much to humans except transform cells; to the other extreme. . . . [Some even said] that these hybrids should never leave [NIH], that [NIH] should provide a place down here for people to work with these agents if and when they wanted to study them.[20]

Although the biohazard question dominated, one question of proper etiquette still remained: Was Lewis obligated to supply samples of the *other* nondefective hybrids, as Watson wanted, even before he published papers on them? Dan Nathans, for one, thought not. Nathans was a future Nobelist for his work on restriction enzymes, and was soon to be a key figure in the rDNA debate, so it is of interest to record that Lewis thought his response was based purely on matters of scientific propriety: "[I] don't recall what Dan was thinking in terms of the biohazard."[21]

But the ND_1 material was already in the literature; so upon return to NIAID, Lewis's immediate problem was supplying a sample of it to Mulder and the people at Cold Spring Harbor. Because of the furor over biohazard at the meeting, Lewis and his group felt that they had to qualify what they were doing with the virus. In a letter to Mulder on 1 September 1971, which was included in the "seed" of the virus, Lewis underlined the safety problem:

[T]hose of us who are working with these nondefective hybrids are concerned about their ability to infect humans and spread through the general population. Due precautions must be taken to ensure that this does not happen. We recommend that you consider the ND_1 virus a potential biohazard. In the interest of public safety, before large scale studies are undertaken we urge that you screen all personnel who will be working with this agent for the presence of Adeno 2 neutralizing antibody [i.e., for immunity to adenovirus]. Only those with evidence of such antibody should be allowed to handle the agent. During such studies, all areas and apparatus used to handle the ND_1 virus should be cleaned immediately after use with an oxidizing agent such as sodium hypochlorite or tincture of iodine; we also use UV lights at night in our laboratory area. All children, especially those under 5 years old, should be permanently excluded from the area where this virus is being used. Personnel working with this agent who develop pharyngitis [sore-throat], conjunctivitis, or viral pneumonia, should be screened for infection with ND_1 virus by both virus isolation and serologic testing for Adeno 2 antibodies.[22]

Lewis recognized that he had no control over Mulder, but he expressed a hope that Mulder would comply:

I hope the precautions listed above meet with your approval. Though no data to date indicate that the ND_1 virus is, indeed, a dangerous agent, I am genuinely concerned about the likelihood of this virus spreading through the general population as we supply the virus to other laboratories for their use. The safety discussions at the Tumor Virus Meeting at your laboratory [i.e., Cold Spring Harbor in August] heightened this concern, as other members of the scientific community shared my reservations. As we proceed with genetic studies on animal viruses, and tumor viruses in particular, it appears that we must be more and more cautious that the mutants we develop do not have an opportunity to leave the laboratory.[23]

Having asked these precautions of Mulder, Lewis decided to include them with all subsequent filled requests for ND_1 from other labs. To ensure that Mulder himself would not distribute the virus without informing others of the concerns Lewis had, he wrote another letter on 15 September, asking Mulder to consider not sending the virus to any other lab, instead referring any individuals who wanted a seed of ND_1 to Lewis. He gave two reasons: By acting as a single supply source, Lewis's lab could safeguard the purity of the seed, providing what they believed to be uncontaminated, biologically characterized virus; and they could maintain adequate records about the material and the advice given to the laboratories. Lewis disclaimed any ulterior motives: "We are not attempting to restrict any collaborative studies in which other laboratories might grow the ND_1 virus under your direction, nor are we asking you not to send [nonhazardous] products of ND_1 such as viral DNA, RNA, protein, etc. What we are attempting to do is to be sure that all workers who are propagating this agent are informed of the safety problems and are acting responsibly in handling the virus."[24] The problem, as Lewis saw it, was that there were no mechanisms to deal with a situation like this. He therefore felt constrained to take it upon himself to establish them. "[A]s you know, we are dealing with many unanswered questions concerning the genetic stability and potentially hazardous nature of these nondefective hybrids. By proceeding methodically with due precautions, we hope to avoid many problems or contain any that may develop."[25]

The two letters, one spelling out the safety precautions, the other asking that no secondary distribution be undertaken by recipients, were then abstracted and included in synoptic form to all who requested stock of the ND_1 virus—a total of about fifteen to twenty labs in the United States and Europe.[26]

All these actions were directed toward dealing only with the ND_1

hybrid, details of which had been published some years earlier. During the year after the Cold Spring Harbor confrontation, work was going forward on the other four nondefective hybrids. Three manuscripts were in preparation. After they were accepted by journals, Lewis would again begin receiving requests for those agents, and he felt that the mechanism established for the ND_1 virus was not sufficiently strong. For one thing, he had no way of knowing whether those to whom he sent the virus were really observing the suggested precautions. He wondered whether his obligations might not be legal as well as moral. "Frankly, we really didn't know what to do. In talking with any number of people, nobody really had too much useful comments. Everybody expressed some concern one way or another either, you know, you're going to do bad things for science, or you're going to do bad things for the public. There was no meeting of minds."[27] The only valuable advice came from one of Lewis's colleagues, Ken Takamoto. Lewis, he said, could not decide a matter of this magnitude on his own—it did not affect just him, or even just the academic community, but the public at large. So Lewis sent a memo to the director of NIAID asking advice. A scientific problem had now become an administrative one.[28]

The memo, dated 27 November 1972, was directed to Dorland J. Davis, NIAID director, but was routed through Wallace Rowe, Lewis's chief at the Laboratory of Viral Disease, and John R. Seal, NIAID's scientific director. The stated subject was a request for an NIAID policy statement concerning the release and distribution of adenovirus-SV40 hybrids.[29] In this document Lewis set out the situation as he saw it:

1. General studies of DNA tumor viruses in NIAID's Laboratory of Viral Disease had isolated five nondefective Adeno 2-SV40 hybrids.

2. Each hybrid carried a variable amount, ranging from 0.5 to 40 percent, of the SV40 genome, as an integral part of the hybrid virus's chromosome.

3. The portion of the SV40 genome involved was from the early region, the region "most closely associated" with the virus's cancer-causing potential. However there was not as yet any evidence that a specific viral gene or genes was responsible for the virus's oncogenic (cancer-causing) ability.

4. Adeno 2 virus is a common human pathogen, especially in chil-

dren. This means that the hybrid, which looks to the body just like an ordinary adenovirus because the hybrid DNA is encased in an adenovirus envelope, *could* infect humans and spread to the general population. This has led to "some concern" about the potentially hazardous nature of these agents.

5. The hybrid viruses were difficult to characterize biologically. Therefore only one of them—the ND_1 agent—had been described in the literature and supplied to other laboratories. This virus contained about 10 to 20 percent of the SV40 genome and induced the newly discovered U-antigen.

6. This U-antigen *could* be responsible for the ability of the adenoviruslike organism to grow in monkey cells, where adenovirus does not ordinarily grow. This means that the acquisition of this segment of the SV40 genome might have increased the range of possible host tissues for the virus. In addition, the role of the U-antigen in tumor initiation was not known, and the labs supplied with the virus have been alerted to the possible hazards and a list of precautions to be considered included.

7. The NIAID laboratory was currently preparing manuscripts for publication that described the other four nondefective hybrids. Two of these had twice as much of the SV40 genome as the ND_1 hybrid, and one of them, designated ND_4, contained the entire early region. That virus was competent for several functions, including T-antigen control, known to be "intimately associated with SV40 tumorigenesis."

8. Despite these concerns, the hybrids were potentially quite useful for genetic studies of SV40. Since SV40 was a tumor virus, these "studies . . . will ultimately define how the virus acts biochemically to produce a tumor."[30]

9. "Once these papers are published, these hybrids will be considered public property and scientific etiquette requires that they be made available to other workers. It is the wide distribution of these agents that generates the greatest risk to the general public."[31]

10. The biohazards in question must still be considered only *potential* biohazards:

a. There was no evidence that any of the hybrids caused human disease, or even infection.

b. None of the hybrids caused tumors in newborn hamsters.

c. However, *all* of the hybrids transformed hamster kidney cells in

tissue culture, and when these cells were transplanted into animals they produced tumors.
d. So far, the ND_4 virus has not been shown to transform human cells. However, nonhybrid SV40 readily transformed human cells.

11. There was little epidemiological data to go on. Two studies, both done in the 1950s, seemed to indicate that there was no danger of an epidemic.
a. Thirty to sixty percent of children under the age of six had some immunity to Adeno 2 virus; while 50 to 80 percent of those over age fifteen had some immunity.
b. Studies on upper respiratory infection in adults showed that Adeno 2 virus was rarely responsible for "colds."
c. On the other hand, little or nothing was known about asymptomatic infection with the Adeno 2 virus.
d. Lewis therefore concluded that

If a susceptible individual were infected with a nondefective Adenovirus 2-SV40 hybrid . . . the frequency of Adenovirus 2 antibody suggests that possibly 1 in 2 to 1 in 10 contacts, depending upon their age, could be successfully infected, thus establishing these hybrids among the reservoir of human pathogens with unknown consequences. Considering the chronicity of adenovirus infection in the adenoids and tonsils of children, the consequences of wide spread infection with these agents are unpredictable. The SV40 genes which are contained within the chromosomes of these hybrid viruses should not be allowed to enter this reservoir.[32]

12. The problems with the Adeno 2-SV40 hybrid, moreover, was not at all unique:

It also appears that the issues raised by these nondefective Adenovirus 2-SV40 hybrids will soon be of fundamental importance to the scientific community. As genetic studies of infectious agents proceed—especially for the production of vaccines and in the field of tumor virology—many more mutants and hybrid viruses of unknown pathogenicity are going to be created in the laboratory. Adequate safeguards should be set up to ensure that these agents remain in the laboratory and that none are allowed to spread uncontrolled through either human or animal populations. Since the nondefective Adenovirus 2-SV40 hybrid viruses isolated in [our NIAID laboratory] seem to represent the first of these potentially hazardous mutants, I feel that the Directors of the Institute must consider the problems these agents have initiated. The implications and precedents to be established by the release of these agents might possibly require a review by the Director of NIH. Hopefully, successful answers to these questions will protect the public interest in scientific inquiry into the understanding and control of malignancies and other genetic disorders while guarding the health of the public by preventing any untoward effects from the potentially hazardous agents such research begets.[33]

13. Given this state of affairs, there were several possible alternatives for disbursing the four new hybrids. Lewis was now asking for advice on how to reply to the expected inquiries. The alternatives, as he saw them, were as follows:

a. He could release the viruses, including a list of precautions and recommendations, leaving the recipient the responsibility of containment. This was essentially the system used for the ND_1 virus.

b. He could release the viruses only to laboratories with proper containment facilities as determined by on-site inspection. "Periodic monitoring of the facility for air, equipment, and personnel contamination, with repeated on-site inspections, would be a consideration."[34]

c. The viruses could be restricted to NIH laboratories; or to laboratories under NIH control. The NIH could then provide noninfectious hybrid products (such as pure DNA) to others on request. Suitable facilities at NIH could be made available on loan for extended periods, if appropriate.

These alternatives concluded the Lewis memorandum. They indicate that even by the end of 1972, the age of government regulation of genetic research was upon us.

Addressed to the director of NIAID, Lewis's memo picked up some additional support on the way. NIAID's scientific director. J. R. Seal, through whom it was routed, referred it to three other senior NIAID scientists and two members of the Board of Scientific Counselors first.[35] In his cover memo,[36] Seal noted that Lewis's memo did not reflect the controversy that erupted over the release of the hybrids at the Cold Spring Harbor meeting. It was obvious, Seal said, that the questions raised were sensitive ones, and there was no clear opinion among scientists on the proprieties to be followed. With new hybrids soon tobe announced, some interim policy was needed.

In the act of soliciting the views of others, Seal gave his own. He believed that specially designed facilities, already used for working with other hazardous biological agents, should be used, but expressed doubt that NIAID could do anything beyond asking for certification as to the adequacy of facilities from requestors of the hybrid; and that in the case of requests from those of unknown competence, some additional information on the experience of the requestor should be obtained. This still left the problem of secondary distribution unsolved, however.[37]

By mid-December 1972 Seal had received three replies. The first was from Sheldon Wolff, NIAID's clinical director. In it he stated his own agreement with Seal's views and he underlined the necessity of verifying the competence and experience of any investigator requesting the virus.[38] Jules Younger, chairperson of the Department of Microbiology at the University of Pittsburgh, and a member of NIAID's Board of Scientific Counselors, put the case even more strongly.[39] The viruses, he said, should be restricted to those laboratories with proper containment facilities, as determined by criteria already set up by NIH's Biohazards Committee. Moreover, all secondary circulation should be strictly controlled and should conform to the same requirements. Speaking as a non-NIH researcher, Younger added that there was already precedent for controlling the circulation of hazardous biological agents in the permit requirements of the US Department of Agriculture's regulations concerning pathogens of domestic animals. In adhering to these regulations, he said, he had never felt hindered in his work. Whatever the legal implications of such a policy for human pathogens, he felt it prudent to institute a similar policy, at least on an interim basis.

But by far the strongest reaction came from James A. Rose, head of the Molecular Structure Section of NIAID's Laboratory of Biology of Viruses.[40] In a correspondence to Seal, Rose pointed out that even the potential benefits cited in the Lewis memorandum had their dark side: "Whether or not Adeno-SV40 hybrids will yield information not obtainable by other currently available biochemical and genetic techniques is a moot point. In my view the hope that they may prove useful does not entirely justify the biological uncertainties resulting from their presence. Indeed, the production of such hybrids could even be stimulated by the fact that they produce new subjects for study in a field which is already crowded and in need of new problems."[40] Even Lewis's contention that a large proportion of the population was immune to adenovirus was not very reassuring. Rather than assuming that immunity gave protection, it might turn out that the immune reaction could "play a role in the adverse biological expression of these agents." (Ironically, it is just such an immune response run amok that serves as a model for the damage done by infection of the mouse by lymphochoriomeningitis virus, the subject of Lewis's first experience with biohazard.)

Rose went on to point out, as had others, that the problem was not confined to Lewis's hybrid virus. He cited four reasons why the Adeno 2-SV40 problem had to be viewed in a much larger context:

1. One of the other adenovirus-SV40 hybrids (in this case a defective hybrid) was highly oncogenic, widely distributed, and being studied in conjunction with similar "human SV40" viruses in many laboratories with little precaution. (Lewis had originally discounted his concerns about defective hybrids because they required normal "helper" adenovirus in order to grow. Rose seemed to be saying that there was enough "helper" adenovirus already in the population to make even defective viruses worrisome.)

2. There was a steadily increasing production in research laboratories of other mutant animal viruses, whose biological properties were not entirely understood.

3. Laboratory manipulation of viruses had resulted in their adaptation to a variety of new host cells, including human cells. The large-scale production of these agents, some of which were oncogenic, could bypass normal barriers to human infection.

His final point spoke directly to the new technology of gene splicing, already visible on the horizon:

4. The biochemical construction of combinations of DNA from different organisms had already been accomplished. No one yet knew what their biological potential might be.

Thus, to be logically consistent, any restrictions in the area of adenovirus-SV40 hybrids should also apply to other potentially hazardous research. Rose's solution to the hybrid problem prefigured almost exactly the latter sequence of events in the concern over rDNA techniques:

I personally oppose the distribution of the Adeno 2-SV40 hybrids at this time. It is possible that an adverse response to such action might be minimized by also requiring that studies with these agents at NIH be discontinued. This decision could be described as a holding action, i.e., one taken while assessing the views of scientists both inside and outside NIH. Alternatively, the existing Adeno 2-SV40 hybrids could be released (with suggested controls) but the further production of hybrids discontinued for the present. I think heavy emphasis should be given to the idea of at least a temporary slowdown in order to acquire better understanding of what now exists and to allow for a more carefully considered and supported approach to the wider range of problems.[41]

NIAID's director Davis received Lewis's memo together with those of Seal, Wolff, Younger, and Rose, sometime during the third week of December 1972. He immediately began to make arrangements for an interim policy, and at the same time tried to set up a mechanism for handling the larger problem. In a memo to NIH's associate director for collaborative research, he asked that NIH's Biohazards Committee and the Public Health Service's Center for Disease Control (CDC) begin to consider the entire problem of the circulation of potentially hazardous viruses.[42] He cited three issues in this general area.

The first was adequate control of agents currently used in NIH laboratories. His concern had been prompted by a recent incident in which an investigator working at the National Institute of Neurological Diseases and Stroke (NINDS) for the summer had, on leaving, taken some biological materials, among them possibly the agent causing the degenerative central nervous system disease kuru. The second issue for consideration was the question of what to do about the large variety of laboratory-created viruses that Rose had been so concerned about, and that had also excited Robert Pollack. The last issue was how to control secondary distribution of the viruses for which Lewis was being asked.

Since there was no central authority in HEW covering these matters, Davis suggested that CDC and NIH confer on a general policy. CDC, he pointed out, had the responsibility to formulate policy on matters pertaining to hazardous biological agents; and NIH, as the funding source for most of the work on preparation and study of these nonnaturally occurring agents (NNOA), had a moral obligation to establish such a policy. Davis volunteered to take the lead in bringing NIH and CDC together, possibly by working with CDC directly in formulating policy.[43]

In the meantime, Davis was asking Lewis for assurance that the individuals to whom he had released the virus had adequate competence and facilities to contain it. Additionally, he wanted assurances from these individuals that they would not distribute the virus to others who were not familiar with its hazards and the measures necessary for minimizing them. He added, "I realize these directions are non-specific and place the burden of judgment on Dr. Lewis. He is however both a physician and a virologist and capable of making judgements."[44]

The same day (20 December 1972), NIH's deputy director for science, Robert Berliner, communicated the new policy to the scientific directors of all the other NIH institutes.[45] He included copies of the memos from Lewis and Seal, and added his own view of the protocol problem: "Contrary to Dr. Lewis' statement that scientific etiquette requires that these hybrids be made available to other workers, the hazards that may develop if they are introduced into the general population are too great to allow for the extention of such courtesy without very careful control." In this instance, then, the confrontation between established convention and potential hazards resulted in the convention giving way. Subsequent events were to show, however, that one person's convention or point of etiquette could be another's principle of free inquiry. Meanwhile, Berliner had extended the NIAID internal policy on the hybrid viruses to all the NIH's constituent institutes.

In early January 1973 arrangements went forward for establishing a more permanent policy on interlaboratory exchange of potentially hazardous viruses.[46] CDC Director David Sencer set up a meeting at his headquarters in Atlanta on 7 March 1973 with several NIAID scientists in attendance, including Lewis and Earl C. Chamberlayne, special assistant to Director Davis.[47] The meeting opened with a summary of the situation by Lewis. A larger context was supplied by W. Emmett Barkley, NCI's biohazards officer, who voiced similar fears about the many oncogenic viruses being studied in numerous research laboratories at that time. Although no definite proof existed that these agents were indeed human pathogens, it was agreed that the possibility of hazard was scientifically plausible and that therefore the risks should not be ignored. The problem, then, was how to establish a mechanism to eliminate or minimize this risk to investigators and the public.

It was the consensus of the meeting that whatever the final result, it should not place the burden of enforcement on the investigators themselves. There should be some system agreed upon that the investigators could follow, but it should be enforced and monitored from the outside. But who should this outside authority be? The suggestion that NIH should administer the control system because of its scientific expertise was rejected on the basis of its lack of authority outside NIH laboratories or those funded by NIH. This left only CDC with sufficiently broad authority to control these agents, an authority deriving from the Public Health Service Interstate and Foreign Quarantine Regulations.[48]

The necessary regulations would take some time to develop, however. In the interim, the group recommended that an alerting mechanism in the form of a Memorandum of Agreement be used to inform requestors about the dangers involved and to receive a commitment from them that the suggested precautions would be taken. Furthermore, their signatures on the memorandum, necessary before they could receive the virus from Lewis, committed them to require the same assurances from anyone else who asked for the agent from them.[49] (This was the forerunner of the Memorandum of Understanding and Agreement that later became part of the control mechanism in rDNA research.)

The dust from the violation of etiquette had apparently not settled, however. In a memo to Seal on 26 March 1973, the chief of the Laboratory of Biology of Viruses, Norman P. Salzman, questioned whether the information to be gained from the use of the hybrids, together with the political cost of doing it, was at all worth it.[50] Judging that much of the work could go forward without the use of the intact hybrid, he suggested that, as a policy decision, studies on these viruses should be discontinued: "The feeling of many scientists outside the NIH is that the real concern is not with safety, but rather that while safety is being invoked the real intent is to allow NIH scientists to use these viruses while at the same time other scientists are denied use of them."[51]

At the same time, CDC efforts to establish regulations were also running into trouble. The first thought was to incorporate the rules into existing regulations covering importation and interstate shipment of disease agents (42 CFR Parts 71, 156, and 72.25).[52] These two sections had needed amending for other reasons, and it was suggested that a new category be designated in the amended regulations to cover "non-naturally occurring agents (NNOA)." But a draft memo, signed by Chamberlayne and appended to a communication from CDC asking for Lewis's thoughts on the matter, already voiced some hesitation. If the requirements for advanced containment facilities already part of the Lewis plan were to be retained for all NNOA, it would essentially stop work on these agents, since few investigators had or could afford such equipment.

The use of the Memorandum of Understanding and Agreement, on the other hand, received strong support; in some cases, such support was conditional on the addition of a mechanism whereby studies could be stopped if it was found that the memo was not being adhered to. But before anything like this could be adopted, it was suggested that wider discussion in the total scientific community be held. One way to achieve this was by having the draft control systems discussed at an appropriate scientific meeting.

Drafts on the Memorandum of Understanding and Agreement (MUA) therefore proceeded, but the proposed regulations that would have put NNOA under CDC permit control only drew more criticism. NCI's biohazard officer Barkley wrote to Chamberlayne in late May indicating his concerns.[53] He pointed out that the worry was not that the adeno-SV40 hybrids were not naturally occurring agents, but that they represented a genetic recombination between a common human pathogen and a tumor virus. Many NNOAs were probably not very dangerous, however, and any attempt to put a blanket control over all of these agents should be discouraged. The decision to regulate these agents should be done on a case-by-case basis; and for those needing regulation, the MUA mechanism, administered through CDC, should suffice.

Apparently this was enough to make the CDC lose interest in the matter.[54] This left the Memorandum of Understanding and Agreement in NIAID hands. As finally drafted, it was composed of two main parts, an Understanding and an Agreement.[55] The Understanding summed up the case for control of these hybrids:

> *Understanding*
>
> The nondefective Adenovirus 2-SV40 hybrids are virions containing varying portions of the early SV40 genome as an integral part of Adenovirus 2 DNA. Thus they are genetic recombinants between a common human pathogen and a simian virus which is not only a well established carcinogen capable of transforming human cells, but is also associated with chronic central nervous system disease in humans. Although there is no known case of human infection with a nondefective Adenovirus 2-SV40 hybrid, there is no reason not to assume that these viruses are capable of infecting humans and producing endemic disease in the general population. Due to the chronicity of Adenovirus 2 infection of the adenoids and tonsils of children and the unknown consequences of such infections by virions carrying segments of SV40 DNA, the noninfective Adenovirus 2-SV40 must be considered potential biohazards.

There followed a detailed Agreement section that specified the technical precautions to be taken.

The legal basis for the MUA mechanism was, in fact, in doubt. In truth, it was primarily a document that was morally, not legally, binding.[56] It became "official" NIAID policy to require that such a MUA be signed on 16 October 1973 and NIH policy about a year after that.[57] But even before these dates, the special problem of the Lewis hybrid viruses was being eclipsed by the sudden awareness by larger segments of the scientific community that the problem was much more general. A memo from the chairman of the NIH Biohazards Committee to Deputy Director Berliner, reviewing its work since its inception, explicitly recognized the connection between the special concerns over the Lewis hybrids and the emerging concerns over the new rDNA technology. Commenting on events of barely a month before, the memo noted, "[A]t a recent Gordon Conference on Nucleic Acids a special session was held and a letter drafted to NSF to raise the scientific communities' concern with the total lack of regulation in the handling of hybrid viruses. One of the scientific outcomes of that meeting was the isolation of a large number of new DNA restriction enzymes which will greatly facilitate the creation of all sorts of new hybrid viruses."[58] The various pathways to the rDNA controversy were beginning to converge with startling rapidity and force.

4

The First Asilomar Conference

In 1971 Congress passed the National Cancer Act as the keystone of an initiative to eradicate cancer within the decade. Many of its supporters expressed a hope that with a full mobilization of skills and capacities in the country's research institutions, we could achieve in the field of cancer what the space program was able to achieve in extraterrestrial explorations. Not only was that expectation not met, but ironically, there was a steady rise in the incidence of cancer of "all sites" from 1970 to 1975 as the funding for cancer research was dramatically increased.[1]

To implement its war against cancer, Congress established the National Cancer Program and designated the National Cancer Institute (NCI) to manage the program. NCI was established in 1937 with the passage of the original National Cancer Act. It was the first of the eleven National Institutes of Health.

For carrying out the objectives of the National Cancer Program, NCI received substantial increases in funding. When it was initiated in 1937, its budget was around $400,000. In 1970, just prior to the passage of the new act, the NCI budget reached $181 million. By fiscal year 1979, NCI had a budget of more than $900 million, around 30 percent of the entire NIH budget.

The Viral Oncology Program (VOP) of NCI was established to investigate the viral etiology of cancer. Early in the program's history, its emphasis was on the search for a human cancer virus. It was initially conjectured that a discrete infectious entity caused malignancies in humans in the same fashion that certain viruses caused tumors in lower animals. As scientists were unable to make definitive connections between viruses and human cancer, the conventional wisdom was modified.

> Virologists are now concentrating upon looking in normal and malignant cells for pieces of viral genetic information which may code for proteins that trigger malignancy, and sophisticated efforts are underway to isolate and analyze these gene products.

Despite the enormous scientific advances made in recent years, the precise mechanism by which a cell becomes malignant remains an enigma. However, there is reason to believe that viruses or viral genetic information may be involved in most malignancies. Where appropriate detection systems have been developed, the carcinogenic activity of chemical and physical agents has often been found to accompany the activation of a latent viral genome.[2]

Thus, viruses were being considered as a necessary but not a sufficient cause of human cancer. In addition to the role viruses were theorized to play in the etiology of cancer, they were also regarded as tools for investigating the molecular processes by which normal cells become transformed into malignant ones. Animal tumor viruses were discovered to be useful probes since they could infect a cell and integrate their DNA into the cellular chromosome.

It was the cell-transforming characteristics of SV40 that made Paul Berg hopeful that it could be used as a vector for bringing new genes into eukaryotic cells. Thus, we begin to see a convergence of the interests in restriction enzymes and animal tumor viruses.

With the substantial increases in funding for work in tumor virology, there was a growing demand from laboratories for animal tumor viruses as model systems or as research probes. For example, from 1965 to 1977, the Viral Oncology Program witnessed a fourfold in-

Table 1
Funding history of the Viral Oncology Program

Fiscal year	Funding (millions of dollars)
1965	15.8
1966	18.6
1967	18.9
1968	19.7
1969	21.1
1970	20.9
1971	36.3
1972	48.9
1973	50.5
1974	57.9
1975	60.2
1976	59.9
1977	60.2

Source: *Annual Report*, Viral Oncology Program, DCCP-NCI, 1977.

crease in its funding from $15 million to $60 million. The largest increases came in fiscal year 1971 (74 percent), and in fiscal year 1972 (35 percent); see table 1.[3]

In 1971 Paul Berg proposed to construct a hybrid molecule comprised of a segment of a bacterial virus (bacteriophage lambda) with some additional bacterial genes attached (genes that code for galactose-metabolizing enzymes) and an animal cell virus (SV40). The hybrid molecule would, it was hoped, enter bacterial cells or animal cells.

Four years after Berg decided to postpone his experiment in which the SV40 hybrid was created, he discussed the reasoning behind his decision in an interview with MIT's Oral History Program.[4] He cited three benefits that the experiment in question would have yielded. They can all be characterized as benefits of an intellectual nature. (1) The techniques being tried are valuable in and of themselves. While this point is not elaborated in this interview, it is implied that there are things to be learned from the successful creation of the hybrid molecules and their implantation into cells. (2) The techniques would enable scientists to determine whether bacterial genes can be expressed in animal cells. (For this achievement, the bacterial genes must first gain entrance to the cells, and second, become integrated into the cells' DNA.) (3) The technique would enable scientists to determine whether, by introducing the animal virus genes into the bacteria, those genes can be expressed.

In his retrospective look at the proposed experiment, Berg distinguished the potential risk for objectives (2) and (3). He reasoned that there is no hazard in introducing bacterial genes into animal cells in culture through the vehicle of SV40 beyond the hazards in working with SV40 itself. Since cells in culture do not propagate by themselves, they could not spread into the environment. However transmitting animal DNA to bacteria is something quite different. Ordinary bacteria can replicate quite easily; and with each progeny strain (clone), the implanted DNA is reproduced.

The principal reason for doing the experiment, according to Berg, was to get the tumor virus genes to express themselves in the bacteria. If the genes could not be expressed, then most of the value for the experiment would be lost. At the time he proposed the experiment, Berg was doubtful about the possibility of such gene expression. He also felt that there was a small but finite probability that the bacteria would take up the tumor virus genes, spread the genes to animals and humans, and then express themselves. Berg, in effect, made a risk-benefit judgment. Referring to his experiment, he recalled the following line of reasoning: "[W]e're not sure it will pay off; it does carry some risk which we can't assess; let's hold off doing the experiment and find out whether the benefit from the experiment is going to be high enough to match that kind of risk."[5]

Thus, Berg gave some attention to the balance of risks and benefits, albeit in a qualitative fashion. Moreover, the benefits he cited were directed exclusively at basic science. The following outline is my reconstruction of Berg's "payoff matrix" decision:

A1. Proceed

B1. Risks: low but finite probability of human infectivity of SV40 cancer virus; cannot assess the risk; qualitative intuitive estimate—low probability of large consequences; all risks are hypothetical.

B2. Benefits: new knowledge; fundamental technique for researchers; some doubts whether benefits can be achieved; benefits minimized if animal genes cannot be expressed in bacteria; secondary benefits more visualizable—ability to clone genes so they could be produced in large quantities.

A2. Delay

B1. Risks: slow research down; no estimates on effect on science; set negative precedent.

B2. Benefits: eliminate small but finite probability of infectivity of population by SV40; raise consciousness of investigators on biohazards; undertake fact gathering on risks; assuage concerned critics who thought the experiment too dangerous to perform.

Once Berg made the decision to postpone the experiment, the next logical step was to acquire some concrete information about the hazards of SV40. Once available, that information could put to rest the hypothetical scenarios that were raised about an SV40 cancer outbreak. Berg and others began thinking of organizing conferences that would address these conjectured biohazards.

[W]hat I realized is that there was a whole question of biohazards that many of us were not facing up to, and they needed discussion. And so I and a group of people decided that we ought to have two conferences to deal with the problem of biohazards; the first one would be a fact-gathering meeting at which we would discuss what is known about the potential biohazards of tumor viruses; then, following that a second conference to discuss the kinds of experiments that people are doing, whether they pose any hazard, how you would find out if they posed a hazard, what we would do while we were waiting?[6]

Berg helped to organize the first meeting, which was held on 22–24 January 1973 at the Asilomar Conference Center at Pacific Grove, California. A second risk assessment conference was held at the same location two years later.

The focus of Asilomar I was on viruses, especially those that are capable of producing infection in humans or tumors in animals. The proceedings of the conference were published by the Cold Spring Harbor Laboratory under the title *Biohazards in Biological Research*.[7] The 100 or so participants who attended were exclusively from the United States.

In addition to Paul Berg, who played a key role in initiating the idea of such a conference, the three individuals who did the organizational work are Robert Pollack (Cold Spring Harbor Laboratory), Michael Oxman (Harvard Medical School), and Al Hellman (National Cancer Institute). The conference was supported jointly by NSF and NCI. Of the many participants who attended, those who would play active roles in the rDNA debate included David Baltimore, Emmett W. Barkley, Andrew Lewis, Herman Lewis, Wallace Rowe, James Watson, and Norton Zinder.

The conference was designed with two goals in mind: to gather the current state of knowledge on the biohazards of viruses, and to develop recommendations on how experimental work should proceed. The conference proceedings made no mention of rDNA work in general or restriction enzymes in particular. Consequently, the idea of breaking up virus DNA and transplanting segments into bacterial cells was not an item on the agenda.

There were two talks given that examined the risks of SV40 on human populations, one by R. N. Hull and the other by Andrew Lewis.[8] According to Hull, humans are susceptible to infection by SV40, but the consequences of that infection could not be assessed at that time. He recommended that SV40 be handled with great care. Lewis pointed out that while there has been no evidence of overt disease in infants or adults infected with SV40, no long-term studies have been undertaken on infected populations. By the year 1963, it became possible to carry out such studies, since millions of people were injected with polio vaccine contaminated with SV40. The risks could not be ruled out, according to Lewis. "Until satisfactory studies evaluate the long-term effects of SV40 infection in humans and clarify the relationship between SV40 and SV40-related agents to chronic degenerative central nervous system diseases in humans, it appears to this reviewer that the laboratory manipulation of SV40 involves some risks."[9]

The value of long-term epidemiological studies was reinforced by another speaker, Philip Cole, who delivered a talk entitled "Epidemiological studies and surveillance of human cancer among personnel of virus laboratories." Cole recommended both retrospective and prospective studies of laboratory personnel who have been or will be exposed to tumor viruses. He cited two factors responsible for the uncertainties about the risks of oncogenic viruses. First, cancer of any given site is an uncommon occurrence; therefore it may be difficult to pick out a rise in the incidence among laboratory personnel. Second, the latency period of the disease among adults is in decades.

Let us examine how the participants dealt with the information on risk that they received. What value issues were at the basis of specific responses to the problem? What types of conflicts were evident during interchanges among participants? What approaches were advanced for dealing with the uncertain risks of oncogenic viruses?

At an evaluation session of the Asilomar I meeting (attended by Baltimore, Berg, Lewis, Pollack, Zinder, and others), the following points were selected as a summary of what was learned from the meeting:

1. Neither the safety nor the danger of the materials discussed has been proved.
2. The laboratory community must be used as an experimental population for epidemiological analysis.
3. A major commitment for safety equipment must be made.
4. There must be a commitment to the education of laboratory personnel.
5. A registry of the health indices and laboratory experiences of all lab personnel should be established.
6. Informal consent statements should be required from grantees and employees.

In many ways this biohazards conference is a precursor of things to come. Notwithstanding the fact that the discussion weighed heavily on viral agents, the form of the problems is similar to what we witness two years later at the international biohazards conference on rDNA molecules. The major issue is how to proceed when risks are unproved and uncertain.

The first Asilomar conference was attended by individuals from a broad range of disciplines within the biological and health sciences. There are epidemiologists, experimental pathologists, pharmacologists, conventional microbiologists, and veterinary microbiologists. Participants were affiliated with cancer research centers, departments of pediatrics, human genetics, biochemistry, and molecular biology. There was also representation from the NIH, the NCI, and the CDC. The final recommendations were weighted heavily in favor of the interests of those who were concerned about the long-term effects of tumor-causing viruses on human populations. There were no recommendations to limit research or delay or proscribe certain experiments. The emphasis was on getting more information, for example, building a registry of tumor virus workers, informed consent, and keeping the decisions within the scientific community.

At one point during the conference, there was a discussion about the rights of laboratory personnel. Michael Oxman argued that once the risks and benefits of a line of research are known, a distinction should be made between the investigator, other laboratory workers including technicians and graduate students, as well as the "uninvolved and uninformed public at large"—including glassware workers, secretaries, and housekeeping personnel. Oxman proposed some limits on informed consent:

> Whereas an investigator may himself decide to assume certain risks, he does not have the right to make that decision for anyone else. In fact, it seems to me that the decision to assume a risk can only legitimately be made by the individual who will be in jeopardy. This principle of "informed consent" can readily be applied to other investigators, graduate students and technicians, as long as they are informed of the nature and magnitude of the risk and are free to decide whether or not they are willing to take it.[10]

The principle stated by Oxman was never recognized for its full significance in this conference. But more important, no principle resembling this is given any serious consideration by scientists as the debate moves from tumor viruses to rDNA molecules. Once the issues reached local communities, several had affirmed their right to informed consent.

The Oxman proposal calls for informed consent on the part of laboratory personnel when the risks are known. But what responsibility do scientists have to the personnel and the larger community when the risks are not known? Of what value is informed consent when there is no assessment? Do the groups that bear the risk deserve some role in the decision-making process? At this conference, these issues were crying out for some attention, but there was no receptive audience. Even the issue of "informed consent" was quite alien to scientific investigators. It might place laboratory technicians in the same category as experimental subjects. Several years after the first Asilomar conference, the issue of informed consent reemerges. When the NIH issued guidelines on rDNA research, critics asked; Who represented the laboratory technicians in the decision-making process? What opportunities did they have to offer comments on the guidelines? Harvard University took the criticism to heart and made one of the dissident technicians the university biohazards officer.

The accountability issue was brought out in another context by the Cambridge Experimentation Review Board—a panel of lay citizens in Cambridge, Massachusetts that was called upon to assess the hazards of the new research techniques to the city residents. Their report to the city manager stated, "Knowledge, whether for its own sake or for its potential benefits to human kind, cannot serve as a justification for introducing risks to the public unless an informed citizenry is willing to accept those risks."[11] The panel of citizens made it clear they wanted nothing less than representation in the decision-making process.

Even the progressive elements in the scientific community involved in the discussions over biohazards were not thinking beyond the laboratory workers. Oxman carefully circumscribed the limits of accountability: "[T]he principle of "informed consent" cannot realistically be applied to glassware washers, secretaries, housekeeping personnel, workers in adjacent laboratories or the public at large. Consequently we must insure that under no circumstances will these people be exposed to risks as a result of our research."[12] Obviously, the idea of informed consent, as conceptualized by Oxman, does not apply to nonspecific, peripatetic populations. To get beyond the confines of the laboratory, a new concept was needed: consent of the community. The distinction between what is inside and what is outside the laboratory does not hold up when we are dealing with self-propagating organisms; if the investigator is at risk, he will share it with the world outside the laboratory. Oxman acknowledged the dilemma: "Yet it may be difficult to limit the risk to the investigator himself, for if he becomes infected with an agent from the laboratory he may transmit it to others. Thus, for example, several of the victims of Marburg virus disease were hospital personnel who acquired the disease from infected laboratory workers for whom they were caring."[13]

Let us formulate the dilemma more concisely:

1. If the laboratory worker is at risk, then the community outside the laboratory is at risk.
2. The laboratory worker should be given the right of informed consent for investigations involving risks to his health.
3. The community outside the laboratory shall not have the right of informed consent.

We can easily generate a contradiction if we add a statement (4).

4. Any constituency at risk shall have the right of informed consent.

One way out of the dilemma is to show that statement (1) is false; investigators may bear a finite risk that is not transmitted to the community. This cannot be ruled out a priori since certain agents may infect by contact and not have a capacity for secondary infections. Of course, without such knoweldge there is no resolution of the dilemma.

Organisms that lack the capacity to infect, to be transformed into a pathogen, or, more simply, to survive outside of a very restricted artificial environment, are said to offer a biological barrier for research. The application of the concept to the uncertainties raised about the creation of rDNA molecules was considered of critical importance in reaching closure on some issues. The idea of such a barrier is said to have emerged from the second Asilomar conference. It was Sydney Brenner of England's Medical Research Council who was the leading spokesman for a disarmed microbe that could host DNA transplants. Brenner had recommended the concept of a biological barrier to the British rDNA committee (Ashby Committee). But while Brenner had gained recognition for his persuasive presentation of this strategy, it was being discussed in a different context prior to the 1975 Asilomar II meeting.

It is noteworthy that the distinction between physical and biological barriers was raised at Asilomar I and did not appear for the first time in connection with gene-splicing research. Two types of biological barriers were discussed at the first Asilomar meeting. When certain cells lack the receptors for specific viruses, those viruses cannot infect the cells. The lack of receptivity to viruses is the first type of barrier. The second barrier mentioned is the immunization of the investigator. Physical containment of organisms was generally conceded to be unreliable. Oxman made a telling point at this meeting that will be echoed many times in subsequent years: "Most investigators experienced in handling agents pathogenic for man have already come to recognize that biological barriers are far more reliable than physical ones."[14] Failing to derive complete confidence from a single containment strategy, scientists were on their way to considerations of biological and physical barriers used in concert.

Both the laws of economics and the laws of physics are inconsistent with the view that something can be created out of nothing. Promoting safety conditions in the laboratory is no exception. When there is a fixed sum for research and greater demands are made for improvements in safety conditions, the results are seen in fewer experiments performed. During one of the discussion periods at Asilomar I, the trade-off of safety against resources targeted for research was poignantly raised by Francis Black of the Yale University School of Medicine: "We are finally discussing the costs of the safety precautions and have come to the realization that they will inevitably reduce the number of grants available and increase the time required to reach our ultimate goal. If we do believe in our mission of trying to control cancer, it behooves us to accept some risk. Even if, as has been suggested, five or ten people were to lose their lives, this might be a small price for the number of lives that would be saved."[15]

Black continued by debunking the notion of absolute safety. Since we have no evidence of disease associated with virus research, he says, we should proceed rapidly to achieve our goal. That remark by Black brought some fiery replies. A participant from the NCI (R. Miller) did not accept the argument that the ends justify the means. He cited the stilbestrol case, where the administration of the drug to pregnant women resulted in the development of cancer in their daughters fifteen to twenty-five years later.

Since both the goal was uncertain (cancer cure) and the risks were uncertain, Black argued, it is, on the whole, rational to accept the uncertain risks for the long shot—a cancer cure. Black was not facing the issue brought out earlier by Oxman, namely, who was making the decision for whom. The issue was brought up again by James Watson: "I'm afraid I can't accept the five to ten deaths as easily as my colleagues across the aisle. They could easily involve people in no sense connected with the experimental work, and most certainly not with the recognition and fame which would go to the person or group that shows a given virus to be the cause of a human cancer."[16]

Watson put his finger on the key dilemma in risk-benefit analysis: Who gets the benefits and who bears the risks? There is a question of fairness that cannot be avoided. Should those who are likely to reap the greatest rewards alone decide on the appropriate course between hypothetical risks and hypothetical benefits? Where are the risk takers in the decision matrix?

Robert Pollack, one of the three organizers at Asilomar I, recalls an interchange between Watson and someone from NCI. The issue was over funding for biosafety. Watson realized that for the safety concerns to have any teeth, there must be sufficient funding. It was an especially sensitive issue for the young investigator who did not have a well-endowed laboratory.

> He [Watson] got into a typical screaming fit with somebody from the NCI, who, of course, was the guy who signs off on the grant to Cold Spring Harbor, where the guy got up and gave a perfectly sound political nonsense talk about how the NCI supports everything that's been said, pro and con. Watson got up and said, "If you're not prepared to back that with fifty grand per grant to a young person and two hundred thousand to an older person, then what are you talking about? Sit down! You know, you're just talking nonsense. You're playing with people's futures if you tell them that they should be working safely and you don't have the money for safety.[17]

Looking several years beyond Asilomar I, when the panel of Cambridge citizens reviewed the potential biohazards of genetic research in the city, the very issue of financing safety was raised. The Cambridge Citizens' Committee recommended to the NIH that with each grant proposal, some funding be allotted to laboratory safety and health monitoring. NIH took the suggestion under advisement.

The one unanimous proposal that came out of Asilomar I was the idea of an epidemiological study. Berg promoted the idea forcefully in his closing remarks. The next step was to follow this meeting up with another that could lay the groundwork for such studies. But as other issues involving the human manipulation of genes across diverse species lines became a principal area of concern, a follow-up conference to Asilomar I never materialized, and neither did the agenda for strategic epidemiological studies.

5

The Gordon Conference Letter

The Gordon Conferences are hardly the place one would expect to find discussions of social responsibility in science. Yet, because of a unique set of circumstances, many of which have already been discussed, and an unusual set of personalities, the Gordon Conference on Nucleic Acids that met in mid-June 1973 was the event that many consider the birth of the rDNA controversy. For example, in its chronology of the events associated with the policy-formation process for rDNA molecules, the US Department of Health, Education and Welfare lists the Gordon Conference as the first citation.[1]

The goals of the Gordon Conferences have been described as follows: "The conferences were established to stimulate research in universities, research foundations, and industrial laboratories. The purpose is achieved by an informal type of meeting consisting of scheduled speakers and discussion groups."[2] These meetings are viewed as an alternative to the more traditional channels of scientific communication. A strong emphasis is placed on informal discussions.

Scientific research progresses through two opposing tensions. On one side we find scientists mutually benefiting from the exchange of ideas both within and among disciplines. Scientists develop informal networks and affinity groups beyond their major professional associations to promote the cross fertilization. These professional groupings within the scientific community have been termed "invisible colleges."[3]

The other tension is the competitive character of scientific work. Research agendas cannot be fully revealed; otherwise, they may be picked up by other investigators seeking to publish first. Ideas are at a premium in science generally, and especially in biological research. Many of the laboratories do not require substantial expenditures of capital, as in, say, physics; new ideas are guarded by investigators since the decentralization of laboratories makes intellectual pilfering a common fear of laboratory researchers.

The organizers of the Gordon Conferences recognize these tensions. A stated goal is that the rights of individuals to guard their ideas be protected. "[I]t is an established requirement of each conference that no information presented is to be used without authorization of the individual making the contribution, whether in formal presentation or in discussion."[4] Thus, the recording of lectures or photographing of slides is prohibited. There are no proceedings resulting from the conferences. However, the formal statements about confidentiality do not reflect what actually occurs at the meetings. Participants take notes and breakthroughs get reported through informal networks rather efficiently, even without published papers. The aura of confidentiality is a means by which scientists can stake out their territories. When scientific results are released at a Gordon Conference before papers are formally published, this provides a signal to others who may have contemplated undertaking a similar study.

According to the official literature, attendance at the Gordon Conferences is limited to about 100 people. A committee is set up for each conference. Its members choose participants from among the applications. The cochairpersons of the conference on nucleic acids were Dieter Söll from Yale University's Department of Molecular Biophysics and Biochemistry and Maxine Singer from the National Institute of Allergy and Infectious Diseases, NIH. The number of registrants for this conference reached 143. There were eleven countries represented, but the overriding majority of registrants were from the United States. The breakdown of registrants by country was as follows:

United States	114	Japan	5	Scotland	1
England	7	Australia	1	Sweden	1
Germany	6	India	1	Switzerland	1
Canada	5	France	1		

The chemistry of nucleic acids was a subject that drew interest from diverse sectors of the scientific community. This is reflected in the preliminary registration list for the conference by academic discipline. (Whereas the preliminary registration list was 143, approximately 130 people actually attended; see table 2.)

The program was organized with two panels on each day, starting on Monday, June 11. By Wednesday, June 13, there was nothing that distinguished the conference from any other scientific meeting. On Thursday morning, June 14, there was a session on "Bacterial enzymes in the analysis of DNA." Daniel Nathans of Johns Hopkins University chaired the meeting. Nathans had been experimenting with restriction enzymes two years prior to the Gordon Conference.

Table 2
Gordon Conference registration list by discipline and affiliation, 1973

Discipline, affiliation	Number registered
I. Department at university, including medical schools and institutes and schools of public health	
Chemistry	16
Biology	16
Biochemistry	23
Molecular biology	9
Radiological sciences	1
Molecular genetics	1
Physiological chemistry	
Bacteriology and immunology	1
Microbiology	3
Pharmaceutical sciences	2
Zoology	1
Molecular biochemistry and biophysics	5
Biophysics	1
Human genetics	1
Institutes	3
Unspecified	9
II. Private industry	8
III. Nonprofit private institutions	10
IV. National or state institutions	17
V. Affiliation uncertain	14

He already completed an analysis of the SV40 genome with the use of restriction enzymes.

Another key personality at the session was Herbert Boyer. According to the account of John Lear in his book *Recombinant DNA: The Untold Story*, Boyer, situated at the University of California at San Francisco, was collaborating with Stanley Cohen of Stanford University. The two scientists had recently developed a way of implanting DNA into bacteria by fusing it onto a plasmid, thereby creating an rDNA hybrid molecule. Lear reports that Cohen and Boyer agreed not to discuss their results until they were published. However, at the Gordon meeting Boyer did discuss the progress of his and Cohen's work. Two restriction enzymes were mentioned in Boyer's talk (Eco RI and RII). He spoke about using restriction enzymes for joining two different drug-resistant plasmids, pSC101 and R6-5.[5]

Some scientists who listened to the talk began to realize that the time had come when the tools of gene splicing had become available to many investigators, not just a few leaders in the field. Edward Ziff and Paul Sedat were working at that time in molecular biology at Cambridge, England. They brought the issues of Boyer's talk to Maxine Singer's attention and expressed their concerns about the use of restriction enzymes to produce hybrid DNA molecules. Singer recalls having some reservations in these early informal discussions that extensive debates might lead to a restriction of the research.[6]

Maxine Singer was interested in the issues of science and ethics before she was invited to chair the Gordon Conference. Both she and her lawyer husband were members of the Federation of American Scientists. Her husband Daniel Singer was affiliated with the Hastings Institute—a private, nonprofit organization that engages in research and develops symposia on issues of biomedical ethics. Daniel Singer had lectured on the subject of the social responsibility of science. In a talk (circa 1970) to the Society of Sigma Chi—the scientific honor society—that was subsequently published in the *Congressional Record*,[7] he had outlined some of the problems society will have to face with the growth of biomedical research and genetics. Scientists, he had maintained, must confront the social and ethical implications of their work and draw the public in on the decisions that bear considerable weight on society:

> Furthermore, whether they like it or not, scientists are also part of the body politic—not above, below, or outside of it. Laymen, standing at a respectful distance, tend still to regard scientists with some degree of awe. At closer range, scientists are very much like people. And while scientists have a near monopoly on talent, they clearly have no monopoly on wisdom, insight, sensitivity, or other human values. But scientists constitute a growing fraction of the articulate population and what they do affects them and the rest of us. They cannot remain aloof and greet each new public problem and misapplication by abdication. They have a responsibility to become involved as citizens in the sometimes unpleasant business of public argument and decision-making.
>
> Last, I think that, for the health of the scientific community itself, scientists must become engaged now in public undertakings. If scientists—who continue to startle us with their damnable cleverness—cop out on the problems they create, the public may turn on them

and/or they will share the fate of physicists who (probably through no fault of their own) have for a generation been wringing their hands and donning hair shirts to expiate their guilt.[8]

Maxine Singer recalls that she was receptive to the issues of bioethics prior to the Gordon Conference. When put to the challenge of either ignoring the concerns of her colleagues or placing them on the agenda of the meeting, she chose the latter. In this instance, a few individuals altered the course of events. There were two scientists who wanted the biohazards issues raised and two chairpersons who could use their discretionary authority to introduce the subject matter in the formal part of the proceedings.

> [I]f Ziff and Sedat had come to Dieter [Söll] and to me and we had been totally unsympathetic, then clearly we could have done nothing. Because it would have been very simple to say, "Well, there's just not enough time for everybody to discuss it. There is no way we can undertake anything like that. It would be wrong." I must say I think there is some substance to that argument. One could make that argument and not be far from the mark. So I think it mattered a lot that both Dieter and I felt that it was a good thing to do.[9]

She and colleague Dieter Söll set aside a half-hour on Friday morning for a discussion of the new techniques of gene splicing.

There was precious little time available for discussion. Nevertheless, three concrete proposals were raised and voted upon. The suggestions were (1) a letter should be sent to the National Academy of Sciences and the National Academy of Medicine to set up an expert panel to study the biohazards; (2) each participant at the conference should do what he or she wished, but that no collective action should be taken; and (3) a group letter should be constructed and signed by all who wished to sign it.

When a vote was taken, seventy-eight people opted for a letter to be sent from the conference. By that time there were between ninety to ninety-two people left for the vote, still a substantial majority of those who attended the meeting. An amendment to the proposal was raised, calling for wider publicity of the letter. Forty-eight people voted for wider publicity and forty-two voted against it.

In a 1975 interview, Singer maintained that she had no reservations about her responsibility and felt comfortable that her first recommendation was the correct choice. However, the discussion of publicizing the letter drew wider polarization among the scientists. On the concerns over publicity, Singer remarked,

> The discussion was, as I recall it, that we had not ourselves discussed the substance of our concerns, that none of us had any way of knowing whether the concerns that we had were real, that none of us had devoted enough thought to them to know whether, in fact, there wasn't some reason why the whole concern was beside the point. And therefore, to raise in the public's mind the notion of possible dangers, when we ourselves had not given the kind of scientific thoughts to those possibilities, was premature and might not be productive. It might make some people worry and concerned about something about which there is no need to worry. And it's very hard to undo such things once you do them. And the notion that, were the public concerned about possible hazards of an unknown kind, of an unestimated kind, there might be restrictions that would follow unnecessarily.[10]

How were the lines drawn among scientists at the Gordon Conference? Maxine Singer had some impressionistic views on how the Gordon conference vote broke down along generational lines. In her July 1975 interview with MIT's Oral History Program, she expressed the opinion that younger scientists were more likely to raise broader socially oriented questions, with the exception of those individuals who were primarily involved in the research under discussion. The older scientists, although somewhat mixed in their reactions, according to Singer, were concerned about the possibility of restricting scientific research.

When the conference was over and Singer had an opportunity to reflect upon the significance of the vote, she conferred with Dieter Söll and they decided to send a draft of the letter to each participant at the conference. This was partly to get input from those who missed the vote. But it was also an opportunity to receive some textual comments. The letter they sent out on 21 June 1973 stated,

> The group voted by a majority of 78 out of about 95 in attendance (142 were enrolled at the Conference) to send a letter, from the Conference, to the Presidents of the National Academies of Science and Medicine. The remainder voted either for a similar letter but with individuals signing if they so chose, or individual action. In addition, a majority of those in attendance voted to try and publicize the letter in *Science* magazine. Because many participants had left by Friday, it was decided to send out a draft of the letter to everyone. It is enclosed. Please send any suggestions for revisions to me by July 15. Also, please indicate below your approval or disapproval and mail to me by the same date. I will assume that anyone remaining silent agrees with the majority and has no serious objection to the draft of the letter.[11]

Sixty-one responses were received from the inquiry, and there was unanimous approval for sending the letter to the presidents of the national academies. But on the question of publicity, the tally was forty in favor, twenty against, and one no opinion.

Once the responses were received, the letter was revised by Singer. With the approval of Dieter Söll, identical letters were sent out to Philip Handler of the National Academy of Sciences and David Hogness, president of the Institute of Medicine. Robert Marston, previously director of NIH and subsequently special consultant to the Institute of Medicine, received one of the copies. In addition, a copy of the letter was sent to *Science*, where it was published on 21 September 1973.

Thus far we have discussed the events at the Gordon Conference leading up to the letter to *Science*. Let us now look at some substantive issues. To what extent did biohazards dominate the concerns voiced by the participants? How did the key actors in the episode interpret their social responsibility? How deeply felt were the concerns that publicizing the letter would place the new scientific research program in some jeopardy? I begin by examining some key passages in the draft of the letter circulated among the Gordon Conference participants, and dated 21 June 1973—exactly three months before the final version of the letter appeared in *Science*. Maxine Singer included in the letter the remarks she made at the conference.

> We all share the excitement and enthusiasm of yesterday morning's speaker who pointed out that the scientific developments reported then would permit interesting experiments involving the linking together of a variety of DNA molecules. The cause of the excitement and enthusiasm is two-fold. First there is our fascination with an evolving understanding of these amazing molecules and their biological action and second, there is the idea that such manipulations may lead to useful tools for the alleviation of human health problems. Nevertheless, we are all aware that such experiments raise moral and ethical issues because of the potential hazards such molecules may engender. In fact, potential hazards exist in some of the viruses many of us are already studying. Other problems will arise with hybrid molecules we are contemplating. Furthermore, these hazards present problems to ourselves during our work and are potentially hazardous to the public.[12]

The previous remarks had identified biohazards as the exclusive issue of concern, and it is clear that Singer believes that scientists have a responsibility to protect the health and safety of their coworkers, other laboratory personnel, and the general public. But what is she getting at when she uses the expression "such experiments raise ethical and moral issues"? (More often than not the terms "moral" and "ethical" are used synonymously.) There is no reference to a moral dilemma, such as when a moral precept is being questioned; nor is there a clearly articulated ethical dilemma such as a conflict of rights. What is put forth as the ethical problem relates to the scientists' responsibility to the public. How do scientists balance their interest and perceived right to free inquiry against the potential laboratory and public health hazards of their work?

In the first draft of the letter to Philip Handler of the National Academy of Sciences, there were two key statements that subsequently drew criticism and comment from participants: (1) "[W]e presently have the technical ability to join together, covalently, DNA molecules from diverse sources. . . . In this way new kinds of viruses with biological activity of unpredictable nature may eventually be created." (2) "Certain of these hybrid molecules are potentially hazardous to both laboratory workers and the public."

The use of the term "unpredictable" is important in this context. As the debate progresses, there will be quite a divergence of opinion on what can and cannot happen when DNA is shifted among different species. Is it possible to generate viruses with substantially greater infective properties? Can a piece of a virus be more hazardous in such experiments than the whole virus?

Another interesting locution is the term "potentially hazardous," which is worthy of more careful scrutiny. The term is ambiguous, and that issue was addressed by written comments. Here are some examples of how the statement "X is potentially hazardous to P" may be interpreted.

1. X can harm P under conditions $C_1, \ldots, C_n$.
2. It is not known that X cannot harm P under some set of conditions or another.
3. There is some evidence that X may be harmful to P.
4. There is a finite probability that X can harm P.
5. There is a posited scenario of events such that X harms P, where the scenario has neither been confirmed nor disproved.

In his comments on the draft letter, Paul Berg wanted to soften the term "are potentially hazardous" and replace it with the expression "*possibility* that such hybrid molecules *could* be hazardous" (emphasis is mine). When the debate reached the public arena several years after the Gordon episode, scientists found it necessary to translate their notion of "potentially dangerous" to the public; some of the above formulations were debunked. In order to make the point, comparisons were stated: for example, the probability of creating a dangerous pathogen is less than that of a meteor hitting the Prudential Tower in Boston. The impact that the generation of scenarios had on the debate will become increasingly evident; many of the scenarios were not easily refutable. The lay public had difficulty determining which scenarios were reasonable and which were extraordinarily unlikely.

Scientists agreed that they could not place any rating on the potential hazards of gene-splicing experiments. But there were others at the Gordon Conference who contended that the hazards were small but admittedly finite.[13]

The full text of the letter that appeared in *Science* was carefully worded and only comprehensible to scientists in the field. Expressions like "overlapping sequence homologies at the termini of different DNA molecules" or "Watson Crick hydrogen bonding" clearly defined its clientele. It was, after all, written to the president of the National Academy of Sciences. The key elements of the letter are *description of the technique in biochemical terms; some things that the technique can achieve* (combining DNA from animal viruses with bacterial DNA, combining DNA from different viruses); *some potential risks* (new kinds of hybrid plasmids or viruses with unpredictable biological properties, hybrid organisms that may prove hazardous to laboratory workers and to the public, large-scale preparation of animal viruses); *some benefits* (advancing knowledge of fundamental biological processes, alleviation of human health problems); *recommendations* (National Academy of Sciences and the Institute of Medicine should establish a study committee).

Edward Ziff, one of the two scientists who urged the chairpersons at the Gordon Conference to set time aside for discussion of biohazards, explained his concerns in the *New Scientist*, 25 October 1973.[14] Ziff's argument in this essay is based upon two ideas: (1) the production of DNA hybrids; (2) large-scale production and isolation of potential cancer-causing viruses. Argument I is built upon the following propositions:

1. There is an unknown risk to humans of certain animal viruses (i.e., tumor-causing (oncogenic) viruses).
2. Creating DNA hybrids from oncogenic viral DNA will increase the risks to laboratory personnel of contracting cancer from tumor viruses.

The cited evidence for (I.1) is that some viral DNAs can transform cells in tissue culture to a malignant state. For statement (I.2), Ziff maintained that DNA hybrids will (a) increase the chance of human exposure or (b) provide new avenues for transmission of viruses. Thus, by grafting a piece of oncogenic virus DNA onto an organism that can reside in humans, the potentially hazardous DNA will have an easier access route to human systems. The second point mentioned is that the new hybrid may be capable of replicating more easily in the environment.

In argument II Ziff contended that the scale of production and the concentrations of the materials used will increase the risks.

1. There is an unknown risk to humans of certain animal viruses.
2. The risk will be increased if large quantities of the virus are produced in concentrated form.

For statement (II.2) Ziff cited the special problems of the formation of aerosols. He also emphasized the long latency period between exposure and the expression of cancer, making it difficult to measure the risks.

Scientists like Ziff were viewed as progressive members of their profession. He was calling for the establishment of better containment facilities for doing the research. This meant additional research dollars targeted to safety. But there seemed to be unanimous agreement about one thing: If there was anything to be done about improving the standards of safety, it was a matter for the scientists and their funding agencies. Ziff reinforced this point in the last sentence of his *New Scientist* essay: "But until such time as reasonable guidelines are established, rigorous discussion and debate within the scientific community will continue to provide the greatest safeguard of the public good."[15]

It is important to note that neither the Singer-Söll letter nor the Ziff article cited the applications of the techniques beyond their use as intellectual tools and their potential clinical benefits to society. Science writer Bernard Dixon spoke about the coming of age of genetic engineering in the same issue of *New Scientist* in which Ziff's essay appeared. Discussing the method of producing DNA hybrids, Dixon said,

> This technique might be used to repair metabolic defects caused by faulty or missing genes. As a means of treating, at the most fundamental level possible, certain hereditary diseases, this would mean the arrival of the long heralded science of "genetic engineering." Other prospects are less welcome. DNA hybridization must also look an attractive proposition for biological warfare researchers (who are, of course, still about their business, despite recent gestures toward biological disarmament). The new technique offers the prospect of fabricating even nastier BW agents, facilitating the combination of "desirable characteristics" that cannot be brought together by conventional microbial genetics.[16]

These remarks by Dixon reflect a trend that will continue as the controversy progresses, namely: the discussions over the impact of the new techniques for the genetic manipulation of man were carried out mainly by people outside of the profession—writers, environmentalists, ethicists and the like, with a sprinkling of scientists whose voices in the distance were ignored.

6

A Voluntary Moratorium

The Gordon Conference letter, cosigned by Maxine Singer and Dieter Söll, was received by Philip Handler, president of the National Academy of Sciences (NAS). Handler referred the letter to the Assembly of Life Sciences (ALS), a section of the NAS. The executive committee of ALS invited Maxine Singer to present her views on the issues raised in the letter. When Singer was queried about individuals who could head a study panel, she offered Paul Berg's name. Berg's role in organizing the biohazards conference at Asilomar in 1973 made him a credible candidate for this task. It seemed appropriate to select an individual who combined scientific competence in the new gene-splicing techniques with some knowledge of biohazards.

When first contacted by NAS to lead a study panel, Berg requested time to think about it. During a visit to Harvard, where he attended a seminar, Berg discussed the NAS proposal with James Watson and John Edsall. He described Watson's response to the follow-up on the Singer-Söll letter as supportive.[1] There was a time when Watson became very concerned about the potential hazards of some virus research being contemplated by a colleague at Harvard. Watson is said to have publicly threatened to get a court injunction to block the experiment.[2]

When Berg had begun to take seriously the concerns raised about the hazards of his proposed SV40-lambda phage experiment, he had called upon David Baltimore for advice. Baltimore's recommendation at the time was to postpone the experiment until there was more knowledge about the risks. Once again Berg contacted Baltimore, only on this occasion to ask him whether he would participate on the NAS study panel. Baltimore already had a track record on social issues in science. He had been an advocate for ending research in biological warfare and supported an international treaty to respond to the problem.

In February 1974 Richard Roblin was preparing to talk for the Biophysical Society's ethics committee. Roblin, a member of the Council for Biology in Human Affairs, was interested in the impacts of biology on society, especially genetic engineering. He already had published an essay in *Science* outlining a responsible course of action for human gene therapy.[3] When Roblin read the Singer-Söll letter in *Science*, he decided to make that issue the basis for his talk. After inquiring into NAS's response to the concerns raised in the letter, Roblin was directed to speak with Paul Berg.

When Roblin was asked during their conversations to suggest people to serve on the panel, the names he came up with were Leon Kass and Jonathan Beckwith. The study panel, he felt, should have some representation on it of the more "radical" and comprehensive views. Beckwith and Kass were primarily concerned about the social uses of science, while Berg's thinking had focused exclusively on biohazards. Eventually, when Berg received the green light from the NAS to convene a study panel, he invited Roblin to participate.

The first meeting of this free-standing committee took place at MIT on 17 April 1974. At that point, the NAS had not yet conferred official status upon the panel. The academy took a cautious approach and waited for the committee's recommendations before placing its own prestige behind it. In addition to Baltimore, Berg, and Roblin, the other participants at the all-day meeting were James Watson, Sherman Weissman, Herman Lewis, Daniel Nathans, and Norton Zinder. Berg outlined the agenda for the panel in premeeting correspondence. The stated goal was "[t]o consider whether or not there is a *serious* problem growing out of present and projected experiments involving the construction of hybrid DNA molecules *in vitro*. If a problem exists, then what can be done about it, both short and long-term actions."[4]

The idea of a moratorium on certain types of experiments was being considered by Baltimore prior to the April meeting. An important concern of Baltimore was whether this small but distinguished group of scientists could effect a moratorium. There were no precedents to go by, and therefore some doubted that other members of the scientific community would embrace the standards of social responsibility set by this panel.

When the meeting finally got underway, Baltimore described the atmosphere as one filled with honest skepticism:

We met here at MIT on April of this [1974] year. And we sat around for the day and said, "How bad does the situation look?" And the answer that most of us came up with was, that the potentials for just simple scenarios that you could write down on paper were frightening enough for certain kinds of limited experiments using this technology, we didn't want to see them done at all. And these in fact were experiments that all of us would like to be doing.[5]

Roblin came to the meeting expecting to get filled in on the details of the rDNA techniques. He soon began to realize that the experiments were far less difficult to do than he had first anticipated. The relative simplicity of the process and the speed in which this new field of research was developing was a major concern for Roblin and others who attended. As a result of discoveries like those of Stanley Cohen, by which plasmid vectors were introduced, gene manipulation had become widely available. Thus, there was some uneasiness expressed by this "scientific elite" about the lack of respect that the ordinary bench scientist might have for the potential hazards of this powerful technique.

There were differences among committee members—particularly on two issues. The first was the question of deferring certain types of experiments. The second related to the general class of experiments in which animal DNA fragments were inserted into bacteria. Roblin recalls that some individuals were opposed to a moratorium purely on philosophical grounds. For them the idea of free scientific inquiry was an absolute, and constraints, whether voluntary or mandatory, were a transgression of this inalienable right. Roblin presented a different view. He was already familiar with the approaches taken to protect the rights of human subjects in clinical research. Thus, there was already a precedent established for a collective responsibility and restraint on some scientific work.

The statement eventually agreed upon by the participants was published in *Science*, *Nature*, and the *Proceedings of the National Academy of Sciences* (*PNAS*). Considerable care was given to the language in the statement. The participants were concerned that their motivations might be misconstrued. According to Roblin's account, some individuals wanted to avoid the word "moratorium" in the letter, on the grounds that it sounded "too officious." They wanted to project the idea that this was a very temporary thing—a pause, which could be best described by terms like "voluntary deferring." But the possibility for a long-term ban on certain work was kept as an option in the final statement, as indicated in the phrasing: "[A]dherence to our major recommendations will entail postponement or possibly abandonment of certain types of scientifically worthwhile experiments."[6]

As the meeting ended, consensus had been reached on the major points. Roblin put together a first draft of the statement. In May Baltimore showed the statement to the ALS committee. It was decided then that the Berg panel should have the full backing and prestige of the NAS behind its recommendation. The panel was given official NAS status and named Committee on Recombinant DNA Molecules, Assembly of Life Sciences.

There was still uncertainty about how the majority of scientists would receive the call for a voluntary moratorium. To test the waters, Baltimore read a draft of the statement at the Cold Spring Harbor Tumor Virus Meeting held in June 1974. One participant at that meeting, Theodore Friedman, recalls that the general response to the statement was not very critical. That was an encouraging sign in view of the fact that tumor virus research was clearly one of the key areas affected by the call for a moratorium.[7] However, there was concern expressed by some participants at the tumor virus workshop over publicizing the issue.

Before the letter was sent to *Science*, four other scientists who heard about the initiative offered to be cosignatories. These were Herbert Boyer, Stanley Cohen, Ronald Davis, and David Hogness. The final list of eleven signers (Herman Lewis was a participant of the April 17 meeting but was not among the eleven signers of the Berg letter) reads like a Who's Who in molecular biology and included those who played key roles in the development of the rDNA techniques.

Three classes of experiments were identified by the Berg committee as warranting special attention. The first two were recommended for postponement. For the third, deferral gave way to a modest "handle with care." Type I experiments were those for which two types of genes were linked to bacterial plasmids. Plasmids are the small circular extrachromosomal DNA that can be implanted in a bacterium and replicate in the bacterial host. The first class of genes is that which codes for antibiotic resistance factors; the second class codes for toxins.

The concern over the spread of antibiotic resistance strains of organisms had been mounting in the public-health community. The insertion of genes that confer this resistance into organisms that normally are sensitive to antibiotics and their release into the environment could exacerbate a growing public-health problem by compromising the effectiveness of clinical antibiotics. Similarly, creating and propagating bacteria with new toxins that could invade the ecosystem also signaled the possibility of a public-health menace.

Any uncertainty about the risks in this category involved the question whether transplanted genes express themselves in the new host, rather than the effect of the recombinant hybrids in human systems once the genes are expressed.

The second category of experiments consists of the linking together of DNA segments from a cancer-causing virus or other animal viruses. At first the concern was over tumor virus experiments. There was still no conclusive evidence pointing to a human hazard in the exposure of animal tumor viruses. Evidence presented to the 1973 biohazards conference at Asilomar showed that SV40 infected humans—antibodies to the virus are found among individuals who have worked with the virus; but there was a vast area of ignorance concerning whether such viruses, over a period of time, cause tumors or other diseases in humans.

In addition to the concerns raised about tumor viruses, this second class of experiments (type II experiments, as designated in the final statement of the committee) was broadened to include all animal viruses. Four months after the statement was released to the press outlining the decision of the study panel, David Baltimore at an MIT seminar provided the rationale for including animal viruses in this secondary category. Note the important point made by Baltimore: Scientists simply do not know when an animal virus can be tumor producing.

> Basically, what we're saying is any animal virus which can cause cancer, and we simply generalize that to any animal virus at all, *since it's not clear when they do and when they don't* [author's emphasis], the genes from animal viruses should not be linked into these bacterial plasmids because you could create a situation where a gene able to cause cancer is then made resident in the human intestinal tract inside a bacterium. And since bacteria [are] constantly lysing and being broken up in the intestinal tract, you could produce a chronic exposure to this kind of gene or gene product.[8]

There was less agreement among panel members on how to deal with type III experiments, which involved linking together, in a random way, fragments of animal DNA to a bacterial plasmid and then implanting the plasmid into the bacterium. The category defines a much wider class of potential rDNA experiments than the other two; consequently, a request for restraint in this matter was more likely to come up against strong opposition. The potential hazards visualized at that time for this third type of experiment can be expressed by the following argument:

1. Many types of animal cell DNA contain sequences similar to cancer-causing viruses.
2. A random recombination of those sequences [shotgun experiment] may include the DNA of a cancer-causing virus.
3. There are potential hazards to humans in implanting the DNA sequences of cancer-causing viruses into bacteria that can take up residence in the human intestinal tract.
4. Therefore, by shotgunning random sequences of animal DNA into coliform bacteria, we introduce potential risks to humans.

Since it was already known that viruses that are innocuous in one host can produce tumors in another, there was some justification for the second premise. But to call for a moratorium on these experiments undoubtedly would have affected so many investigators that the committee believed it would be threatening some of its other goals if it treated type III experiments like the other categories. Once again, Baltimore puts that decision into perspective. "But we felt, strategically, that the recommendations about the type one and the type two experiments were obvious enough that people would accept them without a . . . lot of deep thinking. Whereas the arguments about these other types, which have now become known as type three experiments, . . . were more convoluted, that the types of information talked about were vaguer, that we would have much more difficulty getting the statement accepted if we went strong on those than if we went weak on those."[9]

In Roblin's account of the meeting, the decision taken on the so-called type III experiments was a compromise. Some members of the committee wanted a deferral of those types of experiments as well. The justification advanced for separating type III experiments from the others was based upon two ideas. First, it was held that the probability of getting a dangerous DNA sequence by shotgunning a fragment of animal DNA into *E. coli* was much less than if one were to start out with a known oncogenic virus. Despite all the unknowns about the effect of tumor viruses on humans, rDNA work with these viruses was perceived to be orders of magnitude greater in risk than similar work with random sequences.

The second argument was based upon the notion that our concerns over risks should reflect our current knowledge and not our ignorance. There were some reasonable expectations about how a tumor virus might behave if placed into a bacterium, but that was not so with unknown DNA segments. Consequently, greater caution was warranted for the scenarios for which there was knowledge as compared with those that were more conjectural.[10]

The scientific orthodoxy in the rDNA debate looked for evidence that a particular scenario was plausible. It was not sufficient for these decision makers to pose a hypothetical situation illustrating a hazard that did not contradict any evidence. In a sense, these scientists were saying that it is incumbent on those who believe there are risks to show the plausibility of a hypothetical scenario. This is in contrast to the view that scientists bear the responsibility of proving that what they are doing is safe.

Before leaving this episode, which was to play an important part in rDNA policy, I shall summarize the uncertainties of gene splicing at this stage. (1) If a segment of eukaryotic DNA was linked to the genome of a bacterium by rDNA techniques, could that DNA be expressed? During the discussion of the Berg committee, James Watson raised the point that it is premature to discuss biohazards until it is known whether the foreign DNA is translated in its new host. (2) Do animal tumor viruses pose any threat to humans? (3) In randomly chopping up animal DNA and shotgunning the segments into *E. coli*, was there any risk of picking up a tumor virus?

The final statement of the Committee on Recombinant DNA Molecules of the National Academy of Sciences had four parts. First, it requested that experiments of types I and II be deferred until the hazards are better evaluated. Second, it wanted experiments in which animal DNA is linked to plasmids to be "carefully weighed." It was not clear what they should be weighed against. The statement gave no further guidance to this class of experiments. Third, the Berg panel requested that the director of NIH establish an advisory committee to accomplish three things: (a) oversee an experimental program of risk assessment; (b) develop procedures to prevent the spread of recombinant molecules; and (c) develop guidelines.

From a public-policy point of view, it seems plausible that getting the experimental evidence on risk assessment would precede the development of guidelines. But that sequence was not followed by the NIH. The initial set of guidelines was out before any risk assessment experiments were underway. The final recommendation called for an international meeting, for the purpose of exchanging scientific information related to the problems raised in the statement.

The term "prediction" is used twice in the Berg committee letter. What can or cannot be predicted becomes a key area of dispute in the controversy that ensues. It is, therefore, important to understand what the limits of prediction are in the field of molecular biology. Were the disagreements over what could and could not be predicted a reflection of more fundamental differences about the principles of the science itself? After all, the application of the rDNA technique was producing new knowledge about gene expression in higher-order organisms. Would the new scientific results affect how some individuals viewed the risks?

Closer examination of the phrasing in the letter, which includes terms like "cannot be predicted," lead one to the conclusion that the notion of forecasting and its relation to risks was ambiguous. For example, referring to the plans underway for using the new techniques to create hybrids from viral, animal, and bacterial sources, that letter states, "Although such experiments are likely to facilitate the solution of important theoretical and practical biological problems, they would also result in the creation of novel types of infectious DNA elements whose biological properties cannot be completely predicted in advance." At another point it states, "Since joining any foreign DNA to a DNA replication system creates new recombinant DNA molecules whose biological properties cannot be predicted with certainty, such experiments should not be undertaken lightly."

Among the things one might mean by novelty when discussing the creation of a hybrid DNA molecule are the following:

1. We cannot predict whether the DNA fragment will express itself in its new host.
2. We cannot predict whether the rDNA sequence contains a gene that is hazardous to humans.
3. We cannot predict whether a well-characterized DNA sequence will produce novel properties once placed in its new host organism.

In (3), among the novel properties considered are increasing the survival capacity of the organism; transforming the organism to a pathogenic state from a nonpathogenic state; extending to the organism a capacity to make proteins it was unable to make before the rDNA molecule was inserted; and endowing the organism with new and unanticipated physical properties.

It remains unclear from the context of the letter which one or combination of the above conditions of novelty was being addressed. And yet the explicit nature of the limits of prediction in these experiments plays an important role in how one conceptualizes the risks. For example, condition (3) raises many more uncertainties than either of the other two. Can we assume that a DNA sequence that is well-characterized will not behave in some unpredictable manner once removed from its indigenous organism and implanted into another host? The current wisdom as reflected in the NIH *Guidelines* considers cloning with well-characterized DNA sequences whose product is not known to be dangerous to humans, animals, or plants as a low-risk experiment.

Another problem that has been raised periodically in the rDNA debate concerns the manipulation of viruses, either as vectors (carriers of DNA) or as a source of DNA to be implanted in an organism. Some low-risk experiments require the use of defective viruses. Can a defective virus become transformed into a nondefective virus? Are the conditions of such transformations well understood? Are the outcomes predictable?

For a research program in its early stages, it is highly likely that unexpected results will appear. Science, after all, does not advance through a continuous path of predictable outcomes. But how does the unexpected in science bear on the problem of risk assessment? Science strives for predictive potency. But until the science matures, there will be many incorrect predictions. Some will be based upon conjectures and weakly confirmed hypotheses. A mature science builds its predictions on a theoretical foundation. The lattices in that foundation are adjusted to account for the results of empirical investigations.

The uses of prediction in the assessment of risks for rDNA experiments was, in the main, only quasi-scientific. The predictive power of molecular biology was most effective when it was connected to a frequency interpretation of probability, by which the data of the past is used to illuminate the future. But on many occasions, lacking both a data base and a theoretical framework, scientists applied terms like "prediction" and "likelihood" in a qualitative and nonrigorous manner.

As an example, the relative classification of risks on the basis of the sources of DNA and the host vector system used became the cornerstone of the NIH *Guidelines*. Such assessments implied that scientists could evaluate the risks of cloning frog DNA as compared to insect DNA. In most cases, since the recombinant microorganisms had never been created and the DNA was not well characterized, there was no empirical evidence from which to proceed. This is where the term "prediction" took on political overtones. Those who wanted to see the research stopped or substantially slowed down emphasized the primitive stage of biological prediction when new biotypes were being considered. According to this view, there were too many variables and too many exceptions to make a priori judgments.

A different position on the efficacy of prediction in biology came from those scientists who were concerned about restraints on free scientific inquiry. Their arguments drew heavily on analogies between natural recombinations and what was being planned in the laboratories of molecular biologists. They approached the problem of risk assessment as reductionists. Beginning with a concrete scenario for a hazardous event, they estimated the probabilities of each subevent in the causal chain. According to this method, the likelihood of risk is given by the product of the probabilities of each event in the chain. Those critical of the reductionist hypothesis argued that catastrophes do not conform to a linear process. Furthermore, the reductionist hypothesis treats biology like a mechanistic system and takes no account of emergent events.

Subjective factors also enter into the assessment of risk and the estimate of the probability of untoward events occurring. For example, scientists accustomed to working with certain classes of organisms may underrate the biohazards, because either they have an interest in having no controls placed on their work or they have developed the skills for handling the materials safely. The relation between familiarity and perceived risk may manifest itself at some unconscious level, where the tacit proposition "My work is less dangerous than yours" becomes an important subjective determinant of risk assessment.

Roy Curtiss is a microbiologist at the Medical Center of the University of Alabama (Birmingham). When Curtiss read the Berg letter, he drafted a sixteen-page, single-spaced memorandum on the subject "potential biohazards of recombinant DNA molecules."[11] The memorandum, dated 6 August 1974, was sent to all the individuals who signed the letter, as well as numerous other scientists from whom he solicited comments. Curtiss estimates that his memo reached a thousand people from all parts of the world.

In his statement, Curtiss pledged to cease experiments that would construct bacterial plasmids not then known to exist (described as type I experiments). He also agreed to stay clear of type II experiments by not constructing recombinant molecules comprised of two viral strains or grafting virus DNA to a plasmid.

Curtiss's remarks fall into three categories: (1) concerns about the classes of experiments cited in the Berg letter; (2) institutional procedures for dealing with biohazards; and (3) academic freedom and the regulation of science.

Curtiss proposed to broaden the classification system that was cited in the Berg letter—first, by expanding the three types, and second, by adding an additional category: type IV experiments. Under type I experiments, the Berg letter called upon the scientific community to defer constructing new bacterial plasmids (pieces of circular extrachromosomal DNA) if they were to be used to carry the DNA for toxins or antibiotic resistance. In his memorandum, Curtiss called for extending type I experiments to include the construction, by rDNA techniques, of any new autonomously replicating bacterial plasmids (where "new" meant combining traits not then in existence) and the insertion of the new plasmids into prokaryotic or eukaryotic organisms. To justify his recommendation, Curtiss cited an experiment that he had planned, but, under more careful examination, later abandoned. Curtiss had been investigating the principal bacterial agent responsible for dental caries—*Streptococcus mutans* (*S. mutans*). Following a theory that certain plasmids in the bacteria were responsible for the biochemistry of dental caries, he planned to isolate the suspected plasmids and implant them in *E. coli*. A second approach would be to construct a hybrid plasmid made up of two sources: one piece of the plasmid would come from *S. mutans*, and the other would come from the plasmid normally found in *E. coli*. Curtiss indicated that this could be done with the restriction enzyme Eco RI endonuclease. The potential biohazard he identified was that *E. coli* might gain the capacity of becoming cariogenic (producing dental caries). An even greater concern expressed in the Curtiss statement resulting from transferring genes from *S. mutans* to *E. coli* was that the latter might produce other proteins indigenous to the *Streptococcus* that could prove harmful to humans. The experiments outlined by Curtiss were, according to his own evaluation, within the permissible range under experiments of type I cited in the Berg letter.

The Berg letter identified type II experiments as the linkage of DNA from animal viruses to bacterial plasmids or other viral DNAs (viral-viral and viral-bacterial plasmid hybrids). Curtiss maintained that the classification should not depend upon "whether the viruses come from bacteria, plants or animals and whether the plasmids come from prokaryotic or eukaryotic organisms."

Another significant difference between the Berg letter and the Curtiss memorandum came in the discussion of type III experiments. The Berg group issued a request that investigators work carefully with experiments that splice animal DNA to bacterial plasmids or bacteriophage DNA. The concern was over inadvertently transfering animal tumor virus DNA to a bacterium. No request was made in the Berg letter for suspending such experiments. However, Curtiss called for an extension of the moratorium to experiments that were designed to join chromosomal DNA (from either prokaryotic or eukaryotic organisms) and implant the hybrid molecule into a prokaryotic or eukaryotic organism.

Finally, in Curtiss's type IV experiments, he covered all those cases in which splicing DNA in ways not known to occur in nature "might result in a new pathogenic organism or agent or disease state capable of transmission from one generation to the next."[12] All experiments in type IV should be deferred, according to his recommendation in the memorandum.

Curtiss justified his claims by offering evidence of the following types: (1) Bacterial plasmids can endow bacteria with potentially harmful traits beyond drug resistance, toxin production, or oncogenic activity. Bacteria may inherit the potential of producing certain enzymes that could interfere with the human chemical balance. (2) *E. ciol* is gaining in recognition as an important pathogen and has been identified as the cause of a number of human disease states. (3) Agricultural pathogens are one step away from affecting human populations; therefore plant viruses should be of considerable concern when used in creating DNA hybrids.

The concerns raised by Curtiss extended the range of discourse from known human pathogens to a consideration of systems that humans depend upon, namely, subhuman species and plant life. He portrayed *E. coli* as a very active and very hearty organism that exchanges plasmids (and therefore properties) with a wide population of microorganisms. "*E. coli* also occasionally chooses its mating partners without regard to their taxonomic designations and is known to donate and/or receive plasmids and/or chromosomal information from members of over twenty genera."[13]

The issues began to shift from the laboratory and human populations to the larger ecology. Less than two years after the memorandum was distributed, rDNA research became a *cause celebre* of major environmental groups in the United States, such as Friends of the Earth, the Natural Resources Defense Council, and the Sierra Club. These environmental advocacy groups called for a broader representation in the decision-making process to reflect the range of impact of genes shifted across species barriers.

Curtiss recommended that, in addition to the National Institutes of Health, the World Health Organization was an appropriate agency to deal with the issues on an international scale. After all, what is the value of regulating the research at one site or in one nation, when others have *carte blanche* to proceed as they wish? In addition to the proposal that national and international guidelines be issued, Curtiss also recommended that institutional biohazards committees be created to oversee rDNA research at the local level. Many of his ideas were subsequently included in the guidelines developed by the NIH.

To the biologist, "genetic engineering" refers to the techniques that enable scientists to rearrange or modify genes at the microbial level. To the lay public, genetic engineering is often interpreted as the manipulation of human genes in the person. It conjures up images of the Orwellian society with its special genetic programming of the human ovum. Many scientists deny that their work on rearranging genes of lower organisms has any direct implications for genetic intervention and/or manipulation of human beings. That application, it is argued, is an open question for society to decide.

As a supporter of genetic engineering research on microbes, Curtiss cited the following benefits: understanding of the regulatory mechanisms of eukaryotic DNA—particular reference was made to an understanding of the immune system, cancer, and hereditary genetic defects; use of microbial systems for the production of biologicals such as insulin and growth hormone; use of microbial systems for the production of food stuffs, vitamins, and casein (and thus eliminate the need for cows);[14] endowment of plant species with improved ability of photosynthetic carbon dioxide fixation, and improve capability of plants to fix nitrogen; genetic manipulation of marine algae and blue-green bacteria for breaking down pollutants in the oceans; creation of *E. coli* hybrids that can transform hydrogen into energy; use of bacterial strains for producing safe and effective vaccines to pathogenic organisms.

Even at this early stage of the controversy, the list that Curtiss developed on the possible benefits of the research is impressive. Very little will be added to these generic categories over the next five years. The ambivalence that comes out in Curtiss's remarks is over the safeguards. Will they be sufficient to prevent the creation and escape of a virulent organism? Is *E. coli* the safest organism to use as the host of gene transplants?

Curtiss ended his memorandum with a reminder to his fellow scientists. He asked them to recall the early warnings on the use of radiation that went unheeded and the efforts and time that had been needed to secure regulations for the protection of human and animal subjects. At first, those regulations had seemed unduly restrictive; but once established, scientific inquiry accommodated itself to the uniform standards. Curtiss concluded that with genetic engineering there is yet one more opportunity to establish responsible standards without stifling research.

By the time the Berg committee was formed, the rDNA controversy had been through a lengthy gestation period. The factors that were shaping the events came from many different directions. To some, it may appear that a single incident, like the Pollack response to the Berg experiment, was responsible for the controversy. But the view I have presented shows that many currents were moving in the direction of risk assessment. Pollack could not have made the same impact on Berg were it not for the fact that others in the scientific community were becoming self-conscious about laboratory biohazards.

The attention that some scientists were prepared to give to research hazards was related, in some degree, to political and social events of the 1960s. Those events had brought to the foreground the issue of the responsibility of scientists to the society. Many participants in the rDNA episode had been previously involved in those discussions. By and large, these individuals were more attentive to calls for a collective response to the problems that were raised about new gene-splicing techniques. There were very different views expressed among scientists about where the boundaries of responsibility lie. In this first period, the overriding majority of scientists involved in the issues were thinking of social responsibility exclusively in terms of a professional response to the potential hazards of the research.

All biological development can be viewed both as a process of unfolding potential and narrowing possibilities, and this holds too for the social history of the rDNA story. On the one hand, certain forces derived from the nature of the research, combined with the social and political setting, sent the issue spiraling outward to encompass larger and larger communities of interest. On the other hand, the scope of discourse got narrower and narrower, becoming eventually confined to the single issue of biohazard. It took forces outside the scientific community to stimulate a dialogue on the broader societal problems that would accompany the development of the new techniques.

In examining the events of the early period leading up to the first international conference on rDNA molecule techniques, we can draw the following conclusions:

1. Major efforts were taken to keep decision making within the professional boundaries of the scientific community. There were strong appeals on the part of certain nonscientists, who had been consulted for their views of the social and ethical issues, for broad public input into such decisions. But these voices were heard in vain.

2. The SV40-lambda phage episode, and especially the adenovirus-SV40 affair, were dress rehearsals for the rDNA controversy. In these experiences, many of the pitfalls and problems had already been foreseen and tentative solutions proposed before an active opposition to the research developed.

3. Many of the risks and benefits that were discussed in great detail in later periods were seen with unusual clarity by scientists in the early discussions. The best example of this is seen in the Curtiss memorandum, where an examination of potential biological hazards was offered. But little effort was made to outline the long-term impacts of having rDNA techniques brought into the industrial sector.

4. One of the factors that prompted concern by top-level scientists related to the ease with which gene transplants could be accomplished. When the technique became available to scientists who did not possess exceptional skill, it began to trouble those who had, up until this time, a virtual monopoly on the procedures.

The scientific advances in molecular biology were beginning to reach a wider audience by 1974. The letter that came out of the Gordon Conference and the published letter of the Berg committee were responsible for that publicity. When plasmids were developed as vectors for inserting recombinant molecules into bacteria, it became possible for scientists to gain information about gene structure at a rate that would make all other techniques outdated. Initially, the call to slow down and assess the risks was met favorably by members of the scientific community. They were responding, of course, to the most prestigious members of the profession. Eventually, one could anticipate that pressure would build up for getting on with the work. It is clear that the leading molecular biologists wanted to resolve the issue to the satisfaction of the skeptics in the profession while at the same time make a sound impression on the observers who stood on the sidelines.

Asilomar II and the Rise of Public Involvement

7

An International Affair

The second Asilomar conference on biohazards (hereafter, Asilomar) held in February 1975 was called into being following publication of the Berg letter, which had sought a voluntary moratorium on certain classes of gene-splicing experiments. Asilomar convened an international group of scientists for the purpose of assessing the risks of the new technology and establishing the conditions under which research could or should proceed. No single event has had more impact on the public-policy outcome of rDNA research than this four-day conference. It was a key point in the transition of the consideration of biohazards in gene-splicing experiments from informal scientific to formal governmental channels.

The signers of the Berg letter were all prominent scientists who had become sensitive to the fact that research does not take place in a social and political vacuum. To a greater or lesser extent, they reached the conclusion that scientists have social responsibilities that arise out of their work. Their actions gave legitimacy to the idea that the potential biohazards associated with their research were appropriate targets for peer review.

In the pre-Asilomar period, some scientists, recognizing the broad policy issues and acknowledging that they had no special expertise in matters of social policy, supported wider participation in the decision making. But as the power of the new technique was perceived with greater clarity, the stakes began to escalate and the prospect of losing control over the conditions under which the research would proceed became less and less attractive. In the critical planning period of Asilomar, the controversy over rDNA research was reduced to a set of technical problems related to biohazards in the research laboratory. The objective was not necessarily to close off debate and discussion on other aspects of the research—the social, ethical, and political considerations, for example—but only to separate those issues from the task of establishing the conditions under which the research would be conducted.

The logic of the procedure is another matter. While scientists may be among those best suited to establish safeguards for protecting society against laboratory biohazards, the safeguards they establish may not be appropriate to meet more general social concerns. Yet, although not claiming special expertise in other areas, the Asilomar participants went ahead to make policy as if their technically defined potential hazards were the only ones germane to the question whether work should proceed or not.

The point of special interest about Asilomar, then, is not simply that it brought the debate to its mature stage, but that it represented the development of the issues in a form most congenial to the scientists themselves.

The recommendations that resulted from this conference established the philosophical and practical basis for the guidelines that were issued by NIH sixteen months later. How one judges their effectiveness and the success of the Asilomar conference itself depends on one's fundamental values. The conference has been praised by some as a successful effort to focus intellectual resources toward solving a problem that requires scientific inputs but warrants a public-policy outcome. It has been damned as an assembly of self-interested scientists trying to protect their research programs from external control by nonscientific bodies.

Chapters 7–10 cover the events of Asilomar and explore the competing interests and underlying values that have led specific individuals to assume a particular perspective on its outcome. I shall examine the origin of the conference, its organizational details, its conceptual framework, and the decision-making process that led to the final recommendations. The entire story is a complex one, both from the point of view of the cast of characters and the political setting. The conference became an object of major media attention. Fifteen of the country's top science writers attended, and two of them subsequently published books on the subject.[1] The reasons for this immense interest are apparent. Research that involves cutting and splicing genes has all the prerequisites for a science-fiction thriller: new dread diseases possible; cloning; human genetic engineering; human gene maps; interspecies hybrids (combining genes of plants and animals, toads and humans, etc.). Indeed, one of the difficult problems faced by the scientists at Asilomar was how to distinguish fact from fiction. Many hypotheses were brought into the discussion that had not yet been systematically addressed.

This kind of uncertainty is not unusual in the course of science, which abounds in questions, hypotheses, and tentative explanations as well as empirical tests. This is perfectly normal; there are periods when science proceeds through competition between alternative hypotheses. At such times, long intervals of controversy may precede general agreement among leading scientists. But, with regard to the debate over risk assessment at Asilomar, and in spite of the great variety of hypotheses about the hazards or nonhazards of gene-splicing work and the need for additional scientific evidence, the meeting was marked by a sense of urgency to reach consensus. The result was a negotiated settlement. It involved some science and a lot of politics.

From a public-relations standpoint, the outcome of Asilomar provided some legitimacy to the scientific community for raising and attempting to find a resolution to the biohazard issue. The scientists had even agreed to engineer a safe bug that could serve as the recipient of foreign DNA fragments. Such a microorganism presumably would not survive outside a narrowly constructed, artificial niche in the biosphere. At one level, then, Asilomar is a case of the insiders trying to show the outside world that they are socially responsible and capable of keeping their own house in order.

But on another level, the conference was an extraordinary occasion; scientists representing many parts of the world had been brought together to solve a problem of a political nature, involving social relations among scientists and between science and society. Even at this early stage, many suspected that what was at stake was nothing less than the fundamental values that lay at the heart of science's social framework, namely, self-determination, freedom of inquiry, and peer review. Ironically, despite the efforts on the part of the Asilomar scientists to protect these values, the judgments on the outcome of that meeting would eventually be issued by nonscientists.

The Asilomar meeting by itself did not prove anything about the social relations of science. After all, it was only one event. But it did illustrate a few things. Above all, participating scientists exhibited great variations in interests, opinions, and assessments of risks. Intuition, hunches, and partially educated guesses played important roles in their discussions. The meeting involved a mixture of disciplines, and therefore of disciplinary paradigms, which complicated matters. An additional complication was that there is no rigorous science of risks. (And even if there were, there would still be a role for subjectivity and intuition.)

Before I enter into the analysis of this period of the rDNA debate, I want to emphasize the relevance for understanding the details to follow of certain important generic questions that arise in many situations in which risk assessments are undertaken: How are the problems defined? Who defines the problems? How is information controlled and channeled? How is a resolution arrived at? Can the effectiveness of the resolution be evaluated? Who is at risk? Who benefits from the technology? How was the public involved in the decision, and what was the nature of the involvement? These queries will be addressed in the forthcoming pages on the rDNA debate. I also want to emphasize that this conference was one stage in a long process of risk assessment. It established an important framework upon which scientists and the NIH could build. But, as I shall show, the significance of such frameworks is that they not only establish the boundaries of discourse but also limit them.

The Asilomar meeting drew leading scientists in the field of molecular biology and certain other allied areas from throughout the world. It would be no simple task in these circumstances to find a consensus over such a broad range of issues. The problem of reaching a consensus was further exacerbated by the sheer number of participants—132, excluding science writers. Scientific issues have their own way of circumscribing discussion. But where policy matters are at stake, especially where they could affect the vitality of a new research program, viewpoints multiply.

How did the scientists at Asilomar make the transition from the current state of knowledge in molecular biology to addressing the potential risks of creating novel organisms and, finally, developing policy recommendations for applying DNA techniques? The key to an understanding of the final outcome of the meeting is in the decision process and its determinants.

The issues placed on the Asilomar agenda were organizationally complex. There are at least four reasons for this. First, there are endless ways of creating new hybrid organisms, and no simple grouping of risk experiments could resolve the uncertainties. Second, the scientific questions could not all be framed within a single discipline. Third, there were considerable gaps in knowledge that had a bearing on the evaluation of the risks. Fourth, there were some problems that could not be resolved by an appeal to scientific evidence. Prominent among these were the role of rDNA research in fostering human genetic engineering and whether the prospects were desirable.

The Organizing Committee at Asilomar steered the assembly toward what they viewed as the "solvable" problems. This meant bracketing out certain ethical and social issues. From the perspective of the Asilomar Organizing Committee, to incorporate such issues on the agenda would produce two undesirable effects. First, they would provide the media with copy that might stimulate greater public controversy at a time when scientists were worried about the public perception of science. Second, discussions on the social and ethical impact of the research might cause the meeting to drift away from the biohazard issues, reducing the likelihood that a consensus could be reached.

But a third reason for these exclusions is, in some ways, more significant. To include social and ethical issues as major considerations in establishing the conditions under which work could proceed would automatically open the door to participation by nonscientists. Few, if any, believed that scientists had a monopoly on the understanding and resolution of broader social questions. Thus, restricting the problems to technical issues had the effect of giving the scientists legitimacy as sole arbiters of evidence and makers of policy.

There was considerable concern among Asilomar organizers over the voluntary moratorium established for certain classes of experiments. Scientists had created the moratorium, and now, lest it become permanent, they were determined to end its existence. But at the same time, they wanted to provide reassurance to their colleagues and the general public that gene splicing could be done safely. Therefore, closure was an important consideration in creating the agenda. And the boundaries of discourse had to be firmly established to achieve it.

On the same day that Paul Berg held a press conference on the moratorium letter (18 July 1974), NIH Director Robert S. Stone released a statement that said he would take prompt action in the recommendations of the Berg committee. Very soon after the press conference, Berg, Baltimore, and Leon Jacobs of NIH began a series of discussions on the financial arrangements for running an international conference. (Leon Jacobs was assigned the responsibility of drawing up the contract for Asilomar.) They agreed to limit the number of participants in order to ensure a manageable and productive outcome from the meeting. The size of the meeting was set at between 125 and 150. The obvious impact of this decision was to limit the meeting to invited guests.

Another financial matter that needed resolution before the contract could be written up related to the reimbursement of travel expenses. The issue in question was whether participants would be asked to pay their own costs to attend, or whether those costs would be built into the contract. The outcome of these early discussions was a decision to pay the travel expenses of people who were invited but who could not fund themselves. The policy adopted at that early period stipulated that the selection of guests would be made by the Organizing Committee and that it would not depend upon people's ability to pay. This policy was modified somewhat at a later date when it became apparent how expensive travel costs would be. Speakers and organizers of panel sessions were promised travel reimbursements. Others were asked to find their own funding.[2]

When the NIH wrote the contract for the Asilomar Conference, it designated the National Academy of Sciences (NAS) as the recipient. This meant that the NAS was responsible for transmitting a final report to the NIH. It signified that the Organizing Committee was directly responsible to the NAS. But that left uncertain the official status of the participants themselves. Were they advisory to the Organizing Committee? Was the assembly as a whole going to vote on issues? These procedural questions were left ambiguous, until their resolution was forced from the floor during the plenary session.

The original estimate of expenses for Asilomar was $75,000. Those who drafted the contract neglected to consider the fact that the NAS takes out overhead costs, which, in this contract, amounted to $25,000. To make up the shortfall, Paul Berg called upon the director of the National Cancer Institute (NCI). Subsequently, the NCI allocated an additional $26,000. The National Science Foundation also contributed $10,000 to the cost of the conference.

As chairperson of the Asilomar meeting, Paul Berg had responsibility for selecting an Organizing Committee, subject to the approval of the NAS. David Baltimore and Richard Roblin were chosen because of their participation on the Berg committee. Maxine Singer was asked to serve because of her part in the Gordon Conference letter. At a 10 September meeting at MIT, the nucleus of the Organizing Committee agreed upon two foreign members: Sydney Brenner, an internationally eminent scientist at the Medical Research Council Laboratory for Molecular Biology, in Cambridge, England, and Niels Jerne of the Basel Institute for Immunology in Switzerland. Berg had known Brenner from a previous meeting in London. Jerne was favored because he was chairman of the European Molecular Biology

Organization Council (EMBO). As it turned out, Jerne did not participate in the pre-Asilomar discussions and failed to attend the conference. Consequently, he was dropped from the Organizing Committee when the official report was submitted to the NIH.

During the September meeting, the goals of the February conference were established. Interest was shown among some members of the Organizing Committee to have a detailed report submitted to the NAS on the outcome of the conference, in lieu of a published proceedings. But that idea lost favor when it was discovered there was little enthusiasm for writing such a report.

The NAS had not requested a detailed report of the proceedings, and its officials agreed that a summary statement of the conference would suffice. Conference chairman Paul Berg also agreed to have the entire proceedings taped. According to the plan, the complete set of tapes would be deposited with the NAS, to be made part of a historical archive of the event. But while the Organizing Committee agreed to an official taping of the sessions, they did not intend for those tapes to be released to the news media or the general public in the immediate future. Baltimore recalls that it was the consensus of the committee that the tapes remain in the NAS files at least until the end of the century, at which time they would be available to the public.[3]

There was no compelling argument for this exclusionary attitude toward access to the conference's taped discussions. As it turned out, reporters who attended the meeting used portable tape recorders for their news stories, thereby severely limiting the intent to draw a curtain around the details of the proceedings.[4]

As to the goals, these seemed to be circumscribed at the outset. Of paramount importance was the practical matter of the moratorium initiated by the Berg letter of July 1974.

Paul Berg entered the Asilomar period with the expectation that the conference would lead to an end of the voluntary pause requested for certain classes of experiments. In an interview with MIT's Oral History Program, Berg remarked that he never believed the work should be banned or that the moratorium should be extended until much more information about the hazards was gathered. He acknowledged that Asilomar would not reveal all the risks. It was only a starting point. He reasoned that unless the experiments were carried out, there would be no means for determining where there actually were risks. Furthermore, Berg felt that setting the conditions for doing some experiments too stringently would be tantamount to proscribing them.

Beyond this, the objectives of the conference become less clear and are easier to characterize by what was excluded than what was to be considered. Those exclusions were given public utterance at the opening session of the conference when David Baltimore pointedly noted that two issues were to be kept outside the scope of the meeting.[5]

The first was the utilization of rDNA techniques for genetic engineering. These issues, he held, are replete with values and political motivations. They will obfuscate the technical discussions related to biohazards. The second was the application of gene-splicing techniques to military weapons. At the time, Baltimore believed that the new techniques could conceivably lead to more effective instruments for biological warfare. Nevertheless, he told the participants that this was not an area of investigation for the Asilomar assembly.[6]

The decision to bracket out the issues of genetic engineering and biological warfare at Asilomar had already been made, according to Baltimore, at the April 1974 meeting when the moratorium letter was being drafted.[7] In a post-Asilomar interview, he stated that the public-health aspect of rDNA research was a more important concern because President Nixon had dismantled the machinery of biological warfare.[8]

At the outset, members of the Organizing Committee had considered a conference format similar to the first Asilomar meeting held in 1973. Following that model, selected participants would be invited to contribute papers on subject areas related to the potential hazards. Those papers would subsequently be published in a volume of proceedings. But there was little support for soliciting papers on the biohazards of rDNA techniques. As a result, other formats were considered.

Berg wanted the conference to present a mix of scientific information and biohazards recommendations. Something had to be done to focus the attention of this large assembly and direct it toward reaching a consensus position. Finally, a format was adopted whereby several small groups would be asked to prepare position statements in advance of the meeting. The choice of chairpersons for the three working groups was finalized at the September 10 meeting of the Organizing Committee. Each group was responsible for meeting prior to the Asilomar Conference and for preparing a position paper on a preassigned area. The position statements were then to be distributed to the participants at Asilomar and discussed during regular sessions of the conference. Scientists serving on the working groups were also called upon to organize some of the technical sessions. These sessions were designed to summarize, for the assembly, the current state of knowledge of rDNA techniques, plasmid biology, and gene expression in prokaryotic and eukaryotic cells.

The five subject areas initially designated for planning the technical sessions, and for determining the make-up of the working groups, were (1) ecology of plasmids and enteric organisms; (2) molecular biology of prokaryote plasmids and their use for molecular cloning; (3) synthetic recombinants involving animal virus DNAs; (4) synthetic recombinants involving eukaryote DNAs; (5) ethical and legal concerns arising out of work with synthetic rDNA molecules.

Prior to setting up the working groups, the first two areas were merged for consideration by a single panel.[9] By and large, the technical subject areas reflected the typology of experiments identified in the Berg letter. Maxine Singer was influential in promoting the idea of getting some input from lawyers familiar with biolegal and bioethical issues. Her husband, attorney Daniel Singer, was asked to organize a session of nonscientists.

Three working groups were formed to examine the technical scientific issues and the biohazards. The Plasmid Working Group (PWG), headed by Richard Novick, a microbiologist at the Public Health Research Institute in New York City, was responsible for looking into the risks of plasmid interchange with rDNA molecules. Novick was recommended to head this group because of his knowledge of plasmid transmissibility, especially those plasmids (R plasmids) that confer antibiotic resistance to cells.

The Eukaryote Working Group (EWG) was given responsibility for assessing the risks of transferring DNA from multicellular organisms such as plants and animals into *E. coli*. That group was headed by Donald Brown, a researcher at the Carnegie Institution, specializing in molecular embryology. According to Berg, there were several considerations in choosing Brown.[10] In addition to his expertise in the area of gene expression and control, Berg believed that Brown saw considerable value in eukaryote-to-prokaryote DNA transfers. Berg reasoned as follows: Since Brown was viewed as a scientist with stature in his field, it would carry considerable weight with his colleagues were he to be convinced that certain experiments should not be performed.

The third working group was asked to consider the hazards, if any, in connecting viral genes to plasmids (or other vectors) and transferring the plasmids to bacteria. The choice to head the Virus Working Group (VWG) was Aaron J. Shatkin, who was associated with the Roche Institute of Molecular Biology at Nutley, New Jersey. According to Berg, Shatkin was knowledgeable about tumor viruses, but was not working with them during the Asilomar period. Berg emphasized this point in his choice of Shatkin, since he (Shatkin) would not have a direct stake in the outcome of the experiments with tumor viruses.[11]

The chairpersons of the working groups were responsible for choosing the other members of their respective panels. The number of meetings that each working group had prior to Asilomar varied. The first opportunity that members of the plasmid group had to discuss their roles at Asilomar was in early fall of 1974 at Honolulu, Hawaii, at a meeting of a task force of the American Society for Microbiology (ASM). Established in 1972, the task force was formed to develop a plasmid nomenclature.[12] Novick asked the members of the ASM task force to constitute the Asilomar plasmid panel. A second gathering of the PWG took place in November 1974 at the Public Health Research Institute in New York City. Long hours were spent hammering out the issues for a position paper. The group met again during the weekend before the Asilomar Conference to work on a draft of its statement.

The eukaryote group had one meeting on the weekend before Asilomar for a day and a half. The virus panel also met during that period, but had a preliminary meeting in the early part of November. Once at Asilomar, the working groups met each day reexamining their position statements in the light of the discussions at formal and informal meetings.

Despite the apparent specificity of these working groups, there was still a considerable latitude in how they interpreted their charges. In some cases, specific directions were given to members of the working panels in premeeting discussions and correspondence. For example, in a letter from Paul Berg to E. S. Anderson, the following objectives were outlined:[13]to identify the kinds of experiments scientists would like to do with hybrid molecules; to stipulate the kind of information such experiments could provide; to identify the possible risks involved for the investigator and others; to develop means by which the biohazards could be tested and minimized in order that the work could proceed.

Thus, it appears that when specific objectives were visualized at all, they were organized around assessing hazards of experiments that scientists were particularly interested in doing. From the standpoint of assessing the impact of a new technology, however, even if limited to the question of biohazards, one is also interested in the kinds of experiments that *can* be done. To restrict the field of vision to those experiments in which scientists are currently interested takes no cognizance of commercial ventures or other possible "non-academic" uses or misuses of the technology. This fact emphasizes the limited scope of the Asilomar Conference.

The issue itself, however, was not so easily contained and, indeed, in the deliberations of the conference, the perspective was sometimes broader than the designs the Organizing Committee envisioned. Given the dense network of social and political associations from which the issue emerged, this is hardly surprising. On the contrary, it is a wonder that the containment was as effective as it was. Perhaps a reason can be found in the way participants for the conference were chosen.

What considerations went into the selection of participants? What effort was made to draw in diverse viewpoints in the discussions? After the conference, when criticisms of Asilomar were heard in scientific circles, these questions were raised. A pamphlet issued by the genetics and social policy group of Science for the People, entitled *The Health Hazards of Gene Implantation*, referred to the Asilomar participants as "a group consisting predominantly of research directors," while science writer Nicholas Wade described them as "the paladins of their own special world, the directors of research laboratories and arbiters of scientific fashion."[14] Without characterizing the intentions of the Organizing Committee, one can see that the process of selection depended heavily upon informal networks of a few key biologists. It was naturally skewed in the direction of individuals with similar backgrounds and interests.

Paul Berg solicited names of potential invitees to Asilomar from the program committees and other members of the Organizing Committee. From these responses, a preliminary list of approximately 150 names was drawn up. That list was circulated among the same nucleus of individuals who were then given an opportunity to suggest additional names or recommend that certain names be deleted. Nominations were also solicited from the NIH.

In recalling that process, Richard Novick, chairman of the Plasmid Working Group, explained his interest in having scientists who were (i) involved in the basic foundations of the field and (ii) were also concerned about the consequences of rDNA molecules.[15] Novick gave priority to the seasoned scientists who could contribute to the science, the politics, and the philosophy of the issues.

Considering the charges given to the working groups, this kind of selection procedure seems entirely consistent. On the other hand, there were also people invited who had an interest in rDNA work beyond the potential risks of experiments that "scientists were interested in doing." In particular, Berg, reasoning that one of the principal benefits of the technology was its application to industrial and pharmaceutical problems, supported the idea that the commercial sector should be represented at the meeting.[16] There seemed to be widespread agreement that industrial applications of rDNA techniques could create the most serious hazards—if there were hazards. The reason for this is that the production requirements of biologicals involve the use of large volumes of culture. Containment of organisms for large-scale work becomes more difficult, as does the disposal of wastes and the procedures for dealing with accidental spills.

The individuals attending Asilomar who were affiliated with commercial establishments came from the research divisions or the independent think tanks of their respective companies. The final attendance list shows there were scientists from the Physical Chemistry Laboratory of the General Electric Company, the Merck Institute, the Roche Institute for Molecular Biology, and G. D. Searle & Co. (England).

The Organizing Committee also saw the importance of broadening the participation at the meeting by including individuals who could address the social implications of genetic research. In view of the fact that these issues were deemed out of place, this too seems puzzling. Nevertheless, Jonathan Beckwith, a Harvard Medical School microbiologist and member of Science for the People, was sent an official invitation. Beckwith wrote Berg a letter stating that he would be unable to attend. In his stead, he suggested Dirk Elseviers, whom he identified as a postdoctoral candidate capable of presenting the perspective of Science for the People.

Berg did not accept Beckwith's nomination. In a post-Asilomar interview, Baltimore explained that decision in the following way:[17]

Beckwith was invited because he was considered an important spokesperson for an alternative viewpoint on the social impacts of genetic research, not because he was a representative of Science for the People. Baltimore maintained that the Organizing Committee was not willing to accept individuals to the meeting merely because they held a different point of view.

Two other prominent scientists, Ethan Signer and Jonathan King, both of MIT, had been active in bringing public attention to the social implications of genetic research. Presumably, they too could have brought some alternative viewpoints to Asilomar. Signer had been contacted by Donald Brown, chairman of the EWG, as part of Brown's effort to acquire some information about the potential risks of creating eukaryote-prokaryote hybrid organisms.[18] Signer was asked through correspondence whether he could think of any hazards associated with type III experiments outlined in the Berg letter. In Signer's reply, he wrote that he had more to say about the social impacts of genetic engineering research and had not given serious consideration to the biohazards of rDNA work. He sent Brown a reprint of his 1974 critical essay on current trends in human genetic engineering.[19] Signer's letter was an invitation to Brown for opening up the Asilomar agenda to address the prospects of human genetic manipulation. But Brown rejected the bait. Signer was not contacted again.

When it became known that Beckwith would not be participating at the meeting, Jonathan King was suggested to the Organizing Committee. King had achieved scientific acclaim for his research in the molecular genetics of bacteriophage and gained his reputation as an activist scientist for his participation in the XYY affair. There seems to be some disagreement whether King was invited to Asilomar. Both Baltimore and Berg believe that he was.[20] King maintains that he never received an official invitation and for that reason did not attend. He does recall a passing remark by Baltimore to the effect that he (King) would be invited. In any event, by not having the viewpoints of scientists like Beckwith, Signer and King, the outcome of the Asilomar meeting was more vulnerable to the criticism that its composition was stacked in favor of strong proponents of the research and against those who were concerned about the broad social implications of genetic engineering.

Geographically, the representation was better. There were over fifty foreign participants at Asilomar. Some of these selections were made by the Organizing Committee, and in other cases requests for nominations were sent to the scientific academies of selected countries. Berg avoided inviting official delegations from particular nations.

In the last analysis, the Organizing Committee was after active participation from working scientists, not groups of observers. Asilomar was not planned as an open conference for all interested parties. It was visualized as a working session, consisting of participants selected for their special expertise in the problem areas. This fact alone drew many scientists with a stake in continuing the research.

8

Biohazards

The Berg letter of July 1974 enunciated the most obvious of the biohazards as seen at that early date. With the planning of the Asilomar Conference, it was time to consider the question more systematically.

When the chairmen of the Asilomar working groups were chosen, two of these individuals, Aaron Shatkin and Donald Brown, sent out a net of correspondence to their colleagues in the scientific community. These letters requested information about the possible risks associated with the types of experiments for which the panels in question were responsible. The body of correspondence generated in response to these inquiries during this period is of a very special nature. First, it was not meant for public or media consumption. Many of the scientists who were considering the potential hazards of rDNA activities had not yet anticipated a major public-policy debate. Therefore, one can assume that the concerns expressed in these letters were unencumbered with fears that the issues raised would be misinterpreted or exaggerated. Second, this was the first time that a wide grouping of scientists was asked to examine the risks in a systematic way and the issues were divided up into discrete components.

But the members of the working groups were not the only ones engaged in soliciting and expressing views on the risks of rDNA research in this period. While the groundwork was being set for the Asilomar Conference, members of the Organizing Committee were placed in a position of having to defend the voluntary moratorium as well as possible regulations that might result. They lectured and wrote papers that presented the rationale of the events leading up to Asilomar. Maxine and Daniel Singer jointly wrote an article for the Swarthmore *College Bulletin* entitled "Ethical problems at the frontiers of biological research."[1] Paul Berg participated in a televised debate on the moratorium issue with British scientist Ephraim Anderson, sponsored by the British Broadcasting Company. David Baltimore delivered a paper to the Technology Studies Workshop at MIT entitled "Where does molecular biology become more of a hazard than a promise?"[2]

In one notable interchange of letters, Paul Berg defended the moratorium to Irving P. Crawford, a microbiologist at the Scripps Clinic and Research Foundation in La Jolla, California.[3] Crawford had published a letter in the journal *Genetics* that maintained that the Berg letter and the forthcoming Asilomar meeting was "an open invitation for the establishment of a burgeoning bureaucracy concerned with monitoring and regulating genetic experimentation."[4] Crawford argued that it was unreasonable to think scientists could create assemblages of DNA that were more dangerous than the combinations already tried by nature. He added that the laboratory strains of *E. coli* have, by this time, become "so degenerate from their soft life on nutrient agar slant" that they would not succeed as a viable host in the human intestine. Crawford issued the following proposal as an alternative to the route that the Berg committee was taking: Let each scientist decide at the outset of his experiments whether he would care to expose himself, or better, his child, to the newly assembled "agent." If not, let him learn and use the techniques of containment needed to control such agents.[5]

Berg responded to Crawford in several different ways. He noted that the issues raised by rDNA research were similar to those in other environmental disputes. And just as in those disputes, the individuals adversely affected by regulation were likely to oppose the reforms. Berg noted that from Crawford's standpoint, no system of regulation was necessary in science. Each scientific investigator was normally fit to act responsibly. Yet, the cumbersome regulations on human experimentation research that Crawford so disliked were a good case in point. At least for human experimentation, Berg supported the work and oversight of institutional review boards. The purpose of regulations for research, he said, is for the reckless investigator, not the careful one; and individual investigators aside, Berg questioned the assurances that pharmaceutical industries would be reasonable in the kind of experiments they might be willing to try.

Next, in addressing the scientific arguments offered by Crawford, Berg made several key points. First, the success of predicting the outcome of new scientific programs is not good. Berg asked rhetorically, Should we accept a risk of one in ten for the possibility of a dangerous outcome? Berg referred to the false assurances that were given by the Army that radioactive fallout resulting from nuclear bomb tests would have no "appreciable" risk of inducing leukemia.

Second, Berg questioned the assumption that scientists in the laboratory could hardly make something that nature has not already tried. There is no evidence, he maintained, that genes such as those that occur in SV40 or adeno viruses have ever occurred in combination with plasmids like pSC101 or vectors like lambda phage DNA. Even if these segments had been in contact in some past age, how can we know what their selective advantage will be in our present environment?

Finally, Berg raised the possibility that the laboratory strains of *E. coli* could be transformed into a pathogen. At the time (late 1974), Berg felt there was not sufficient information to make a conclusive judgment on these potentialities.

One of the best sources for understanding the early concerns scientists had about the risks of the research is in the correspondence of the working groups. These and other letters have been deposited in the Recombinant DNA Controversy Collection at MIT's Institute Archives and Special Collections. They were contributed by major actors in the rDNA affair. It is important to understand that many of these letters were written in response to a question of this general type: Could you imagine what might go wrong if we do . . . ? Those who thought of scenarios in which hazards might result were drawing conclusions from a set of worst case hypotheses that, most would agree, had not been confirmed.

The correspondence written between September 1974 and February 1975 has been divided according to the following major scenarios of risk: (1) creation of plant pathogens; (2) *E. coli* with cellulose-degrading genes; (3) troublesome proteins in *E. coli*; (4) natural recombinations versus artificially created DNA molecules; (5) immunological hazards; (6) tumor viruses; (7) perturbation of biochemical pathways; and (8) drug resistance genes in *E. coli*.

Plant geneticists were optimistic about the impact of rDNA techniques on their field. Plant growth could be studied in a unique manner by isolating and amplifying plant genes. These advances might lend themselves to a revolution in plant genetic engineering. They might also improve our ability to extract drugs from certain plants. But, according to the view of some biologists, if genes that code for plant toxins were inserted into bacteria, it could prove hazardous to plant and animal life.[6]

Arthur Galston of Yale's Department of Biology, for example, wrote to Donald Brown (November 1974) outlining the potential hazards of creating bacterial hybrids with plant DNA.[7] Galston offered three scenarios that I have reconstructed for purposes of illustration. For case I, the reconstruction takes this form:

1. Some plants are susceptible to specific toxins that can be traced to a genetic sequence.
2. Assume that the gene responsible for the toxin is implanted in bacteria.
3. Assume that the bacteria with the plant toxin spread to the susceptible plants.

---

4. Under the above conditions, there is some likelihood that the susceptible plants can be wiped out.

For case II the reconstruction is

1. Select a certain species of plants with a specific tolerance to atmospheric oxygen (i.e., plant grows better at certain oxygen concentrations than at others).
2. Genetic components in certain microorganisms control the oxygen tolerance for the species.
3. Suppose that the genes that reduce oxygen tolerance in microorganisms are introduced into plants.

---

4. There is a likelihood that the plant yield can be reduced or, worse yet, the plant can be killed off.

In his final example, Galston discussed the possibility of transferring to plants a toxin that has been characteristic so far only of microorganisms. He questioned whether recombinant techniques would allow one to build into certain species of plants essential for human nourishment the genes for botulinus toxin.

The final paragraph of Galston's letter highlights the frustrations of those scientists who expressed their concerns during this early period that everything capable of being done should not necessarily be done. The problem was how to prevent a laissez-faire attitude and yet save science from enslavement:

> I am personally fearful that one day an experiment in this area is going to produce unexpected and unfortunate results. I do not think one can prevent such experiments from being done by voluntary declarations of the Berg type. On the other hand, I would hate to see a world in which there was an enforced prohibition against such experiments. Basically, this leads me to a pessimistic outlook for the

future. Conferences such as yours [referring to the forthcoming Asilomar meeting] may help in some way to forestall or delay the inevitable.[8]

Not all plant pathologists were as concerned as Galston. Some had reasoned that even if an agricultural mishap resulted from the use of the new gene-splicing techniques, it could be detected and reversed. R. H. Burris, a biochemist at the University of Wisconsin–Madison, in a letter to Donald Brown[9] discussed the scenario of transferring nitrogen fixation (nif) genes from bacteria to plants. If plants could express the genes, conceivably they could fix their own nitrogen from the air. Could a plant species become too prolific? Could the nif gene inadvertently reach plants that are pests to society? Responding to the prospect that the nif gene could find its way into the wrong species of plant, Burris still saw no catastrophe. Plants that grow too fast or plentiful could be handled by modern weed control. Thus, in his view, we presently possess a technology that could help us reverse the hazards of this type of experiment. Burris also could not see any risk to plant life in constructing plant DNA-bacterial hybrids. Even for the unlikely case in which there was destruction to certain plant types, others could easily be substituted. It was unimaginable to him that an entire species of plant would be wiped out. If there was a danger in mixing plant and bacterial genes, it might appear, according to Burris, as a depressed immunological reaction of plants to invading pathogens.

These differing approaches to risk are in many ways typical of the evolving DNA debate. On the one hand, Galston speculated freely on what might happen in the worst of all worlds, however unusual those possibilities are in our present experience; and he drew the conclusion that voluntary restrictions were inadequate. But Galston also feared the consequences of enforced regulation.

Burris, on the other hand, saw the problem in terms of our present experience and could not see the new technique as having altered anything fundamental in that experience. Should new problems arise, we are not without resources to handle them. This was more than a difference in professional judgment. It was a difference in style, temperament, and faith in our ability to control run-away technologies.

Another notable exchange of correspondence took place between A. M. Chakrabarty and Roy Curtiss III. Chakrabarty, then a researcher in the physical chemistry laboratory of the General Electric Company, was among the many hundreds of scientists around the world who received a copy of the Curtiss memo distributed in early August 1974.

In responding to Curtiss, Chakrabarty described an experiment of his own. He was planning to construct a plasmid with the genes in *Pseudomonas* that coded for enzymes that can degrade cellulose. Furthermore, Chakrabarty was interested in cloning the plasmid with the cellulose-degrading genes in *E. coli*.

Curtiss first responded to Chakrabarty's letter on 16 September.[10] He told the GE scientist that his experiment was a valuable one and he did not see any biohazard associated with it. Curtiss issued a word of caution for other genes in *Pseudomonas* that might prove hazardous if cloned in *E. coli*. He suggested a test to Chakrabarty. Inject the *E. coli* with the plasmid into a mouse and observe whether it was more toxic than the *E. coli* strain without the plasmid. Thus, in first thinking about Chakrabarty's proposed experiment, Curtiss was not troubled by the transfer of the genes in question, only the unknown additional baggage that could accompany that transfer and the effect it might have on *E. coli* strains. Four days after his response to Chakrabarty, however, on 20 September, Curtiss had a change of mind.[11] He sketched out two possible hazards in implanting cellulose-degrading genes into *E. coli*. They were based upon the dietary importance of unreduced cellulose as roughage in healthy bowel activity.

The following are two reconstructed arguments based upon Curtiss's remarks. For argument I the reconstruction takes this form:

1. Cellulose in the diet satisfies people's appetite without adding calories.
2. If all cellulose were degraded during digestion, it would be considerably more difficult for people to reduce their caloric intake.
3. If *E. coli* with active cellulose-degrading genes occupied the human gut, people would find it difficult to reduce their caloric intake.
4. Greater caloric intake means there would be greater obesity in society, leading to a reduction in the average life span.

---

5. *E. coli* with cellulose-degrading genes in the human gut can result in a reduction in life span.

Curtiss was merely suggesting the above argument as a possibility that warranted serious consideration. In a second, less developed idea, Curtiss mentioned the importance of roughage in the digestive tract.

For argument II the reconstruction is

1. In the human diet, roughage is primarily unprocessed cellulose.
2. Roughage passes through the bowels undigested or degraded.
3. Some evidence links low-roughage diets to higher incidences of bowel cancer. (It is conjectured that the undegraded cellulose helps to recycle bacteria, thereby reducing the potential for disease.)

---

4. If the resident bacteria in the human gut degraded the cellulose in the diet, thereby eliminating the usefulness of roughage, there could be a higher incidence of bowel-related diseases, including cancer.

When Chakrabarty responded to Curtiss's warning, he explained the potential benefits of his experiment.[12] He wished to determine whether animals with cellulose-degrading *E. coli* could benefit from the calories of cellulosic foods. If this type of experiment was successful, the implications for a new Green Revolution were substantial. In countries where there has been substantial food shortages, Chakrabarty saw great promise in a technology that enabled people to utilize high-cellulose-content foods for their untapped calories. Vast populations in underdeveloped countries could have the flora of their intestines transformed so they could digest and obtain calories from vegetation that is inexpensive and plentiful but possesses little value as food.

Chakrabarty was convinced by the potential hazards sketched out by Curtiss. He admitted his primary focus was on people in developing countries for whom starvation was a serious problem. He had not given thought, he said, to the effects the cellulose-eating *E. coli* would have for people in other parts of the world.

There are a number of extraordinary things about this exchange, not the least being the apparent agreement these two scientists reached about a fairly far-fetched scenario involving an area neither had special expertise about (chronic-disease epidemiology and the effects of diet and nutrition on mortality, and morbidity from large bowel cancer). But more important is the scope of Curtiss's doomsday scenario, which is more than matched by the vision of Chakrabarty's scheme to enable the third world to nourish itself on otherwise inedible and nonnutritious foods—in effect, to give them a totally new and qualitatively different nutritive capacity. One is immediately struck here by the breadth and scale of the argument and how it came to be resolved with so little contention. The contrast with the previous example is stark. Somewhere between them is the issue of the troublesome proteins.

When Donald Brown began the preparation for the meetings of his working group, he sent out a number of inquiries that requested information about the type III experiments outlined in the Berg letter. In his correspondence, he posed a scenario and asked the recipient to respond.[13] The following is a reconstructed version of Brown's scenario for a biohazard:

1. A gene from a eukaryote is placed in bacteria.
2. The gene codes for a protein, say P_x.
3. The bacteria colonizes the human gut and produce significant quantities of P_x.
4. P_x is released into the blood stream.

If the above sequence of events were to occur, it is possible that P_x could produce untoward effects. No one was sure whether animal DNA could even be expressed in bacteria. The translation of DNA to messenger RNA and finally to a protein product requires certain promoter sequences. It was not understood at the time whether the spliced DNA contained these promoter genes or whether the host itself had promoters that would work on the genes of the eukaryote. Brown questioned whether inserting the entire viral genome into a bacterium would yield the complete virus protein as a product. Along these lines, one would want to know how easy it is to get such products; is it a very skillful engineering feat, or can it occur quite easily and unexpectedly?

Brown also polled virus experts on the potential risks of cloning viruses in *E. coli*. Many experts were given two alternatives and asked to grade their relative riskiness. The alternatives were (1) studying a virus like SV40 by growing it in tissue culture cells and purifying the living virus, and (2) studying SV40 by inserting its genome in *E. coli* and obtaining a large amount of the virus by cloning. According to Brown, most virologists believed that the recombinant form of the virus would be safer.[14]

Aaron Shatkin also had sent out letters to scientists, in which he inquired about the effects of animal tumor viruses on humans. That discussion would certainly loom heavily at Asilomar, since the proposed implantation of SV40 into *E. coli* had been a key factor in starting the controversy. In a response to Shatkin's letter, Frederick Robbins, dean of the School of Medicine at Case Western Reserve, summarized what he knew about the investigation of 1,000 schoolchildren who had been vaccinated with polio virus contaminated by SV40.[15] The investigators, who were focusing exclusively on excess mortality, reported that in the fourteen years since the study began, no excess mortality had been discovered.[16]

In the last paragraph of his letter, Robbins praised the initiative taken by Berg and Baltimore. Then he rather gingerly touched on the subject of creating new pathogens. He said that when one tries to create a new recombinant, a possible outcome could be the production of a defective virus, which is analogous to what is done by radiation. He then reminded Shatkin that the manipulation of these viruses could produce an oncogenic product regardless of whether the parent organism is oncogenic. Robbins raised an issue of considerable importance. In creating recombinants, can scientists produce a product that is more dangerous than any of its component parts? The implications of an affirmative answer are extremely dismal for rDNA research involving animal tumor viruses or viruses in general. If we cannot venture an intelligent guess about which hybrid viruses will cause disease in humans, then all such viruses are candidates for high-containment handling.

Renato Dulbecco, an eminent scientist at the Imperial Cancer Research Fund Laboratories in London, England, put forth several arguments in a letter to Shatkin that had led Dulbecco to the view that working with rDNA viruses required no greater precaution than working with ordinary transforming viruses.[17] He based his conclusion on evidence indicating that the dangers from recombinant molecules were small. The evidential support included these observations: the plasmid or phage DNA containing a transforming gene would have a probability of infection similar to that of SV40 DNA in the absence of adjuvants; the laboratory strain of *E. coli* entering the intestinal tract would have no selective advantage and would subsequently disappear, and a plasmid carrying bacteria would disappear even faster; if a cell is infected by DNA in the gut, it will not be the source of reinfection.

In his conclusion, Dulbecco remarked that the danger of working with plasmids of transforming genes in bacteria is less than the hazards associated with work on SV40 or Rous sarcoma virus. He did admit that unexpected things can happen. One of these is that DNA released from bacteria will infect cells with far greater efficiency than expected. But all in all, he figured, the precautions taken in working with transforming viruses are sufficient. In conclusion, Dulbecco saw no justification for banning recombinant research of any type and placed on those who would limit such research the burden of justifying these limitations.

Howard Temin, a biologist at the University of Wisconsin, concurred that the risk of spreading "cancer genes" was negligible as long as one was not working with cancer cells. In correspondence to a biohazard subcommittee at his university, Temin maintained that the probability of induction of cancer was no greater in having eukaryotic DNA present in bacterial cells than there is in having it in the eukaryotic cell.[18] According to Temin, there was no evidence that implanting animal virus DNA in bacteria would make the virus more dangerous than it was originally. Temin added that animal virus DNA and fragments of it were less pathogenic than the virus itself.

On the other side of the issue, those who disagreed with Temin were concerned that latent viruses might be propagated in bacteria, thereby opening up new routes of infection for humans. The viral DNA might replicate as fast as the bacteria. If there were a viral oncogene, its insertion in a bacterial host and subsequent replication could establish that gene in an entirely new ecological niche with a possibility of having its effects amplified. Those concerns were expressed by Paul Berg in a letter to E. S. Anderson during the Asilomar organizing period.[19] Berg felt that the conference should give some thought to the following issues: (1) Might a hazard result from exposure to *E. coli* replicating a viral DNA segment joined to a plasmid? (2) What is the probability that SV40 DNA introduced into the gut via *E. coli* would infect the brain? (3) Could a herpes virus DNA fragment, when joined and amplified, establish a latent infection (or induce a malignancy) in man?

Ecological niches were also on the mind of Allan Campbell, a Stanford University biologist, and one of the scientists Donald Brown had written for some views on the hazards of transferring nitrogenase genes into exotic species. Campbell's response focused on the potential hazards of disturbing biochemical pathways.[20] Suppose we introduced the nitrogenase (nitrogen-fixing) genes into organisms that are lacking them. Campbell maintained that there would be an extremely minute chance of producing a catastrophic result. But since such an improbable but finite catastrophic result exists, Campbell recommended that containment procedures for such experiments be established and followed. The burden of proof, according to Campbell, should be on the investigators to show that possible ecological risks have been explored. Next to the problem of a possible disturbance to the biochemical cycles, Campbell wrote, the worry that rDNA research might cause a few thousand extra cancer deaths seemed less significant.

Another area of concern during the pre-Asilomar period was the risk of spreading antibiotic resistance properties to pathogens through rDNA technology. Some individuals, however, argued that the risks of increasing bacterial resistance to antibiotics through genetic manipulations in the laboratory were negligible next to what was occurring outside of the laboratory under natural conditions. One of these scientists was the English bacteriologist Naomi Datta. When the Berg committee's letter appeared in *Nature*, Datta and her colleagues at the Royal Post Graduate Medical School (Hammersmith Hospital, London) discussed the experiments classified as type I, and concluded that such experiments were not hazardous. Datta communicated these findings to Paul Berg in October 1974, while Berg was steeped in the organizing efforts for the Asilomar meeting.[21]

Datta's arguments were based upon two assumptions. The first holds that what can be done in the laboratory will have been done more significantly under natural conditions. According to this view, bacteria of all types are constantly exchanging plasmids, some of which will bear the resistance genes for certain antibiotics. Thus, people coming into contact with a myriad of bacteria will eventually pick up some with the antibiotic resistance genes. Datta argued that this effect will be a more frequent occurrence than what could happen in the laboratory.

The second argument states that even if novel antibiotic resistance strains are produced, this in itself cannot be dangerous, since these strains will not be disseminated. Moreover, Datta pointed out, there was plenty of room for hairsplitting in the interpretation of the Berg moratorium. Suppose someone wanted to create a plasmid with two antibiotic resistance genes. Assume further that this plasmid is not found in nature, and therefore falls under the moratorium. Since, to the bacteria, the situation is equivalent to having two plasmids, each with one of the resistance genes on it, Naomi Datta argued that the "new construction" does not add to the natural hazards.

Others considered the issue more serious. In the fall of 1974, biologist Carolyn Ruby of the Merck Institute for Therapeutic Research in Rahway, New Jersey, requested a special strain of bacteria from Martin Gellert, a molecular biologist at the National Institute of Arthritis, Metabolism and Digestive Diseases. Ruby explained that she planned to use the bacterial strain for implanting drug resistance genes. Gellert refused to comply with her request.[22] He wrote back to Ruby, identifying her work as "a dangerous game with totally unpredictable results." Furthermore, Gellert believed that her experiment would be barred by a reasonable interpretation of the rules of the NAS (Berg) committee.

A particularly interesting example of a conjectural risk is provided by the case of immune response. It was raised in a letter to the head of the Eukaryote Working Group, Donald Brown. The correspondent was Sumner C. Kraft, an MD and professor of Medicine at the University of Chicago. Brown had asked Kraft to provide any information about the potential hazards of implanting foreign protein into bacteria. Kraft wrote back in late October 1974.[23] He made several points: Enteric bacteria can release proteins giving rise to immune responses. Proteins do not have to be released into the circulation system to bring about such a response. Many unknown and diverse factors may be at work in deciding who will be allergic to a particular protein. A radical transformation of an individual's bacterial flora could have untoward effects on that individual.

Among all the biohazard scenarios raised at this time, the immunological effects of rDNA molecules had the least visibility. Nearly six years after Kraft framed the problem, a formal risk assessment conference was organized by NIH to examine whether immune responses to active polypeptides in rDNA molecules warrant serious attention.

Another issue raised in the pre-Asilomar period was whether anything created in the laboratory with rDNA techniques could "short-circuit" evolution. Can man do what nature has seen fit not to? One of the strongest consistent proponents of the view that rDNA techniques would not introduce anything that nature has not already created is Bernard Davis, Adele Lehman Professor of Bacterial Physiology of the Harvard Medical School. Davis presented this line of reasoning in a letter to Paul Berg in September 1974.[24] His reconstructed argument takes the following form:

1. DNA recombinants are arising all the time in nature—including in the human gut.
2. Whether the new recombinants reproduce and proliferate into the

environment depends upon their selective advantage with respect to other organisms with which they are competing (Darwinian selection).
3. The new recombinants developed in the laboratory do not have a selective advantage in nature, otherwise they would already be widespread.

---

4. Therefore, the only people at risk from the laboratory creation of DNA recombinants are the workers, who may risk getting an infection.

Davis recognized that his first premise was a conjecture for which he believed there was some evidence, although it certainly was not conclusive. His ideas were more carefully developed in the next several years. They represent a unique attempt to draw into the risk analysis a broad theoretical framework, namely, evolutionary theory, to examine the potential hazards of the research.

My purpose in this chapter has been to explore the types of arguments that scientists were using in response to the possible hazards of creating hybrid organisms with rDNA molecules. The analysis of the pre-Asilomar correspondence illustrates the range of problems considered by scientists during this early period. Many of the arguments were simply sketches; considerably more knowledge was needed to determine whether the hypothetical risks were something more than logical or theoretical potentialities. Each issue touched upon several disciplines. For example, scientists would begin with a bacterium; but then the DNA of an animal virus would be inserted into it; and finally, it would enter the human intestinal tract. Since the individuals discussing the potential risks generally did not have expertise in all relevant fields (bacteriology, virology, and infectious disease epidemiology), it was incumbent upon the decision makers to view the problems across scientific disciplines.

The early correspondence illustrates a degree of polarization among some experts; it also brings into question the very notion of where the expertise lies.

9

Building a Consensus Position

The heart of the Asilomar Conference was the work of the three groups that had been convened to review the potential hazards in the areas designated by the Organizing Committee: (1) plasmids and bacterial DNA; (2) eukaryotic DNA; and (3) viruses and viral DNA. As with the pre-Asilomar correspondence, these three groups exhibited varying styles and temperaments, and each interpreted its tasks somewhat differently.

By far the most wide ranging and ambitious report was produced by the Plasmid Working Group (PWG), chaired by Richard Novick of the Public Health Research Institute, City of New York and on which were four other members.[1] The analysis and recommendations of this panel were delivered in thirty-five single-spaced pages. The document contained six main sections. The introduction, in a section entitled "Scope and purposes," stated that the report planned to go beyond the specific concerns of the Berg letter. The Plasmid Group explained that it had included in its focus "*all* genetic manipulations involving the introduction into a prokaryote species of genetic material that may or may not be native to that species and may be unlikely to be acquired by it in the natural environment."[2]

While the group nominally framed its recommendations with prokaryotic organisms in mind, its members supported the position that the report's conclusion should be extended to eukaryotic microorganisms as well.

The proposal was an extraordinary effort in many respects: the desire to achieve comprehensiveness; the emphasis on detail, its clarity; and its specificity in classifying different types of experiments in which biohazards may arise. Beyond these elements, what stands out in this document as a tribute to the broad environmental concerns of its authors is a set of six principles forming the philosophical basis of the proposal:

1. Since man has some measure of control of [sic] his actions, there

is an operational dichotomy between the activities of man and the processes of the natural world. The distinction between "man-made" and "natural" is therefore meaningful and control of the former is both worthwhile and possible.
2. It is possible to modify profoundly the genome of a (micro)organism by artificial means involving the *in vitro* joining of unrelated DNA segments. Such modifications may find expression in the organism's phenotype as well as its genetic constitution.
3. Modified (micro)organisms may behave in an unpredictable manner with respect to the expression of foreign genes and to the effect of this expression upon their ecological potential (including their pathogenicity).
4. The genetic effects of these manipulations may be different from anything that ordinarily occurs during the natural process of evolution.
5. Historically, unforeseen ecological effects of technological developments have been more often than not detrimental to man and his environment.
6. The release of a self-replicating entity into the environment will prove to be irreversible should that entity prove viable in the natural environment.

According to the document, some of these principles were regarded as assumptions. And as assumptions, they entreat us to behave cautiously. It places the burden of proof clearly on those who claim that the recombinant hybrids have already been fashioned under natural conditions. It draws upon our experience of past technologies to convey that we should expect adverse consequences from rDNA research. And it links self-replication and irreversibility; once an organism is released and survives, there can be no going back to square one.

The plasmid report echoed the risks that had been discussed at the Gordon Conference and in the correspondence that the chairman of the working group had received. The report cited the potential biohazards of microorganisms carrying "novel combinations of genes that have never existed before or are very unlikely to arise in the course of natural evolution."[3] Modified organisms might be able to occupy a novel ecological niche or function differently than they do in their present ecological setting. One possible outcome mentioned in the report was the creation of microorganisms that can increase human disease or inhibit the treatment of disease—such as bacteria resistant to treatment by antibiotics. The analogy was made with cases in which organisms were introduced into new environments with results both unpredictable and uncontrollable. The report cited fire ants, killer bees, *Xenopus* toads, and the organism responsible for

Dutch elm disease as cases in point. These species, when brought into new environments lacking their indigenous ecosystem's population inhibitors, create significant and irreversible perturbations that cause widespread havoc.

Four advances are cited on the benefit side of the new research: (1) fundamental knowledge of gene structure, organization, and function; (2) genotypic modification of plants or animals to improve their usefulness to man; (3) the instruction of bacteria that could produce valuable biological substances such as insulin and growth hormone; and (4) curing human hereditary diseases.

The PWG report went on to provide the following approach to the problem:

1. Determine the biohazard potential of the experiment.
2. Match the biohazard potential with a system of containment that limits the frequency of the escape of organisms.
3. Where the escape of a small number of organisms is deemed hazardous, the experiment should not be performed.

The expression "biohazard potential of experiments" was introduced to discriminate among different types of gene-splicing activities. Eight factors were identified for establishing some relative weightings for "biohazard potentials": increase in pathogenic potential of organism resulting from genetic manipulation; extent to which modified organism can spread into the environment; extent to which a modified host organism will alter its natural ecosystem; degree of persistence in an ecological niche; extent to which foreign DNA is expressed in host; extent of characterization of the organism; purity and characterization of DNA used in creating hybrid molecules; scale of the experiment—large-scale experiments are to be considered more potentially hazardous than small-scale experiments.

For most of these factors, their connection with biohazards is fairly straightforward. A brief elaboration of a few points is offered for explanatory purposes.

The fifth point asserts that there is an increase in the "biohazard potential" if the foreign DNA can be expressed in the host, as compared to the case where such expression is unobtained. The sixth point distinguishes between random fragments of DNA and well-characterized segments. If it can be shown that a segment of DNA is well characterized and expresses a single product and that the product is known *not* to be harmful, then presumably the risk of a biohazard is greatly mitigated, if not zero. The reason for this is that a given sequence of DNA is believed to code for only one protein. To say that the code is universal means that a sequence that codes for rat insulin in rat cells will not code for any other protein when implanted in *E. coli*. Thus, a well-characterized sequence of DNA is much less of an "unknown" than random gene fragments.

The Plasmid Group classified experiments into six types, starting with class I experiments, for which it is known conclusively that the biohazards are insignificant, and ending with class VI experiments, for which the biohazards are judged to be of such great potential danger that the gene transplants are not to be undertaken under any type of containment conditions. The six containment classes as presented in the PWG report are as follows (emphasis added):

Class I. Biohazards can be assessed and *are known to be* insignificant; genetic exchange occurs naturally.

Class II. Biohazards can be reasonably assessed and *one expects them* to be minimal; genetic exchange occurs naturally.

Class III. Experiments in which the biohazards usually *cannot be totally predicted*, but (1) the donor DNA will not significantly increase the pathogenicity of the recipient, nor significantly alter its ecological potential; (2) pathogenicity of genetically altered microorganisms or its parents is minimal; (3) genetically altered microorganisms do not escape treatment of infections caused by it.

Class IV. Experiments in which the biohazards are *usually unknown* and cannot be accurately assessed; transplanted DNA is judged *potentially significant* in affecting either the *ecologic potential* or *pathogenicity of the recipient organism*.

Class V. Experiments in which the biohazards are *usually unknown*; transplanted DNA could severely affect the pathogenicity or ecological potential of the recipient organism.

Class VI. Experiments in which the biohazards are judged to be of such *great potential severity* as to preclude performance of the experiment, under any circumstances at the present time.

Each class of experiments was matched with a specific level of containment that included rules of microbiological practice around the work area and certain types of laboratory equipment. Class V experiments called for the most technically advanced facilities for handling infectious microbiological agents. The experiments placed in class VI were deemed potentially too hazardous for any containment conditions.

The final section of the PWG report was devoted to implementation of the guidelines. It proposed that local biohazards committees be established at every institution, both commercial and academic, where work is undertaken with rDNA molecules. The committees would play a critical role in assessing the biohazards of the planned experiments. Investigators would be required to submit a proposal for each rDNA experiment, describing the purpose of the experiment, its explicit benefits, an assessment of the potential biohazards, and the kinds of precautions to be taken. The proposal and other documents related to the assessment of the project would all be kept on record at the institution. The Plasmid Group also suggested that proposals of each project become available to editors of scientific journals, who could require, as an additional standard for publication, that results be obtained according to a set of agreed laboratory practices.

Despite the review by an institutional committee, a principal investigator still would bear the major responsibility for the assessment of the project. The proposal said nothing about an overseeing role for the NIH or any other federal agency. It seemed sufficient to this group of scientists that a peer review mechanism be established. No penalties were cited for investigators who violate the guidelines: "We believe that a combination of scientific integrity combined with peer pressure generated in the face of public availability and scrutiny of the documents mentioned above, will result in strict adherence to the guidelines, while at the same time avoiding the extremes of the approval of [a] hazardous series of experiments by a poorly informed local committee or a veto by a local committee of experiments which would be generally accepted as valid and worthwhile under the conditions of containment that have been proposed."[4]

The authors of the Plasmid Working Group took on a kind of missionary zeal in promoting adequate microbiological training for those contemplating work in rDNA. Their concern was that DNA chemists beginning work with microorganisms would take a naive reductionist approach to the issue of risk. The report asks rDNA investigators to start thinking in microbiological terms, that is, beyond the genetic parts of an organism to its whole being in the environment. A memorable phrase captures the intent: "The microorganism is not simply a 'warm body' to house a recombinant DNA molecule of interest."[5]

Near the end of their report, the Plasmid Group inserted a strong political statement about the government's potential exploitation of the technology: "We believe that perhaps the greatest potential for biohazards involving alteration of microorganisms relates to possible military applications. We believe strongly that construction of genetically altered microorganisms for any military purpose should be expressly prohibited by international treaty, and we urge that such prohibition be agreed upon as expeditiously as possible."[6]

Notwithstanding the expressed concerns of this working group, the discussion of rDNA and the military never materialized at Asilomar. It was left for individuals to pursue the issue after the conference prepared its final recommendations.

Chaired by Aaron Shatkin, the Animal Virus Working Group (AVWG) consisted of eight members.[7] Its report took up all of one page—typed single spaced. An additional half-page was appended to the report containing a minority statement by Andrew Lewis. The AVWG report recommended that work with rDNA proceed under the same guidelines used by the National Cancer Institute (NCI) for work with oncogenic viruses. NCI has three categories of risk: low, moderate, and high. According to the AVWG, recombinants containing animal virus genes should be classified in the moderate-risk category, and it recommended that investigators follow the precautions stipulated by the NCI for that category of experiment. For purified segments of viruses that are not associated with pathogenicity, the report called for low-risk containment. And finally, for work with highly pathogenic viruses, high containment was recommended. Most experiments would fall into the moderate-risk category.

It was a very brief statement that simply extended what was already in place at NCI to the entire spectrum of bacteriological work with rDNA molecules. The report implied that there was nothing qualitatively new about the risks of rDNA technology. To implement these classifications, the AVWG suggested institutional peer-review committees similar to those suggested by the Plasmid Group.

Andrew Lewis took a minority position. At that time, Lewis believed, the risks from some rDNA work exceeded the benefits. He called for significantly slowing down the research until there was more data on the risks. The burden of proof, he maintained, falls on those who assert, without the validating data, that the risks are negligible.

The Eukaryote Working Group (EWG) was chaired by Donald Brown of the Carnegie Institution and consisted of seven other members.[8] The panel delivered a five-page report with a three-page appendix. Brown communicated with the Plasmid Group while his own committee was preparing its recommendations. The EWG report cited specific sections of the Plasmid Group's report. It also utilized the latter's six-stage classification scheme for experiments.

The EWG conceptualized three types of potential hazards with DNA molecules from eukaryotes: (1) A eukaryotic gene when placed in a bacteria could function and produce a toxic product. (2) Eukaryotic DNA could enhance the virulence or change the ecological range of the bacterium in which it was cloned. (3) Eukaryotic DNA could infect some plant or animal, integrate into its genome, and modify the cells of the organism.

The report acknowledged that while current rDNA work involved the insertion of prokaryotic or eukaryotic DNA into *E. coli*, future developments would surely take such work into animal cells. It went on to say that if gene therapy is to be pursued, eukaryotes must be found as host-vector systems. At the same time that the EWG cited the importance of eukaryote systems, they had some reservations about their safety: Will foreign DNA sequences inserted in eukaryotes be more hazardous than when the same DNA sequences are inserted in prokaryotic systems? The report conjectured that propagating DNA fragments with the same biohazard potential would be more hazardous for eukaryotic systems than if the same fragments were propagated in prokaryotic systems. This hypothesis was reflected in the group's assessment of experiments according to potential biohazard. For any particular source of DNA, the EWG generally graded the risks higher for animal or plant vectors than for plasmids and bacteriophage used for similar purposes.

The EWG also ranked what we may call "estimated potential biohazards" of specific classes of experiments. These assignments of relative risks were derived neither from some algorithm nor some rigorous analytical method. Much of the evidence necessary for making such assessments was lacking. The classification scheme was grounded on assumptions deemed plausible to the EWG. The report stipulated some principles that guided the decision.

One principle sets a lower boundary on the risks of an experiment involving several biological components. The principle can be stated as follows: If a eukaryote virus gene is implanted in some host, then the risks of the experiment are not less than that for the donor virus itself. According to this view, the estimated potential biohazard in cloning a combination of entities is not less than the estimated potential biohazard of its most hazardous fragment.

Another assumption in the report is that DNA from higher mammals has a greater potential for human hazard than the DNA from lower animals. The reason offered is that there is greater likelihood of unleashing a human pathogen by working with species closer to *Homo sapiens* than by working with species further removed on the phylogenetic scale. The upshot of this view is that, for example, mammalian viruses have a better chance of infecting humans than do insect viruses.

In contrast to the Plasmid Group, the EWG had nothing to say about implementation. But they did end their report with comments about the kinds of knowledge that would help in risk assessment, framed in the form of questions: Will a eukaryote gene function properly in a bacterium? (The enzymes necessary to start the coding process may not be available for DNA in bacteria.) Can DNA and RNA infect animals and plants? (A closely related question was, Can hybrid DNA molecules or their transcripts find a transmission route from bacterial cells to plants and animals?) Will bacteria containing eukaryote DNA recombinant molecules have an altered ability to colonize any eukaryote?

The eight-member eukaryote panel agreed that there were many significant questions that remain to be answered before the risk assessment could be empirically validated. However, they still felt confident in assigning containment values for a wide range of experiments. Those values fell within classes II through V of the plasmid panel's taxonomy. The EWG did not classify any experiments in class I or class VI. Three basic factors helped determine the estimate of potential risk: where the DNA came from; the vectors used to implant the DNA; whether the DNA in use was purified (well-characterized or random sequences).

The class assignments for different experiments was the result of a consensus from members of the EWG. At the high-risk end of the scale (class V), the following types of experiments were cited: shotgun experiments with primate DNA using animal viruses as a vector; purified DNA expressing known toxic products attached to any vector—eukaryotic or prokaryotic. At the low end of the scale, requiring class II containment, only one class of experiments was listed—purified eukaryotic sequences are implanted in a host in which they are not expressed when the vector is a prokaryote type.

One noteworthy feature of the EWG report is its relatively narrow scope. Where the Plasmid Group took every opportunity to broaden its concerns, the EWG passed each of these opportunities by. The issue of gene therapy, for example, was an open invitation to make at least a passing reference to the fact that this was a controversial use of the new technology. Instead, only the issue of biohazards was acknowledged.

Although the three panels were responsible for reviewing different aspects of the problem, there is a basis for comparison. Next to the eukaryotic and virus panel statements, the report of the plasmid panel was monumental in scope. Unlike the other two panels, the Plasmid Group set up a category that proscribed certain cloning experiments. Two examples were given of such experiments. The first was the insertion of certain sequences of the diphtheria genome into *E. coli*; the second was the insertion of the genes that code for botulinum toxin into *E. coli*. There were also important differences between the AVWG and the EWG position statements in how animal viruses would be handled. The Virus Group was satisfied with employing the practices and containment provisions established by the NCI, while the Eukaryote Group favored a new containment strategy following the recommendation of the Plasmid Working Group. Finally, both the Virus and Plasmid Groups supported the idea of institutional committees advising the principal investigators on the appropriate safeguards for experiments.

10

A Framework for Genetic-Research Policy

The sessions of the Asilomar Conference were organized around the themes of the working groups, with additional time set aside for the presentation of the ethical, legal, and public-policy perspectives. The introductory talk was given by David Baltimore, who summarized the events leading up to the meeting and presented the goals of the conference. Baltimore told the assembly that the new techniques may allow scientists to outdo evolution by establishing new combinations of events that nature has not seen. The issue of creating novel organisms lay at the root of many of the early concerns raised by scientists. Could molecular biologists construct an assemblage of genes that would behave differently than any of its parts? If that were the case, it would be difficult to build a risk analysis of the hybrid organism exclusively from the individual components.

On the other hand, if it could be successfully argued that new gene assemblies are being formed constantly in nature at a much higher frequency and in much more diverse circumstances than would be achieved in the laboratory, then the so-called novel biotypes may have already been tested for their persistence and virulence.

During his opening remarks, Baltimore also placed the Asilomar meeting in a policy context. He stated that the scientists had been called together to balance the benefits and the hazards of the research and to produce a strategy that will maximize the benefits and minimize the hazards. The conference agenda, however, did not reflect such a risk-benefit approach, except possibly in the narrow sense in which benefits were viewed primarily as breakthroughs in basic knowledge. There was no sign, during the meetings, that a general framework for a risk-benefit analysis had been considered. Individual participants may have, consciously or subconsciously, struck personal balances based upon an interpretation of the current state of knowledge, subjective criteria, and personal values. But they did not play an active role in shaping the outcome of the meeting. For the most part, they were passive or reactive, leaving the bulk of key decisions to the working groups and the Organizing Committee.

Paul Berg followed Baltimore's introduction with a discussion of the techniques then available for joining DNA from different species and implanting the recombinant molecules into a bacterium. Following the general orientation, technical talks were given on topics related to the subject area of the working groups; after each report was completed, the appropriate working panel presented its recommendations.

For three full days, the lectures, panels, and working sessions ran from 8:30 A.M. to 10 P.M. Periods were set aside after lunch for informal or spontaneous meetings. That was time put to good use as several unscheduled workshops took place at those hours. One notable workshop was organized by Sydney Brenner, who examined the problems of designing safer cloning vectors (plasmids and bacteriophages) and safer bacterial hosts.

But there was also activity at the conference reflecting the ethical, legal, and social problems of genetic research, despite the fact that these matters were not considered by the Organizing Committee as germane to the question whether there was risk in the work. Some of these discussions about social impact were scheduled into the conference. More will be said about this below. But there was also an unsolicited position statement to all conference participants from a group affiliated with Science for the People, and it bears some examination.

In 1973, a study group had formed out of the Boston area chapter of Science for the People (SESPA) around the issue of genetic research. It was called the Genetics and Society Group (GSG). Study sections like GSG in Science for the People are loosely formed voluntary associations of individuals who share a similar perspective about the role of science as a social institution. At the minimum, this usually amounts to the views that science is shaped, to a significant degree, by social and political events in society; scientists are people with special interests, sometimes class interests; and the activities of science, including the choice of research and the advancement of theories, are not politically neutral activities.

The GSG study group initially had chosen to address two types of issues: (1) social or psychological theories grounded on genetic evidence, that is, the genetic determinants of IQ; (2) scientific research programs designed to generate evidence that social or political behavior was genetically determined. Two members of GSG, Jonathan King and Jonathan Beckwith, had received international attention for their campaign against research that sought a connection between an extra Y chromosome in males and antisocial (criminal) behavior.

After the Berg letter had been published in July 1974, several scientists in the Boston area, including graduate students and postdoctoral fellows belonging to GSG, initiated informal discussions on the social impact of rDNA research. By December 1974 some members of GSG began to hold meetings to which other interested scientists were invited for the purpose of examining the broader issues of gene-splicing work. The working group that was formed from these meetings called itself the Genetic Engineering Group of Science for the People (GEG/SESPA). GEG learned of the upcoming Asilomar meeting and drafted an open letter to the participants which was marked to each attendee and sent to the conference location.

The letter addressed both the potential biohazards and the social consequences of the new gene-splicing techniques. It made a connection between rDNA techniques and other technologies such as cell fusion and in vitro fertilization. The combination of these technologies was viewed as converging toward human genetic engineering. The letter stated that the public should be informed about rDNA research and that decisions about who benefits and who bears the risks should not be left in the hands of scientists. The GEG statement emphasized that science was a value-laden activity. Genetic engineering, it held, does not arise out of general social needs; scientific interests are not always synonymous with social interests. For example, research into the cure for disease takes precedence over research into prevention. Foremost, the statement called for broad public participation in the decision-making process at Asilomar. Responsible self-regulation of science, while an ideal according to the GEG, was not possible on this issue at this time.

There were five concrete proposals made in the GEG statement: (1) Involve those most immediately at risk—technicians, students, custodial staff, and so forth—in collective decision making on safety policy for the laboratory. (2) Integrate into the curriculum of biology and medical courses the social implications of present and future biomedical research. (3) Require social and environmental impact statements on the means and goals of biological-research projects. (4) Continue examinations of these matters at public sessions of scientific meetings. (5) Expand participation on the proposed rDNA advisory committee to the NIH; the NIH might serve as the structure through which the involvement of nonscientists in decision making could be implemented.

In retrospect, these are very modest proposals from an organization that has been branded "radical." Ironically, two years after GEG/SESPA drafted these recommendations, most of them were supported by politically moderate environmental groups like Friends of the Earth and the Environmental Defense Fund. By January 1979 three of the five proposals were adopted by NIH (first, fourth, and fifth); the third was only partially adopted with the issuance of an Environmental Impact Statement for the NIH *Guidelines*.

The GEG/SESPA statement was signed by nine Boston-area scientists. It received no official attention at the Asilomar meeting. Representatives of the group were not in attendance. There was no recognition given to members of the GEG for their input to the meeting.[1]

Despite this disregard of the GEG statement, there were two sessions at Asilomar devoted to social questions. The first was scheduled on the first day of the meeting after the luncheon break. The speaker was attorney Harold Green of the National Law Center at George Washington University. The title of his paper was "A public policy perspective." A second session was set aside for Wednesday after the dinner break. The three lawyers who participated in this session were Daniel Singer (affiliated with a Washington, DC, law firm and with the Hastings Institute), Roger Dworkin (a law professor at Indiana University), and Alex Capron (a law professor at the University of Pennsylvania).

Green brought the attention of his audience to the responsibility of scientists when their work may put people at risk. He pointed out that it is easier to make public-policy decisions when the risks are real and known. Asilomar, he noted, presented a very different situation. There was uncertainty whether adverse consequences would result from experiments with rDNA molecules, nor was there any knowledge about their extent. According to Green, society must assess the risks of new technologies at their inception. Otherwise, the risks are built into our way of life and it becomes more difficult to evaluate them in an objective fashion. He used the analogy of nuclear power to illustrate the limitations of the risk-benefit model. As more nuclear plants are made, their benefits increase since society becomes more dependent on their products. But by that time it may be more difficult to exercise reasonable controls because of the opposition by entrenched economic interests.

Slowing down the use of a potentially hazardous technology can postpone certain benefits; Green considered that outcome, however, to be a sensible policy. He stated, "[S]ociety can more easily tolerate a postponed, or even a lost benefit than it can actual injury."[2]

Green commended the participants for taking a responsible role in examining the risks before any harm has been done. In addition, he believed that the participants represented a broad range of expertise from the scientific community (a claim that was disputed by environmental organizations after the conference ended). He also praised the organizers of the conference for including participation by nonscientists such as himself.

In his summation, Green made an important distinction between balancing risks and balancing benefits: "[B]enefits are always relatively obvious, immediate, intensely desired, while risks are usually relatively remote and speculative. There is, moreover, usually no constituency for the risks."[3] In Green's view, more attention must be placed on risks, as the benefits have their natural constituencies.

As organizer and participant of the Wednesday panel, Daniel Singer approached the legal and ethical issues more like an applied philosopher than a lawyer. He cited two underlying assumptions of the Asilomar meeting that could be challenged. The first was that the risks are remote when compared to the scientific and medical benefits of the research. The second was that it was morally wrong as well as politically dangerous to place restrictions on intellectual activities. Scientists at this meeting, he said, would be making ethical and/or social judgments, and they should not convince themselves otherwise. In this regard, the scientists should act responsibly by considering the circumstances in which one human being may subject another to nontrivial risks. Once those circumstances are established, the next question concerns the responsibility a purveyor of risks has with respect to the victim. One individual may not knowingly subject another to risk unless there is a clearly determined right, as in the case where society has made such a collective determination.

Singer ended his remarks with three prescriptions to the Asilomar participants. First, it is up to the proponents of the research to prove that it is safe, not merely to themselves, but to the broader public. Second, the principal investigator has a responsibility to inform fully all those in the laboratory environment of the potential risks (to them and the outside community). Third, scientists must be willing to face the prospect that some experiments, though elegant and intellectually satisfying, may be morally reprehensible.

Alex Capron was introduced by Daniel Singer as a professor of law at the University of Pennsylvania, where he teaches Torts and Law and the Life Sciences. Capron's talk was entitled "Legal analysis and response to rDNA hazards." He noted that while the major responsibility for safety has been in the hands of individual investigators, these informal codes of responsibility are too vague and generally inadequate. He told the Asilomar scientists that they could no longer ignore public participation. Capron distinguished two levels of participation. First, one must consider the participation of those who work in the environment of the laboratory. They must be advised of the risks and given options for avoiding them. The second level of participation, which, he acknowledged, could be more difficult to satisfy, involved the broader public. At this time, supporters of public participation, like Capron, were hard pressed to form a clear idea of how to achieve it in an area involving hypothetical hazards in scientific research. What are the rights of communities living in the vicinity of institutional laboratories? Can participation work at some national advisory level, where decisions involve a body of esoteric knowledge?

As far as the law is concerned, Capron said, the scientists should understand that freedom of thought does not imply freedom to cause physical injury to others. Where there are such potential outcomes with rDNA work, prior restraint is a reasonable response whenever "irreversible harm is threatened." The values of science and the values of the public may differ, in Capron's view, and the risk-benefit outcome of the scientific community should not reign supreme: "In legal systems which are based on public sovereignty, it is the right of the public, through the legislature, to reach erroneous decisions—while it remains your [the scientists] right, and is probably your obligation to lobby against such decisions but to abide by them in the meantime."[4]

Capron addressed the issue of having guidelines for research. The administration of "soundly drafted regulations" could be left to the scientific community. Here he was referring to the determination of physical-containment requirements for particular experiments. But at some higher stage, where revision of the guidelines may take place, Capron supported the involvement of nonscientists.

Roger Dworkin reached Asilomar much on his own initiative.

When he got wind of the conference, he contacted Paul Berg directly, requesting an invitation to speak. At the University of Indiana Law School, Dworkin taught in the fields of biology and the law and ethical values and legal systems. He had been a visiting professor at the University of Washington School of Medicine during the Asilomar period.

Dworkin's talk took a very practical turn in comparison to the other lawyers. Foremost on his agenda was an examination of liability. The issue of liability, he said, will often be determined by nonscientists and independently of the issue of intention or negligence on the part of an investigator. Negligence, in the case of rDNA work, can result either from the way the work was done by the researcher or from doing the work at all. In judging liability, the conduct of the researcher will be evaluated by nonquantifiable factors; for example, how likely was the hazardous event, how serious was the injury. Dworkin alerted the scientists that even guidelines would not protect the investigator from being found negligent in a liability case.

How did the talks given by the legal experts affect the scientists at Asilomar? The evidence (from oral interviews of Asilomar participants, questions raised during the discussion period, and from the written accounts of those who attended) suggests that the social issue of greatest concern to the scientists was their personal liability emphasized in Roger Dworkin's talk. Dworkin's account of the reaction to his talk is found in his post-Asilomar recollections. "What a legal audience would have regarded as commonplace, elementary, and obvious, struck the distinguished scientists as novel, shocking and frightening. Calling the researchers' attention to their potential liability induced a fear in them akin to a layperson's fear of virulent bugs crawling out of a laboratory."[5]

Another participant at Asilomar, Robert Sinsheimer, also felt that the scientists responded to the lawyers strictly from their concern about suits and not because of any moral response to issues raised. Sinsheimer was disturbed about this lack of ethical fiber among the scientists: "If somebody gets cancer because you've done . . . [an] experiment, the point isn't that you're going to pay them a million dollars to recompense them for it. And yet that kind of attitude was expressed."[6]

As a consequence of Dworkin's presentation, scientists began to understand that their laboratories could fall under the jurisdiction of the Secretary of Labor. The Secretary has the responsibility of implementing the Occupational Safety and Health Act (OSHA). John Lear gives an account of the interrogative period after Dworkin's paper:

> The question-and-answer session that followed Dworkin's talk left no doubt that he had made the deepest impression of the evening. The vision he had conjured up of inspectors from the U.S. Department of Labor swooping down on research laboratories without warning and slapping fines or jail sentences on slovenly experimenters was too much to contemplate with serenity, so alien was it to the permissive regulations of the NIH, which are promulgated at least as much by the researchers themselves as by the officials who are ultimately responsible.[7]

Richard Roblin recalled that many scientists in the room listened attentively when the OSHA laws were discussed. Roblin observed that there was little awareness among scientists that the Secretary of Labor and not the Secretary of HEW was responsible for the safety of the workplace.[8] As a consequence of this session, some scientists were willing to proceed more cautiously with their work and accept a modicum of regulation. Moreover, they preferred that any oversight come from their benefactor, NIH, and not from "alien" sectors of government.

The most significant business of the conference, of course, was to reach a resolution on the open questions so that the scientists could get on with their work. The Organizing Committee continued to play a central role during the conference, as it had done earlier in establishing the goals and determining the makeup of the meeting. There were two outcomes that the committee wanted to be sure were achieved: (1) the general moratorium should be lifted, and (2) the oversight of potentially hazardous experiments should remain, as much as possible, within the scientific community. How this kind of resolution was to be achieved by the conference, however, was not at all clear.

From the outset of the Asilomar meeting, there was no clear policy on reaching a consensus on final recommendations. The issues broke down into so many parts that entertaining a vote on each point could have taken considerably more time than was available. The Organizing Committee took a wait-and-see attitude. It was difficult to assess in advance how much dissonance would result from among those assembled. Berg described the process as that of "feeling out the consensus of the participants."[9]

When Baltimore completed his introductory remarks at the opening session, a participant from the audience asked him how consensus would be determined. His cryptic reply was that the way in which consensus would be determined would depend upon the extent of the consensus. He told the audience that if there were significant differences, voting would not help.[10]

Without some form of voting, how could the assembly reach a conclusion? The answer lies in the special relation the Organizing Committee had with the NAS. The committee's mandate did not include using the assembly as a legislative body. Its job was to produce some recommendations based upon input from the participants.

As the Asilomar meeting progressed, Berg began to sense that there might be some difficulty in getting consensus among the diverse interests. The major obstacle to consensus was the concern among scientists that limitations on research might affect their own work. Berg commented, "People, I think, were being very self-serving . . . as I'd discovered two years earlier, everybody [would] like to draw a circle around their work and stamp it as pure and unadulterated, and it's what you're doing which is nasty and needs to be proscribed."[11]

On the third day of the conference, members of the Organizing Committee became increasingly anxious about producing a document. Starting on Wednesday evening, they worked into the early morning hours with members of the three working groups, passing revisions back and forth for final editing. The committee discussed how it would present its provisional statement to the participants. According to Maxine Singer, Organizing Committee members initially decided against a vote because they thought it might be turned down, and they wanted some recommendations to come out of the meeting.[12]

At the plenary session held on Thursday morning, there was some confusion among participants as to the proper role of the Organizing Committee in preparing recommendations. There were also questions raised about the significance of the vote taking. Was the committee obliged to represent the views of the assembly? Should they also present minority positions? Berg did not see a clear pattern of consensus emerging in the first three days of meeting. Consequently, the Organizing Committee chose to interpret the final report as their statement and not a statement of the entire Asilomar body.[13] Berg discussed his strategy for clarifying the relation between the Organizing Committee and the other participants in his post-Asilomar interview: "I tried to put forth the view that this was *our* paper, and we were obliged to make a report to the Academy of some recommendations, and that these were our understandings of what we think are plausible recommendations and the way of proceeding; it comes from having listened to the debate and the discussions."[14]

In this way, the failure to achieve a consensus position would not immobilize the committee from issuing its recommendations to the NAS. Of course, the Organizing Committee might have delivered a report stating that consensus was not reached. But that could have meant relinquishing the decision-making responsibility to some other body, resulting in a loss of control over the issues. That warning was issued by Baltimore at the outset of the conference in his introductory remarks. Unless this assembly provided the leadership and guidance, he said, there would be no one else to whom to appeal, and the mission of Asilomar will have ended in failure.[15]

That final session of the Asilomar Conference took place on Thursday morning, 27 February. It was devoted to the discussion of the preliminary statement prepared by the Organizing Committee that summarized the outcome of the conference. The participants had approximately a half-hour to read and digest the contents of the five-page document, which was divided into six sections. Paul Berg chaired the session. He began by restating the charge to the Organizing Committee from the National Academy of Sciences. The committee was responsible for filing a formal report to the NAS on the progress achieved by the conference, including any recommendations. The NAS would review the report, make any changes it saw fit, and then release it for publication.

Berg informed the assembly that the recommendations they were given were provisional. The purpose of the Thursday morning session was to get feedback from the participants, mainly to clarify ambiguities or confusions in the document. But Berg made it clear that the Organizing Committee would accept full responsibility for the statement and that it should be viewed as a statement of the committee and not the assembly. Recalling how the Organizing Committee viewed the role of other participants, as the plenary session drew near, Berg said, "We want to get their views on it [the provisional statement] and where their views are acceptable and useful to us, we'll incorporate them."[16]

At this point, after three days of intense discussion and debate, and after staying up until dawn the evening before putting together the provisional statement, Berg and others on the committee were counting on achieving some convergence. When comments from the assembly were sought, some interrogators wanted to know whether there would be additional opportunities to revise the provisional statement. There was some discomfiture expressed over the idea that the document would be reported as a recommendation of the entire assembly, while the Organizing Committee had the final authority on its contents. Berg contended that the report represented the assessment by the Organizing Committee of the existing consensus. But his view was not left unchallenged. How would it be determined whether there was consensus over the provisional statement if a vote was not taken? And if there was going to be a vote on the statement, how would it be carried out—line by line, paragraph by paragraph, or section by section? Berg reaffirmed the view of the Organizing Committee that a vote had not been intended. A consensus would be determined, he held, without having a vote. Some participants remained skeptical of Berg's ability to determine a consensus without a show of hands. What was the point of eliciting comments if they were not recorded? Participants were assured by the Organizing Committee, particularly by Baltimore and Berg, that their comments would be considered in the final draft of the Asilomar statement. But which comments were adopted in the report would still be left to the discretion of the committee.

Finally, Sydney Brenner requested a vote, and Berg conceded to having a show of hands, beginning with the first paragraph of the statement. After an overwhelming acclamation was given to the principles behind the statement, it was clear that Berg and others had underestimated the degree of support the assembly was prepared to offer. Maxine Singer recalled, "And it wasn't at all clear at that point if they voted our statement down, what we would proceed to do, although, we had said we would send it in by ourselves."[17] But not all sections of the provisional statement achieved the same kind of support, especially in the consideration of the potential risks of particular classes of experiments. Nevertheless, as one observer noted, "at every point the Organizing Committee won overwhelmingly with never more than five or six hands raised in opposition."[18]

The great majority of participants at Asilomar did not play an active role in drafting statements or contributing input into the contested points. One participant estimated that somewhere around thirty individuals provided most of the input that went into the working panel reports, including the writing, revising, and critiquing of the final statement.[19] Nevertheless, the votes taken showed that the Organizing Committee had gained the confidence of the vast majority of the assembly. However, among those most concerned that the final statement was too weak for the problems at hand were some of Asilomar's most informed and involved individuals.

What, then, was the result of this process? The Plasmid Working Group had proscribed certain rDNA experiments by placing them in their class VI. The other two working groups had not placed such limits on experiments, and the provisional report of the Organizing Committee followed their recommendations.

When the issue of proscribing certain areas of research was raised before a vote of the assembly, Berg formulated the notion as follows: (1) Some experiments should be done only at the highest containment facilities that presently exist. (2) Some experiments should not be done even in the highest containment presently available. The vote was taken, and the majority supported the second position, overriding the warnings of some that the prohibition of experiments could set a dangerous precedent.

The Organizing Committee, following the recommendations of the virus panel, established a three-tier classification for experiments: low risk, moderate risk, and high risk. Some criticism was expressed in the discussion, particularly from the microbiologists, on the use of the triadic classification scheme. The model in the provisional statement for handling all recombinant organisms was based upon procedures adopted by the National Cancer Institute (NCI). Critics argued that the viral procedures were not necessarily appropriate for bacterial work. Thus, there was a standoff between those who supported a model resembling a classification of etiological agents established by the Center for Disease Control and those who supported a model based upon NCI standards for research with viruses.

The questions raised during the plenary session underscored the fact that many loose ends still existed. Why was a line drawn between warm-blooded and cold-blooded animals rather than mammals and nonmammals in distinguishing levels of risk? The Eukaryote Working Group wanted to divide the vertebrates in the risk assessment, whereas the virus panel did not want to make such a division. Why should frog DNA require lower containment than monkey DNA? One participant poignantly remarked, "There are tumor viruses in frogs."

There was disagreement over what it means for DNA to be well characterized or rigorously purified. And if it is well characterized, then does that ensure that it is safe? The fact that the loose ends were not resolved to everyone's satisfaction was anticipated, to some degree, by the process itself. This outcome, after all, was merely the trial run for a much longer deliberation that would be carried on by the NIH and its advisory bodies.

When the consensus was reached on the final section of the statement, Berg asked for a show of hands for support of the entire document. He called it a fair consensus, without unanimity. The document could now be appropriately called the provisional statement of the proceedings of the meeting.

Some of the most intense opposition to the consensus statement came from members of the plasmid panel. Stanley Falkow recalled the events in an interview with a member of the Oral History Program at MIT.[20] He felt that, for the most part, the plasmid panel was ignored by members of the Organizing Committee when the provisional statement was drafted. Despite the fact that the plasmid group put significantly more time and effort into their report than the other panels, Falkow contended that the final recommendation reflected the interests of the virologists: "And even though we had gone to great extremes, it seemed that the most stringent prohibitions were being put on the people who worked with microorganisms. So we . . . felt that we had done a lot of work, and they had overlooked a lot of what we had done, and we were very angry about that."[21]

Falkow recalled that the plasmid panel had voted to adopt the provisional statement; nevertheless, they all dissented from the conclusion. He rationalized his vote by reasoning that some gains had been made. After all, the provisional statement was somewhat stronger than the Berg letter. But the input of the plasmid panel did not end at the Thursday session. After Asilomar, Falkow revised the plasmid section in the provisional statement. It was passed around to the members of the Plasmid Working Group, and finally sent to Berg for insertion in the final version of the report.[22] Novick had pressed to have the plasmid report published in its entirety because he felt the summary of it that appeared in the provisional statement did not do justice to its scientific findings. The Organizing Committee rejected that proposal. They were concerned that publishing it might give the impression it was a minority report. This has special significance in view of Baltimore's early remark: If the scientists cannot reach consensus, the issue will be taken out of their hands.

The resolutions voted on by the assembly were incorporated into the final position statement of the Asilomar conference. By late May 1975 the consensus statement received approval from the Executive Committee of the NAS. And in June the official Asilomar position was published in the major scientific journals (see appendix B). Important changes were made in the provisional statement as a consequence of the discussions and votes taken in the plenary session. Appendix B examines the impact of the plenary session on the final document.

The consensus statement of Asilomar laid the philosophical and practical basis for the guidelines that followed. Briefly, it called for the use of physical and biological containment commensurate with the potential hazards of the biological agents used. It proposed four general containment levels for gene-splicing experiments and gave examples for each level. Following the consensus of the assembly, it asked that some experiments not be done using the containment facilities then available.

There was a call to the scientific community for the development of safer hosts and vectors as well as experimental investigations on the risks of recombinant organisms. Under a section entitled "Implementation," the final statement referred to the development of codes of practice for the conduct of experiments. But unlike the provisional statement, no mention was made of institutional review committees to whom principal investigators would be accountable on safety issues. This left a significant area of public policy unresolved, namely: How are scientists to be monitored and who will enforce the code? But there was still much to be said about the Asilomar scientists as architects of public policy.

The Asilomar meeting had two stated objectives, namely, to review the progress in the field of rDNA molecule research and to address the potential biohazards of the work. It is frequently described as a conference devoted exclusively to biohazards. Some participants who were given an introduction to this area of scientific investigation had an opportunity to consider whether it could be useful for their own work. It is in the context of both the promise and the perils of the technology that the biohazards were considered. It is fair to say that the scientific participants had a strong interest in seeing the research flourish.

The success of the conference can be judged on its initial objectives, but its unintended impact was far reaching. Asilomar emerged as a model for setting science policy, and it is this outcome which has stimulated as much debate as the recommendations themselves.[23]

Let us examine, briefly, the outcome of Asilomar from three perspectives: (1) risk assessment; (2) as an instrument for policymaking; and (3) in furthering the values and goals of science.

In comparison with earlier efforts, such as those of the Gordon Conference and the Berg committee, Asilomar discussed biohazards broadly and intensely. The Berg letter had emphasized caution for DNA recombinants comprised of virus DNA or the genetic determinants for antibiotic resistance and toxins that are implanted in new bacterial hosts; at Asilomar, scientists went further and began to consider the potential hazards of shotgun experiments in which unspecified fragments of a genome are cloned into bacteria. The pre-Asilomar correspondence showed some concern about environmental hazards resulting from the release of novel biotypes into the ecosystem, for example, the disturbance of biochemical pathways. But this issue had not received much attention at the conference compared to the hypothetical scenarios of *E. coli*, with an assortment of eukaryotic genes, inhabiting the gut.

A strong consensus emerged from the meeting for engineering a strain of *E. coli* with sufficient genetic mutations to render it safe for cloning—that is, it would be unable to survive in the human intestinal tract or in the environment outside the laboratory. As already indicated, the summary statement provided a framework for matching experiments with levels of physical and biological containment. However, this initial framework did not grow out of a coherent and systematic set of principles. It was a negotiated settlement among scientists incorporating some science and considerable conjecture and intuition.

The meeting has been evaluated as an instrument for developing policy in science. Figure 1 shows the stages of the decision process from the initial organization of the meeting to the published recommendations. First, I shall examine some positive features of the Asilomar meeting as a model for setting science policy. Second, I shall summarize its limitations.

The conference brought together some of the leading international experts in molecular biology. Most were seasoned scientists from major institutes and universities in this field. This conference provided an efficient and fruitful vehicle for problem-solving. Meetings of this type, where the interactions are unrelenting and intense, can provide benefits that long and protracted discussions may be unable to yield. The international scope of the meeting had two effects. It underscored

NAS contracted by NIH to organize biohazards conference

NAS chooses chairman of Asilomar Organizing Committee (OC)

Chairman OC selects other members

OC creates heads of three working groups, selects participants, determines goals and objectives, organizes agenda

Heads of working groups choose members, solicit information from scientists in their communications network; provide suggestions of participants

Three working groups hold meetings, prepare provisional statement on risks and strategies for contending with biohazards

Working groups organize technical talks and panels at conference with oversight of OC

Working group reports presented at assembly; comments made at scheduled meetings and at informal sessions

Revised statement of working groups submitted to OC

OC develops provisional statement from all inputs

Asilomar assembly reviews and votes on the contents of provisional statement of OC; some changes are requested

OC revises provisional statement of conference and submits to NAS for approval

NAS makes minor changes in Asilomar statement and authorizes publication

Asilomar statement published; serves as the principal policy document for NIH rDNA Molecule Advisory Committee

Figure 1 Decision flow diagram, Asilomar Conference, 24–27 February 1975.

the point that scientific inquiry has no national boundaries and that some degree of international cooperation was necessary for the rDNA issue. As one observer noted, "Germs don't need visas." On the other hand, the foreign guests reminded the US-dominated meeting that the policies that individual nations might choose to address concerning the potential biohazards are unlikely to conform to what worked well in the United States.

The meeting was open to selected representatives of the press who could ensure that opposing views received public attention. The journalists in attendance played an important role subsequent to the meeting in alerting other sectors of the scientific community and the general public about the state of the controversy. The mixture of working groups and a large assembly of scientists also had some advantages. Since the issues could not be reduced to a simple set of questions, small clusters of scientists with expertise in one of the key areas could bring together the relevant knowledge more quickly and efficiently than might be accomplished by a large and more diverse assembly. For a problem of a purely technical nature, the decision process appears well suited. A large body of scientific experts served primarily in a reactive mode, while a smaller body of scientists assembled the information and outlined a strategy.

As a policy-making model for the rDNA debate, Asilomar was severely limited in the following ways: selection of participants; clarity of the decision-making process; boundaries of discourse; public participation, and control of dissent. In the matter of selection of participants, the meeting was subsequently criticized for drawing its scientific expertise from too few fields. Most participants were from disciplines that would draw benefits from the new research program. There was little representation from the health sciences and no participants representing environmental interests. An alternative way to convene a conference of this nature might have included: contacting the professional associations that had the expertise in a relevant area (infectious disease, medical microbiology, immunology) for nominations of participants; an open request for papers on the potential risks of using certain hosts in the cloning experiments; contacting environmental organizations for their expertise in problems of monitoring agents that are disseminated in the environment; soliciting participation from agencies and organizations concerned about occupational health. Once these inputs were available, working panels could have been established with a wider range of representation from the scientific community and public interest groups. However, opening up the process in this way posed the risk of losing control of the issues. This concern was preeminent among the Asilomar organizers.

From the outset of the meeting, it was unclear how a consensus would be reached. Participating scientists questioned the meaning of a vote, when the Organizing Committee was given responsibility for making the final recommendations. The Organizing Committee had considerable discretion in setting the parameters of the meeting. Thus, although the issues surrounding this area of scientific investigation had broad policy implications, the Asilomar decision-making process was controlled by a relatively narrow sector of the scientific community. This is yet another reason why it has limited value as a model for setting public policy.

The Organizing Committee had placed specific boundaries on the types of issues it would review. Problems associated with willful misuse of the new technology, the role of rDNA in genetic engineering of humans, and the commercial applications and misapplications were bracketed from the discussions. Thus, the international scientific assembly could, at best, provide one set of the inputs into the policy. While the biohazards were an important input, they did not fully represent the problem.

No one seriously considers that Asilomar had a public-participation component, although it did have lay participants from the field of law and public policy who provided some alternative viewpoints on the responsibility of the scientific community. Participation in the issues by environmental groups came considerably later, when some avenues of discussion had already been closed.

The final report issued by the Organizing Committee did not reflect the level of dissent that existed at the meeting. Since this report was widely circulated among scientists, it may have provided a false impression about the degree to which scientists at Asilomar, especially members of the working panels, had supported the document. The Plasmid Working Group's report was the most carefully developed of the three. Many of its recommendations were rejected, yet there was no opportunity for the public or the scientific community to judge how that report differed from the Asilomar report.

Disciplinary autonomy emerges as a key value at Asilomar. Its viewpoint may be expressed as follows: Let those who will use the new research techniques bear the major responsibility for its safe and intelligent deployment. But Asilomar does represent an important shift from individual to collective responsibility in this area of biological research. Scientists had alerted the world that they had come upon a field of inquiry that warranted the development of a peer-review system for issues of biosafety. Some scientists saw in this an expanded role for study sections to make recommendations on grant applications. But others had pointed to a conflict of interest in having the same individuals review research quality and biosafety conditions.

A major tension developed during the meeting between those who refused *on principle* to proscribe any experiments and those who considered it appropriate for some rDNA experiments to be deferred. The latter group had been in the majority and the Asilomar statement reflected its position. No attention was given, at this meeting, to potentially *verboten* areas of research, either on ethical grounds or because the risks of misapplication were too great and the outcome too ominous.

To secure the goal of disciplinary autonomy, the organizers of Asilomar had accomplished two objectives: (1) they defined the issues in such a way that the expertise remained the monopoly of those who gain the most from the technique, and (2) they chose to place authority for regulating the use of the technique in the agency that is the major supporter of biomedical research in the United States. As the controversy developed, these objectives came under attack from persons both inside and outside the scientific community. The scientists who played the principal roles in organizing and implementing Asilomar had only waged the first battle. Two major struggles lay ahead: the development of workable guidelines and the threat of direct public intervention into research policy through federal legislation.

11

Formation of RAC

In its widely publicized letter,[1] the Berg committee called upon the director of the National Institutes of Health to consider establishing an advisory committee charged with developing guidelines for scientists using rDNA techniques. This was the final stage of a three-part process for erecting a policy that had begun with the formation of the National Academy of Sciences' committee, led to the Asilomar Conference, and finally resulted in an NIH advisory committee for translating the recommendations of Asilomar into guidelines. Few people considered how arduous the path would be from Asilomar to the NIH *Guidelines*. And yet from the time Asilomar ended (February 1975), it took NIH sixteen months to deliver its guidelines. Still, this is not an unreasonably long period in comparison to the time it takes OSHA, EPA, or FDA to regulate a single substance, for in the matter of rDNA molecule regulations, the rules being considered had to be applicable to a potentially vast domain of research and technology.

The formation of the advisory committee as the final piece in the policy process was also a critical element. While officially only advisory to the director, the committee would bear the burden of constructing the policy framework and the content of regulations. How would this body be constituted? What latitude would it have to depart from the Asilomar recommendations? In choosing the committee, what efforts were made to win broad public confidence? These questions are a key to understanding how scientists were able to maintain control of the issues despite attempts by new participants in the controversy to steer the issues away from the technical biohazards.

Secretary of HEW Casper W. Weinberger signed the charter officially establishing the Recombinant DNA Molecule Program Advisory Committee on 7 October 1974, seventy-three days after the Berg letter appeared in *Science*.[2] The authority for such action comes from section 301 of the Public Health Service Act (42 USC 241), which gives the secretary of HEW (now Health and Human Services) the right to conduct investigations into the causes of disease and its treatment. In outlining the potential hazards of the research, the charter cited the need to investigate the broad environmental issues in addition to the laboratory biohazards. The approach followed the Berg recommendations "to evaluate the potential biological and ecological hazards." The wording in the charter emphasized the notion of controls and the duality of positive and negative benefits growing from the research: "[T]he use of this technology [rDNA] has various possible hazards because new types of organisms, some potentially pathogenic, can be introduced into the environment if there are no effective controls. The technology is also capable of producing microbial organisms which can be useful or harmful to agriculture or industry, and thus secondarily affect human health."[3] The objectives of the charter resembled closely in wording and content the recommendations of the Berg committee.

Within its boundaries there was considerable leeway for the committee to act. However, those boundaries excluded some important impacts of the technology. The charter made no reference to the broad social and ethical issues; it defined the twelve-member committee exclusively in scientific terms, to be comprised of authorities knowledgeable in the fields of molecular biology, virology, genetics, and microbiology. In spite of its emphasis on environmental hazards, the charter neglected to see a role for ecologists, infectious-disease experts, or specialists in occupational health. It reflected a belief that the potential hazards of escaping microorganisms could be handled at the laboratory level.

Over a month elapsed after the charter was signed, and still the advisory committee had not yet been chosen. Paul Berg was disturbed by the delay. He, as well as others at Asilomar, was concerned whether scientists would continue their support of the voluntary suspension of certain experiments. Berg wrote to Robert S. Stone, director of NIH, requesting that Stone move quickly and decisively in naming the advisory committee defined by the charter.[4] The timing was critical. A delay in the establishment of this advisory body could have threatened support for the moratorium. An effort had already been under way to form such a body, but changeover in administration and the imminent departure of NIH director Stone had contributed to the delay.

When the Recombinant DNA Advisory Committee (Initially named The Recombinant DNA Molecule Program Advisory Committee, subsequently shortened to Recombinant DNA Advisory Committee, or RAC) was convened, it began its life as a technical body of restricted scope. In its evolution the committee experienced two important transformations. First, there was the addition of two nonscientists to RAC within a year's time. Second, the size of the committee was almost doubled in 1979, giving it one third public representation, including some prominent critics of NIH policies. The following section examines the considerations that went into selecting the first NIH Recombinant DNA Advisory Committee (RAC).

DeWitt Stetten received his baccalaureate in biochemical sciences from Harvard College and an MD from Columbia College of Physicians and Surgeons. After a two-and-a-half-year internship at Bellevue Hospital, he returned to Columbia in biochemistry, where he received his PhD in 1940. Stetten's association with NIH dated back to 1954. He was dean of a medical school in New Jersey for eight years from 1962 to 1970 and returned to NIH, where he became director of the National Institute of General Medical Sciences. Stetten then moved into a high-level administrative position in NIH as the deputy director for science, which oversees intramural programs. He was chosen by Robert Stone to head RAC.

As chairman, Stetten played a key role in selecting the other RAC members. According to his interview with MIT's Oral History Program, Stetten relied heavily on the Division of Research Grants for nominees to RAC because of their direct contacts with scientists.[5] The advisory committee was chosen after discussions had taken place between Paul Berg, Leon Jacobs (NSF), Stetten and Steve Schiaffino, the deputy director of the Division of Research Grants. Stetten described the type of people they sought: "We wanted people who are energetic, competent, well-informed and, in some degree, diverse."[6]

Some members of RAC were selected because of their expertise in a particular area, for example, Stanley Falkow for his knowledge of *E. coli* and Waclaw Szybalski because of his expertise in bacteriophage. Stetten recalled that three members were active in the field of rDNA research (David Hogness, Stanford University; Charles Thomas, Harvard University; and Donald Helsinki, University of California).[7] There was also an effort to obtain representation on the committee of women and minorities, such as Jane Setlow of Brookhaven National Laboratory and Ernest Chu of the University of Michigan. Roy Curtiss III, University of Alabama, was selected for his background in bacterial mutations and bacterial selection. Edward A. Adelberg of Yale University had worked in plasmid biochemistry with *E. coli*.[8]

At the very first meeting of RAC, the issue of public members on the committee was raised. Stetten recalled: "I . . . was interested and a little surprised to find out at an early meeting that the members of the committee wanted to have at least one nonscientist present."[9] Stetten was concerned that choosing the "wrong" person might create a disruption in the process. In his interview with MIT's Oral History Program, he explained his views about having nonscientists participate on RAC: "[I] have reservations as to the contributions which a nonscientist is likely to make in considering an issue which I regard as a scientific issue."[10]

Stetten maintained there are two stages to the resolution of the problem, the technical stage and the policy stage. Scientific judgment alone belongs in the technical stage:

> Once the experiments have been decided upon, once a description at least of the risk and of the benefit can be given, even though we can't prove in precise, annotated terms, then the problem becomes a political issue as to what to do about it. But we weren't at that stage yet. . . . The person who never worked in a containment situation, and a person who [has] never handled obnoxious microorganisms . . . can have very little sense of what's right and what's wrong. [A] lay judgement about a surgical procedure, is usually not very relevant.[11]

On more than one occasion Stetten commented on the extraordinary difficulty of coming to some determination of the risks of rDNA molecules. Notwithstanding his views that the nonscientist had little to offer the process of risk assessment, he also acknowledged the nonscientific aspects of the problems. A week after Stetten was interviewed by MIT's Oral History Program, he wrote about these feelings to Richard Goldstein and Harrison Echols.[12] The context of the letter is the criticism by forty-eight scientists to an early draft of guidelines prepared by a subgroup of RAC. "It has been an interesting experience for me to watch experts in the field struggle with the intrinsically insoluble problem of estimating hazards in the total absence of actuarial data. This is, as I see it, the task before the members of our Committee. Conclusions which must be reached cannot be reached by scientific judgements alone. They are reached certainly in part intuitively, in part by analogy, in part by hunch, and in part by consensus. Certainly, none of these are the method of science."[13]

In responding to the request of some RAC members to invite a nonscientist, Stetten sought such an individual by contacting the American Civil Liberties Union, Common Cause, and the staff of the

President's Advisory Council on Human Rights. These inquiries did not lead to any candidates. Finally, an acquaintance of Stetten suggested the name of Emmette Redford, a professor of government at the University of Texas at Austin. Redford attended his first RAC meeting on 4 December 1975, at La Jolla, California. (He was not an official member of the committee at that time, but participated as a consultant.)

A second nonscientist sought for the committee was philosopher-theologian LeRoy Walters, director of the Center for Bioethics at the Kennedy Institute at Georgetown University. There was some difficulty at first in getting Walters's appointment accepted. A possible explanation of the delay was his affiliation with the Kennedy Institute during a Republican administration. It is also a curious fact that a bioethicist would be chosen, since no interest had been shown in expanding the discussions to the bioethical domain. But Walters, like Redford, provided his strongest input on procedural issues. In recalling his appointment to RAC, Walters cited his work with the National Commission for the Protection of Human Subjects as his preparation for the rDNA episode, which involved studying the social control of some phases of biomedical research.[14] Walters conjectured that his appointment was spurred on by the first public hearing on the proposed guidelines (February 1976), where there were discussions about getting participation on RAC from people in the field of bioethics.[15]

Other appointments to RAC were made to fill in gaps in scientific expertise. Wallace Rowe, a virologist and MD with the National Institute of Allergy and Infectious Diseases, was added. To deal with issues involving plant pathogens, Peter Day of The Connecticut Agricultural Experimental Station was also brought on the committee.

There is an interesting story behind the appointment of Elizabeth Kutter. RAC member Waclaw Szybalski had discussed the issue of broadening the participation on the committee with DeWitt Stetten. Szybalski advised against having lay representation because of the technical complexity of the subject matter. But he did see the need to draw into the committee a biologist who was not affiliated with the major research institutions.[16] Szybalski had such an individual in mind. He recommended Elizabeth Kutter of Evergreen State College in Washington. Kutter not only filled the bill as a woman, but she was from a small liberal arts college, had been engaged in active research in bacteriology and bacteriophage, and also had teaching interests in the social and ethical aspects of biology. For some members of RAC, Kutter's appeal may have been chiefly as window dressing. But Kutter played a vital role in this critical period, as a skeptic, as a link to other scientists who had been on the periphery of the discussions, and later as one of those given major responsibility for redrafting the guidelines.

The second meeting of RAC was held in May 1975 at Bethesda, Maryland. It was decided at that time to have a subcommittee work on a draft of the guidelines. Selected by Stetten, the subcommittee consisted of David Hogness (chairman), Donald Helsinki, Waclaw Szybalski, and Ernest Chu. Reflecting on that choice, Elizabeth Kutter felt it was a tactical mistake to have someone who was involved in the research chair the subcommittee writing the guidelines.[17] Stetten later described his choice of Hogness: "I think it was a good selection although he was . . . fairly partisan on the side that the dangers were not all that great, and that really these experiments were perfectly safe. Still he was in Paul Berg's laboratory; it was well known that he and Paul were intimate; and we thought we would get some input from Paul Berg, which we certainly did."[18]

The early RAC meetings exhibited some of the same skepticism and diversity of viewpoints found at Asilomar. And with Hogness taking on a leading responsibility for the first draft of the guidelines, a strong link was established with the Asilomar scientists and their recommendations. A major setback to that process came after the Hogness draft was distributed and discussed by the full RAC at the Woods Hole meeting, 18–19 July 1975.

Several days before going to Woods Hole as an invited consultant to RAC, Elizabeth Kutter stopped off in Boston to meet with some phage (bacteriophage) scientist colleagues. While in Boston she also talked to some members of Science for the People. Kutter, who had already received the Hogness draft, shared its contents with her Boston colleagues and some members of Science for the People. She delivered a talk at the Harvard Medical School on the progress in developing guidelines. She advised those (like herself) who were disappointed with what the Hogness subcommittee produced to attend the RAC meeting at Woods Hole and make their voices heard.

When the meeting finally took place, there were few observers; there was no representation from Science for the People. More important, some key members of RAC were absent from the meeting, namely, Stanley Falkow, Wallace Rowe, and Donald Helinski. During this period both Falkow and Rowe had taken a conservative stand (in this context, a cautious approach that placed the burden of proof on those who maintained there were no risks) on the research; their presence would have influenced the course of that meeting.

At the meeting, discussions centered around (1) weaknesses and gaps in the draft guidelines, and (2) composition and responsibility of local biohazards committees. Elizabeth Kutter supported the idea of having strong local committees. But their role, she believed, should not be merely advisory. She also supported broad representation on the local bodies, which would include technicians and graduate students. Other RAC members, opposed to the idea of local (institutional) committees, raised the objection that nonscientists could turn out to be obstructionists.

After a vigorous discussion of the Hogness draft, it was favorably voted on by the committee.

Once a year in August between 200 and 400 specialists in bacterial viruses (bacteriophage) gather at Cold Spring Harbor to hear lectures and share current information with colleagues. The meetings last about a week and address technical subject matter almost exclusively. In 1975 the meeting was held near the end of August.

During the May 1975 RAC meeting, Elizabeth Kutter's suggestion that she organize a discussion of safe cloning vehicles at the upcoming Cold Spring Harbor meeting was received favorably. By the time the phage meeting took place, attention was turned from safer cloning to the recent draft of the guidelines.

One session of the phage meeting was set aside to discuss the Woods Hole proposal. Kutter estimated that over 100 people attended the discussion that she moderated. Copies of the Woods Hole guidelines had been available to participants a day before the session. While the discussion showed there was considerable opposition to weakening the Asilomar recommendations, no consensus was reached on what action should be taken. For this, a second informal meeting was scheduled a few nights later. Scientists who played key roles in the critical discussions at this point were Harrison Echols (University of California, Berkeley), Richard Goldstein (Harvard Medical School), Mark Pearson (University of Toronto), and David Botstein (MIT). At the informal meeting it was decided to draw up a petition that would outline some areas of disagreement with the proposed guidelines. The petition was posted and signed by forty-eight of the CSH participants. It was sent to RAC chairman DeWitt Stetten, and copies were sent to *Science* and *Nature*. *Science* did not publish the letter, nor did it provide coverage of the action. *Nature*, on the other hand, did provide some coverage.[19]

The statement accompanying the signatures made three points: (1) It urged that the most hazardous experiments be curtailed until there was an experimental determination of the risks. (2) It expressed a concern that in the current draft of the guidelines, mammalian DNA, including animal viral DNA, could be cloned in low physical containment. It recommended limiting shotgun experiments of mammalian DNA to the highest physical containment until vectors that were proved safe became available. (3) It called for broadening the scientific representation on RAC to include areas such as animal virology, plant pathology and genetics, and epidemiology.

The letter also requested that there be stronger representation of scientists on RAC who were not involved in cloning work. In addition, the letter requested public representation on the committee.

On 9 September DeWitt Stetten responded to the petition with letters to Goldstein and Echols (the first two signers). In his letter, Stetten revealed the sense of frustration he was having on the advisory committee. There was not enough empirical data to estimate the hazards, according to Stetten. Many of the decisions that RAC reached were arrived at by hunches and intuition.[20]

The Cold Spring Harbor petition had evoked strong reactions on both sides. RAC committee member Jane K. Setlow of the Brookhaven National Laboratory (who was named chairperson of RAC several years later) took umbrage with the CSH statement. She responded to the allegation that there were vested interests in the advisory committee. Of the committee members who voted on the Woods Hole draft, Setlow maintained that five were not involved in cloning (Chu, Curtiss, Littlefield, Setlow, and Szybalski), while three persons were (Adelberg, Hogness, and Thomas). As for lay representatives on the committee, Setlow was dubious about the contributions such individuals could make to the committee's deliberations. But as a public-relations measure, Setlow could see the point.[21]

The criticism of the Woods Hole draft of the guidelines came from inside RAC as well as from scientists who had attended Asilomar and many who had not. It was an attack leveled on many fronts. Those who played major roles at Asilomar were disturbed that a smaller group of scientists had considered downgrading the recommendations. The Asilomar organizers carried considerable weight at NIH. It is for these reasons that the initial support of the Woods Hole guidelines gave way to a more conservative approach.

Let us examine more closely the nature of the opposition. At least four members of RAC had voiced their disagreement with both the content and the philosophy of the provisional guidelines. They were Roy Curtiss, Donald Helinski, Elizabeth Kutter, and Stanley Falkow. The CSH petition had been greeted favorably by Falkow, who was sufficiently fired up by the Woods Hole draft (although he had not been there to vote on it) to generate a stream of correspondence.[22] RAC members did their lobbying outside the committee to muster support for their positions. Paul Berg, the dean of Asilomar, wrote a six-page, single-spaced letter to Stetten expressing his disagreement with the draft. Berg's comments centered on the following points: the draft departed from a principle in the Asilomar statement that the levels of containment should be greatest initially and subsequently modified when the risks were assessed; the requirement for weakened host-vector systems was too low; the containment levels for shotgun experiments were set too low; there were no provisions dealing with experiments that create organisms carrying antibiotic resistance determinants not known to occur in the organisms naturally.[23]

In this letter, Berg maintained that the effectiveness of physical containment was overrated; even the moderate (P3) containment level was, he felt, "reassuring to the psyche" but not something scientists should rely upon foremost. (Physical containment refers to laboratory safety procedures and the physical construction of the facility. The four recommended levels were noted as P1 (lowest) to P4 (highest). Biological containment refers to weakened strains of hosts and vectors that were less likely to survive outside a laboratory. The three recommended levels were noted as EK1 (lowest) to EK3). He argued that for physical containment subject to human error, exposure can be reduced but not eliminated. He revealed that in his own P3 facility at Stanford designed for virus research, almost everyone who entered that lab acquired substantial antibody titres to SV40 (a sign of infection) after one-half to a full year. Berg supported biological containment as a first line of defense. He went on to argue that physical and biological containment are not equivalent (it would be illogical for them to be interchanged). While Berg thought it reasonable to trade off physical containment (lower the P value if the EK value is increased), the reverse should not be permitted (P3 + EK1 is not an acceptable alternative for P2 + EK2 for many experiments). Berg said,

"To suggest that a 'shotgun' experiment with warm-blooded animals can be done with an EK1 vector even under P3 conditions is careless."

While many scientists had directed their criticism to the relaxation of the Asilomar recommendations, some were questioning the assumptions of Asilomar. One of the most basic of these was the use of *E. coli* as a host for rDNA experiments. Paul Primakoff (formerly at the Department of Microbiology and Molecular Genetics, Harvard Medical School) wrote Stetten that *E. coli* was the wrong microorganism to use in rDNA experiments. The preferred host, he argued, should be "ecologically distant" from human beings, one that occupies an ecological niche that would keep it away from human systems and does not exchange genetic information with bacteria that inhabit humans. If this is ignored, Primakoff predicted that foreign DNA segments will, in time, appear in wild-type *E. coli* that interact with humans.[24] Another important actor in the controversy as it developed, Wallace P. Rowe, of the Laboratory of Viral Diseases, NIAID, in a response to Primakoff's letter, averred that a non-*E. coli* host should be a "high priority goal."[25]

The criticism of the Woods Hole draft was overwhelming. Of all the letters NIH received, only two complained that the restrictions were too great.[26] A new working group of RAC appointed to develop another draft of the guidelines included Elizabeth Kutter, Donald Helinski, Wallace Rowe, and Stanley Falkow. The weight of the criticism had influenced the makeup of the working group, consisting of those who had expressed serious reservations about the Woods Hole draft. Jane Setlow was annoyed at how the criticisms of the Woods Hole guidelines were handled. She wanted to know why the whole committee was not informed about the formation of the new subcommittee, why three out of four subcommittee members were not present at the Woods Hole meeting, and why Elizabeth Kutter, chairperson of this new subcommittee, had not yet been officially appointed to RAC.[27] The official letter appointing Kutter was dated 31 October 1975, the date Setlow's letter was written.

This episode illustrates that Asilomar had indeed identified the center of gravity of safety precautions. When a shift from that center of gravity was attempted (as in the Woods Hole draft), the scientific forces were sufficiently mobilized to bring the guidelines back to its "equilibrium" position. Thus, we have an example of a decision process that reversed its position. The stages were as follows: (1) David S. Hogness served as chairman of an RAC subcommittee to draw up guidelines labeled "provisional." (2) A revised set of guidelines (modification of the provisional) was adopted by the full RAC (although not all members were present) at Woods Hole (18–19 July). (3) Strong reaction was received from members of RAC, Asilomar attendees, and other scientists. (4) A decision was made by RAC not to submit the Woods Hole draft to NIH for publication and comment.

At this time criticism of the developing research policy came from within the sciences. Concurrently, however, efforts were under way to alert the public and broaden its involvement in the decision-making process. The first congressional hearings that addressed rDNA research focused almost exclusively on the public's right to decide the policy outcome. And once the new draft of the guidelines was approved by RAC, NIH held its own public hearings, soliciting comments from diverse interest groups.

12

Early Public Hearings

In the spring of 1975 the Senate Subcommittee on Health under the chairmanship of Edward Kennedy (D-MA) held a series of hearings on topics in scientific research and the public interest. One session of that series in April was devoted to genetic engineering and the aftermath of Asilomar.[1] Kennedy introduced the hearings with a battery of questions that struck at the heart of a key issue—the role the public should play in the development of the new research techniques. Kennedy called upon four individuals to provide statements; two of them played important roles at Asilomar. The guests were Stanley Cohen (Stanford University), Donald Brown (Carnegie Institute of Washington), Willard Gaylin (Institute of Society, Ethics and the Life Sciences, Hastings-on-Hudson, NY), and Halsted Holman (Stanford University).

Stanley Cohen's testimony is of interest on two counts. The first concerns the significance of the new research tool. The second relates to the boundaries of public decision making vis-à-vis science. In Cohen's abbreviated description of the new technique, he emphasized that "the method involves the joining together of hereditary materials—genes, into biologically functional combinations that do not occur in nature."[2] Moreover, Cohen used the term "natural barriers" to describe the separation of genetic material among dissimilar life forms. In an ironic turn of events, several years later Cohen and his colleagues tried to demonstrate that natural barriers may not exist. The point to be emphasized is that the conception of natural barriers between divergent species at that time was an orthodox view, not one held only by those critical of the research. The phrase "natural barriers" was used on two other occasions in Cohen's writings in 1975.[3]

Kennedy questioned Cohen about the kind of public involvement that took place in the Asilomar decision making. "The fact remains that the public involvement is really after the fact, is it not?," Kennedy asked, adding that the public was being drawn in "after decisions had been made by scientists." The Massachusetts senator had put his finger on what was to many scientists a very sensitive point. Few scientists would disagree with the principle that public accountability was warranted in this situation. But there was much greater disagreement on the nature and timing of public input. Cohen acknowledged the right of the public to gain assurance that rDNA research is carried out safely, but, he added, it is not in the public interest for a public body to direct the course of rDNA investigations.

Donald Brown reinforced Cohen's view that scientific research must be kept free of public control. Some decisions, according to Brown, should be left to the scientific community, as if by some social contract. The public, Brown maintained, can have no grounds for "directing the pathways of research." He also told the committee that scientific research should not be evaluated on the basis of a risk-benefit equation. Brown's argument can be reconstructed in the following manner:

1. All research has a potential for good and a potential for harm.
2. No weighting of the benefits and risks can be achieved ahead of time.
3. Some estimation of benefits and risks is essential in a risk-benefit approach for determining whether the research should proceed.

---

4. A risk-benefit approach for evaluating scientific research is untenable.

Brown was then faced with the next logical question in his line of reasoning: How should risk-laden research be handled? He concluded that "the risks of science to society must be reduced to a minimum no matter how great the anticipated benefits."[4] This is rather a remarkable conclusion. It implies that the foremost and overriding obligation is to minimize risks that are generated from a scientific research program. Some have argued that the application of such a principle could have delayed, considerably, the discovery of a polio vaccine. One could argue that the risks to polio researchers in the 1940s were not minimized.

But until it is clear what minimizing risks means and who makes the decisions on these matters, Brown's conclusion remains open-ended. What assurances are needed to satisfy the *de minimis* condition? Even at this early stage of the debate, some scientists claimed to have all the evidence they needed to satisfy themselves that cloning in *E. coli* K12 was a very low risk experiment. Other scientists were converted at a later period. The conversion process itself involved factors that extend beyond scientific rationality. Notwithstanding some of the aforementioned problems, Brown did offer a guiding principle for evaluating the rDNA issue: The scientist's first responsibility is to do no harm.

Halsted Holman is a professor at Stanford University's Medical School and a specialist in immunology and internal medicine. In his testimony before the Health Subcommittee, he drew an analogy between rDNA experiments and human-subject experiments. The latter, he said, should have the informed consent of an individual, while the former should have the informed consent of society. Holman cited three principles that guide human subject experiments that he would apply to genetic research: (1) The purpose of the experiment or treatment must have high importance unattainable in other ways. (2) Probable gain must clearly outweigh probable risk for the subject. (3) The subject must be fully informed of the nature of the experiment or treatment, assent to it freely, and be able to terminate it at will. For genetic-engineering experiments, Holman contended, the risks were not known, and the benefits would have to be great to justify the experiments. Even where great benefits may be foreseen, in his view, it would have to be shown that rDNA experiments were the only path to achieve the end result.

Holman offered a very different view of the social control of science than Cohen's or Brown's. Brown would minimize risk regardless of the potential benefits. Holman was willing to accept minimal risks if the benefits are unique and significant. Where the principal justification of the research is in the acquisition of knowledge, there is yet another important dividing line between the views. Brown and Cohen would exclude public input into the development of knowledge, whereas Holman maintained that "decisions concerning the development and uses of knowledge are a public concern, and a joint responsibility of both the public and those with the knowledge." Extrapolating beyond rDNA research, Holman maintained that the direction of research in general is a matter for public oversight. Holman's conception of the scientific endeavor as a partnership with society added yet another perturbation to the policy path of the rDNA episode. Regardless of whether the risks are there or not, should the research be directed, in part, by nonspecialists? Are we better off with an anarchy of elites setting the courses for research programs? How far can we take the "informed consent" analogy? Which person(s) or group(s) can represent society at large?

The views expressed at the Kennedy hearings foreshadowed the struggle in Congress between policymakers and scientists. They illustrate a fundamental difference regarding the distinction between science and technology and the policy implications following therefrom. But a more subtle distinction, emphasized by Donald Brown, is that between nondirected and directed research programs.

Physician-bioethicist Willard Gaylin, the fourth panelist at the hearing, criticized the position that accountability to the public may be warranted for directed research but never for nondirected research. Gaylin, then president of the Hastings Center, divided up the issue of public involvement in science into three parts: (1) Does the public have the right to regulate scientific pursuit? (2) Can the public intelligently regulate science? (3) Should the public regulate science, and if so, what are the appropriate methods of control?

Gaylin pointed out that the conventional wisdom would approve regulating scientific activities if they impose a threat to society. But he presented a position close to that of Halsted Holman. The knowledge engendered by science is a social product because of its historical roots, its public resources, and "because it has become an indispensable part of our common culture."[5] We have a right to control science not because of its failure, but because of its success. This is a fundamental departure from the laissez-faire conception of scientific pursuit underpinning the rationale of progress and the liberal view of intellectual freedom. Gaylin's position has been associated with radical groups like Science for the People, but is rarely heard in the liberal circles of the Hastings Center. His testimony before the Health Subcommittee upset some members and associates of Hastings. Maxine Singer wrote a five-page, single-spaced critique of Gaylin's remarks point by point.[6] Daniel Singer was concerned that Gaylin's testimony would give Hastings an antiscience reputation.[7]

From this brief congressional hearing, one begins to witness the emergence of several positions on the nature and involvement of the general public in science-policy issues. These positions are represented in figure 2.Movement to the left represents a broader or deeper involvement by the public in scientific activities (risk assessment, funding, etc.), while movement to the right represents a narrowing of involvement. Gaylin and Holman supported public participation in decisions covering scientific activities for which no hazards are recognized, for nondirected research (basic science), at a level of the peer-review committees. Brown, on the other hand, regarded non-

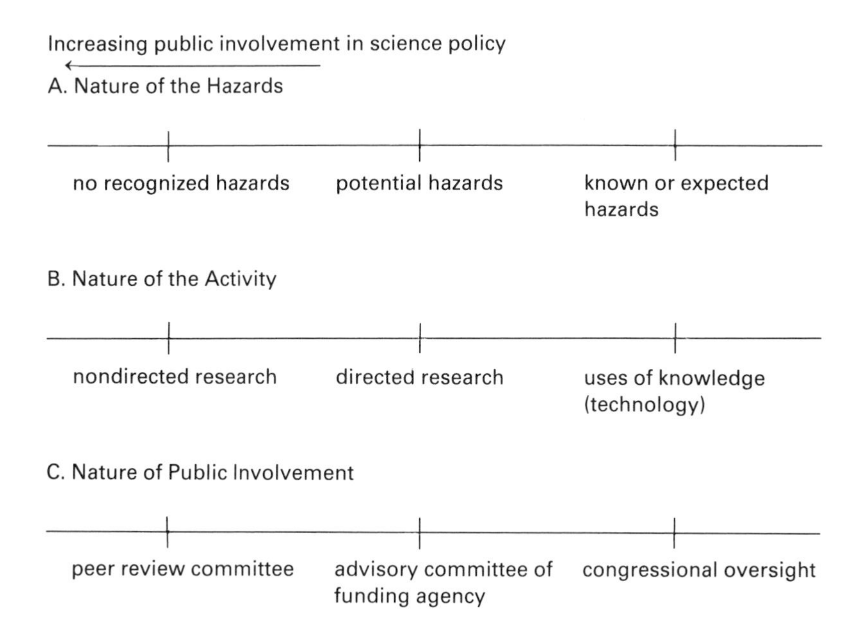

Figure 2 Access points for public participation in science policy.

directed research as outside the province of public involvement, except perhaps through the indirect route of the support of funding through Congress.

Public participation in the rDNA policy process had its debut, at least in a symbolic sense, at the meeting held on 9–10 February 1976 of the NIH Director's Advisory Committee (DAC). This public advisory body was established by the director of NIH on 9 May 1966 (by authority of Section 222 of the Public Health Service Act). The DAC is chaired by the NIH director and is comprised of sixteen additional members appointed by the secretary of HHS for overlapping terms of four years. The purpose of the committee as stated in an NIH publication is to advise the agency "on matters relating to the broad scientific, technological, academic, managerial, and socio-economic setting in which the continuing development of the biomedical sciences, research training, and biomedical communications must take place."[8] The scope of this public advisory body is characteristically broad. It may become involved in issues such as resource allocation, administration of NIH, evaluation, and policy development.

By having DAC hold a public meeting on RAC's newly proposed guidelines arising out of the La Jolla meeting, NIH director Donald Fredrickson saw a way to revitalize the committee and at the same time demonstrate publicly that NIH was receiving a broad range of inputs for its decision. At the time of the meeting, DAC consisted of the NIH director and six other individuals, less than half of the committee's full membership. Fredrickson compensated for the dwindling membership by inviting three former DAC members and eleven consultants. This latter group was made up of a mixture of lawyers, consumer advocates, bioethicists, and scientists.

In his interview with the MIT Oral History Program, DeWitt Stetten discussed some of the factors leading up to the meeting.[9] He placed some significance on the public hearings that had been held by the secretary of transportation during the controversy over the landing of the supersonic transport (SST). According to Stetten, those hearings were cited as an example to Fredrickson on how to handle a highly technical controversy with major public-policy implications. During this period, Fredrickson was receiving advice from Stetten, Maxine Singer, and a federal circuit judge, David L. Bazelon (also appointed to DAC). At the same time, some members of RAC wanted him to open up the decision-making process to nonbiologists.

With these considerations in mind, Fredrickson planned the meeting around laboratory biohazards, in conformity with the objectives of RAC and the National Academy of Sciences. He invited some scientists previously involved in the debate to provide the technical background and explain the significance of the newly drafted guidelines. Following these technical presentations, there was a group of thirteen "public presenters" consisting principally of scientists, only a handful of whom were there to challenge the guidelines. In preparing for the meeting, NIH notified seventeen public-interest groups. Most of the organizations contacted did not have technically skilled people on their staffs who could address the problems of rDNA technology. The exception can be found in the environmental groups that depend upon ecologists, biologists, physicists, and the like to evaluate the environmental impact of public policies. Of all the public interest groups contacted, only the Environmental Defense Fund sent a spokesperson.

Elizabeth Kutter, one of the newest members of RAC, believed that NIH did not go far enough in its public-participation effort. She wanted participation from ordinary folk, such as an autoworker or a cabdriver. Kutter also felt that travel expenses should have been available for individuals representing a broader constituency.[10]

While the DAC meeting drew no cabdrivers or autoworkers, as Kutter had proposed, there is an irony to the choice of her examples from the transport sector. Genetic transfer at the molecular level was indeed the main topic of discussion. Could the chimeric rDNA organisms be kept in the laboratory? Would the debilitated strains of *E. coli* pass on their genetic information to more hardy strains?

At the meeting itself, public presenters were limited to a ten-minute talk, with five additional minutes for discussion. I shall now examine the principal arguments, both critical of and in support of the guidelines. They are subdivided into the following categories: (1) pathogenicity of *E. coli*; (2) gene transfer; (3) emergent events; and (4) grading the risks.

Strong appeals came from some participants at the DAC meeting to find an alternative host to *E. coli* for rDNA molecule experiments. *E. coli*, it was argued, is a normal inhabitant of the gut and pharynx of mammals, including, of course, man. An experiment that could transform *E. coli* into a pathogen is, therefore, capable of infecting the human population. Those who advocated using a different host organism were not persuaded that the physical or biological containment discussed at Asilomar would protect laboratory workers or the public if a pathogenic strain of *E. coli* were produced. They argued that containment provided a statistical barrier, that is, protection against large numbers of escaping organisms, but not an absolute barrier. Low escape frequencies, in their view, must be understood in terms of the total amount of culture used, the number of experiments carried out, and whether investigators are forthright in complying with the safety standards established.

In response to the claim that the laboratory strain of *E. coli*, named K12, was sufficiently enfeebled not to warrant greater concern, proponents for an alternative host cited three counterarguments: (1) Contamination and reversion of the crippled strain into a hardier strain was possible. (2) The crippled strain could transfer its recombinant molecule into a strain that could infect humans. (3) Experiments in which random segments of DNA are transplanted into *E. coli* K12 (shotgun experiments) could introduce a DNA sequence that would change the characteristic of the strain.

One presenter at the meeting, Allen Silverstone, a representative of the Boston Area Recombinant DNA Group, called upon NIH to ban shotgun experiments in mammals and avian species. He maintained that "DNA derived in a shotgun experiment could produce enzymes that would create various kinds of biological havoc."[11]

The distinguished NIH virologist Wallace Rowe disagreed with those who dismissed the risks as purely fictitious. He identified the risks of deliberately putting toxigenic genes into *E. coli* as real. If such transfers can be made deliberately, Rowe argued, then one must consider the possibility of accidental occurrences.[12] At the time, Rowe was concerned about using certain phage vectors for cloning that might carry genes for toxins, like botulinum, diphtheria, or streptococci. Overall, he believed that science was in a poor position to evaluate the risks. Rowe supported the idea of undertaking some key risk-assessment experiments in addition to the one he already was planning, namely, seeing whether polyoma DNA inserted into a cloning vector can be released and induce infection in animals.

Some scientists, who were less concerned about the transformation of *E. coli* into a human pathogen, emphasized the possible transfer of genetic matter from *E. coli* to a hardier organism that would prove more hazardous. Roy Curtiss, already a veteran of the rDNA affair, told DAC that normal wild-type *E. coli* K12 was capable of exchanging plasmid DNA from over forty genera of bacteria, including some that are pathogenic. Thus, an investigator placing gene segment X into *E. coli* K12 might find that segment transferred to a robust strain of *E. coli*, which then might gain residence in a human gut.

One of the most captivating philosophical issues of the meeting concerned a question of emergent properties. Should one be concerned that a recombinant product will be more hazardous than its individual components? There were two lines of thinking on this matter, one characterized by a reductionist and another by an organismic view of molecular biology.[13] Reductionism is the term used to describe a metaphysical or epistemological position in which the properties of the whole can be accounted for, explained, or predicted from the elemental parts. Organismalism, a term coined by the zoologist W. E. Ritter in 1919, holds that the organism in its totality is as essential to an explanation of its elements as its elements are to an explanation of the organism. Organismic biologists believe that the "form" of the single whole organism is a factor in its embryological development, behavior, reproduction and physiology. According to the organismic view, the additivity principle does not apply to the organism. The parts of an organism are mutually determining and interdependent.

The relevance of these philosophical distinctions to the molecular biology of rDNA risks is clear. Some scientists argued that a new hybrid organism would not behave radically different than its original form. For example, David Baltimore told DAC that for innocuous recombinants, "[E]ven if those recombinants got out of the laboratory the hazard they pose is no more severe than the hazard posed by the microorganism that contains the recombinant DNA."[14] A variant of this view is that the risk of genetic experiments will be no greater than the risk associated with the DNA transplanted into the host organism. This led many scientists to the view that there were no intrinsic risks in rDNA techniques themselves; the concern was over the genes added or the hosts and vectors employed.[15]

An interesting approach to getting some boundary conditions on the risks of rDNA hybrids was taken by Donald Brown. His analysis was based upon information he had received and discussions he had held while serving on the eukaryote panel of the Asilomar Conference. Brown believed that one had to examine the potential hazards from a comparative framework. He raised the question, Which is safer, to grow large amounts of tissue culture material and purify large amounts of virus from tissue culture material, or to clone the viral genome in bacteria?

Brown posited a set of conditions under which the cloned DNA could be harmful to an investigator. That scenario was then compared to nonrecombinant methods for achieving similar results. He recognized that discussions about the hazards of gene-transplantation experiments were not grounded in a body of experimental evidence. His approach was based upon the idea that the estimation of risks is a contextual issue. For new activities, one should compare the estimated risks to more conventional behaviors for which the risks are better understood.

Brown reduced the prospect that an investigator could become ill from a recombinant experiment to the following series of steps:

1. A researcher would have to drink a large amount of the bacteria containing the DNA.
2. The bacteria would have to colonize or transfer their infectious plasmids to other bacteria.
3. The bacteria would have to liberate the poisonous plasmids.

4. The naked DNA would have to withstand the hostile environment of the gut.
5. The naked DNA would have to enter the gut cells to produce an infection.

Brown concluded that these steps are not easy to reach. Biological containment is built into the process. But even in the case in which the final step is achieved, the resulting infection will be no worse, according to Brown, than the donor (virus) is by itself.

There are two aspects to Brown's approach to risk assessment: (i) the reductive, and (ii) the comparative. For the reductive part, a final stage S_f is analyzed into a set of sufficient conditions, $S_1 \ldots S_i$. The probability of S_f is $p_1 \times p_2 \ldots \times p_i = p(S_f)$. The risk of S_f is the product of its probability of occurrence and its hazard.

When the potential hazards cannot be explicitly quantified, the comparative approach is frequently used as an alternative. In this case, Brown compared the method of working with SV40 in tissue culture with that of working with SV40 in recombinant organisms. However difficult it may be to assess the risks of the latter, Brown told DAC he was confident that those risks were smaller than the so-called conventional work with SV40. The effectiveness of his argument hinges upon two factors: first, the so-called baseline of risk is acceptable; second, the subjective comparison is acceptable.

Another facet to this approach is the use of a worst-case scenario, which gained support at Asilomar despite some early skepticism. If it can be shown that the case in which the potential for human infectivity is greatest represents a negligible risk, then it stands to reason that one need not consider the spectrum of other scenarios. The logic of this approach assumes an ordering relation for the vast number of possible gene-splicing experiments that are potentially hazardous. Let the ordering of experiments be designated by E_i with E_w signifying a worst-case situation. For any two experiments, let E_i precede E_j in the ordering when $i < j$, and let the risk of $E_i <$ the risk of E_j. If the risk of E_j is negligible, then the risk of E_i is negligible for all $i < j$. Since E_w is the upper bound of risk for all experiments, if the risk of E_w is negligible, then the risks associated with the class of experiments represented in the ordering are also negligible.

One of those leading the opposition to the reductionist approach to risk was Robert Sinsheimer. His disagreement with those who claimed there were no additional risks with rDNA techniques beyond that of working with the organisms themselves fell into two areas. Sinsheimer expressed more concern about random DNA sequences than the DNA sequences that are known to be hazardous. He raised the possibility that novel pathogens would be generated through the introduction of eukaryotic genes into prokaryotes. Viruses indigenous to prokaryotes, he observed, ordinarily do not affect eukaryotes because they lack the means of expressing themselves in eukaryotes. However, he argued, once the recombinants of the viral DNA are made, the genetic control mechanism can be transplanted with the DNA fragment.

If Sinsheimer's view is plausible, then the factors that determine expression of DNA in an organism are transferable between divergent species. If there are any natural barriers to expression of eukaryotic genes in prokaryotes that cannot be overcome in this manner, then most of the hazards of such gene transplants are zero. But it is precisely the expression that scientists were anticipating and that allows the technology to revolutionize the field of molecular biology.

In a second line of argument, Sinsheimer responded to the claim that the laboratory DNA transfers have occurred at *some* frequency in nature and consequently are safe. The transfers in nature, if they indeed have occurred, were at a very low frequency. He implied that a quantitative difference will result in a qualitative effect. He gave DAC an analogy with nuclear risk: "It is like saying ' . . . the human species evolved in the presence of background radiation, and therefore we don't have to be worried about radiation.' "[16]

One of the most interesting interchanges during a discussion period of this meeting took place between Donald Brown and a consultant to DAC, Joseph Melnick, a virologist at Baylor University. Melnick focused on a special type of effect that he postulated could result when linking virus DNA to enteric bacteria: "Dr. Brown raised the question as to whether working with the virus itself might not be more dangerous than working with the DNA recombinant. And while this may be true . . . SV40 will not spread from person to person, whereas if it gets into *E. coli* and is let loose in the community, theoretically it may be spread from those people infected with *E. coli*."[17]

Melnick's point is that the virulence of the material one works with is not the only factor; one also must consider whether new routes are created for potentially hazardous genes to become established in the environment. Brown replied to Melnick that extending the host range of SV40 is not a problem so long as one stays with *E. coli* K12. His reason: K12 will not multiply in the gut. Since its range is limited to begin with, we cannot extend the field of influence of a donor DNA when it is implanted in K12. Many scientists at the time were satisfied that K12 could not be made a robust organism by inserts of DNA and that its debilitation was accepted for the broad range of human bacterial flora, except perhaps when an individual was on a regimen of certain drugs.

The approach taken at Asilomar and thereafter in the construction of guidelines was to establish a gradation of potential risks where each level could be matched to a containment strategy. Differences were expressed at the DAC meeting on the rationale for some of the distinctions used in classifying those risks. Richard Goldstein cited as unjustified the separate classification of embryonic tissue from other cells in an organism. His reasoning was based on a purity argument, namely, the embryonic tissue will not be fully detached from follicle cells from which they are obtained.[18] The reason given for a lower risk estimate for embryonic tissue by Wallace Rowe was that it had not been shown to harbor exogenous viruses. One of the presenters to DAC, Stephen Wiesenfeld of the National Jewish Hospital and Research Center, opposed such a distinction. He placed shotgun experiments with adult tissue equal in risk to those with embryonic tissue.[19]

Elena Nightingale of the National Academy of Sciences threw another wrench into the risk-classification scheme. She questioned the distinction between cold-blooded vertebrates and all other eukaryotes. Why, she asked, should cloning insect DNA into *E. coli* be less risky than cloning vertebrate DNA? Richard Goldstein believed that invertebrates may just as easily carry human disease as vertebrates:[20] "[T]he dangers inherent in shotgunning invertebrates are just as dangerous as any vertebrate, mainly because of the unknown factor [namely] . . . all those genes you are shotgunning."[21]

Other critics also warned against shotgun experiments with invertebrates because some, for example, insects, bear potentially harmful viruses. The fact that invertebrate DNAs code for hormones that may exhibit activity in human systems was also cited. But it was not only the pharmacologically active agents of animal systems that were a matter of concern. Questions were raised about cloning a human hormone like insulin into *E. coli*. John Sedat of Yale University, who was active in raising the consciousness of his colleagues at the 1973

Gordon Conference, addressed this issue to DAC: "[C]an we take any risk of letting the population accidentally have a bug that will produce, say, insulin in their intestinal flora?"[22] Sedat saw the beginnings of a rapid rise in cloning experiments if restrictions were not in place. The research benefit for much of this work was to amplify genes for the purpose of detecting sequences.

On this issue, Sedat was confident that alternative techniques for amplification were on the horizon. One need not rush into the work. Falling on the same side of the issue as his professional colleagues from Science for the People and other eminent scientific critics, Sedat encapsulated the "go slow" approach: "We have several billion years of evolution behind us. Who is to say that we should just march in and start changing these things?"[23]

The new technology had clearly opened up a cornucopia of risk scenarios. In response, scientists divided themselves into two camps, the offense and the defense. The goal of the offense was to find a scenario that was plausible. The function of the defense was to discredit the scenario with some scientific artillery. But even in those cases where the potential hazard seemed remote, the offense frequently brought up some mitigating circumstances.

One of the clearest examples of how those arguments developed is in the scenario of the insulin-producing *E. coli* in the gut. In reconstructing the argument, viewpoints are presented from several independent dialogues.

Offense: One can imagine a dangerous situation in having a resident in the human intestinal tract that is (a) self-propagating and (b) harboring a functional gene that codes for human insulin.

Defense: *E. coli* with an insulin gene in the intestinal tract would be harmless because insulin has to be injected into the blood stream to be pharmacologically active. (Diabetics cannot benefit from insulin taken orally.)

Offense: Suppose an investigator has a stomach infection and the *E. coli* produces the insulin at the juncture of the epithelial cells.

Defense: The guidelines exclude individuals with such illnesses from performing the experiments. One also has to consider that even if a few bacteria produced insulin that found a pathway to the blood stream, if the quantities were insignificant compared to the body's own normal production, there would be no risk. Since *E. coli* K12 would not colonize the gut, this is an additional safeguard against such a scenario.

The strategy of the defense was to show that the offense was exaggerating the hazards by pointing to a boundary limitation in the scenario that it argued, demonstrated that the results could not be attained, or, if they could, that the effects would be insignificant. The offense focused on the novel and unexpected aspects of biological research, such as contamination, the establishment of new niches, unknown viruses, or the gross errors or sloppiness of investigators.

The events covered in this period of the debate illustrate two trends that build up momentum in the next few years. First is the emergence of organized dissent within the scientific community. This is a dramatic shift from the informal lines of communication and professional meetings with which scientists felt at home in expressing their criticisms of rDNA policy proposals or in venting their feelings about hazards.

Second is an organized opposition among scientists willing and ready to take the issues outside professional circles. Those critical of the approach taken by NIH in developing guidelines moved on four fronts. They wrote articles in journals and magazines that eventually reached broader sectors of society. They communicated with congressional staff and addressed legislative and oversight hearings to influence national policy. Some scientists took the issues back to their local communities where they supported ordinances regulating research in academic and commercial laboratories. Finally, scientists formed alliances with environmental groups that brought technical expertise and advocacy skills under a single initiative.

The DAC meeting was the starting point of organized opposition from the most influential environmental public-interest groups in the country, namely, Friends of the Earth (FOE), the Environmental Defense Fund (EDF), the Sierra Club (SC), and the Natural Resources Defense Council (NRDC). The first three organizations were invited to attend the meeting. These groups either had their own technical staff people who could understand the issues and critique the proposed guidelines or were in close contact with scientists willing to play such a role.

The policy process, starting with the Berg letter, passing next to the drafting of guidelines, and finally leading to the DAC meeting, was moving smoothly with only minor perturbations due to attempts to shift the locus of control away from the scientific community. In the aftermath of that meeting, the strength of nonscientific opposition to the research guidelines began to mount. With this came the greatest threat to disciplinary autonomy that biology has ever witnessed; and the threat of public oversight and control stimulated strong reactions. Many scientists began to close ranks and abandon their earlier criticism. For some, the "real dangers" of public participation were far greater than the "hypothetical dangers" of rDNA research. The public-interest groups had shifted the debate off the course of biohazards by raising procedural issues, concerns of legal liability, and requests for an environmental-impact statement. By this point, a new orthodoxy among scientists began to emerge. The biological air was becoming filled with self-righteous indignation. "We acted responsibly and now the public is overreacting."

NIH Director Donald Fredrickson issued the guidelines four and a half months after the DAC meeting, on the same day that the Cambridge (Massachusetts) City Council was holding hearings on a proposed new facility for rDNA research at Harvard University. Cambridge was one of a dozen places that debated the issues on its own terms. Nothing could be more threatening to the goals of scientific inquiry than to have a variety of local and state codes regulating gene-splicing experiments.

The production of the NIH *Guidelines* represented an enormous quantity of human labor. The magnitude of documents produced, including the letters and memoranda that originated at NIH, is testimony to that. But how did that effort affect the quality of the product itself? The next chapter examines the rationale behind the *Guidelines*. The architects of the *Guidelines* acknowledged that much of its edifice was built upon untested suppositions. Those more critical of the process viewed the final outcome as weighted heavily in the direction of the principal users of the technology. Nicholas Wade, a staff writer for the journal *Science*, put it this way in his book *The Ultimate Experiment*: "Because most of the decisions about appropriate safety levels involved questions of judgement rather than hard scientific fact, the charge of unwitting preference—no one has accused the Committee of deliberate bias—was hard to rebut."[24]

Wade's point must not be taken lightly. There are two kinds of bias, one that is overt and explicit, another that is subtle and tacit. Where there are matters involving subjective choice and conjecture, the latter bias is more difficult to identify. It is a relevant datum that 30 percent of the US participants at Asilomar were involved with the use of rDNA techniques within a year after the meeting. Also, three members of the original RAC were engaged in gene-splicing experiments while the guidelines were being framed. It is fair to say that scientists were not about to regulate themselves out of business. But it is also the case that the rDNA affair provided the biological community with the opportunity to complete an unfinished agenda that they had begun at the first Asilomar meeting in 1973—a systematic examination of the hazards in biological research.

Instead of using the rDNA debate toward that end, the biohazards were examined exclusively in terms of gene-splicing techniques, foregoing a more comprehensive approach that would also include conventional viral and bacteriological work.

To the extent that the guidelines were designed to diffuse the controversy and diminish the polarization, they represent a negotiated settlement. But this settlement was not without its scientific basis. Scientists learned to play a dual role. They compromised when necessary; but they also brought to bear on the problems the technical armory that they were accustomed to when they approached a purely scientific issue.

13

The Logic of the First NIH Guidelines

The guidelines issued by the National Institutes of Health were the culmination of a process that has been traced back to 1971 when Robert Pollack at the Cold Spring Harbor Laboratory began to raise the safety issues associated with recombining genes. The following five years of formal and informal meetings and discussions can best be characterized as an adventure in inductive risk assessment. By this I mean that the final outcome did not result from a set of basic notions or principles; rather, the risks of rDNA experiments were evaluated on the basis of the experiences of participants engaged in viral or bacteriological work. No general framework was advanced for assessing the potential risks. The first attempt to establish such a framework came nearly two years after the issuance of the US guidelines (23 June 1976). In the spring of 1978 Sydney Brenner, a participant at Asilomar and distinguished scientist at the Medical Research Council's Laboratory of Molecular Biology in Cambridge, England, tried his hand at such a task. Brenner produced a paper for the Genetic Manipulation Advisory Group (GMAG), the British counterpart to RAC, that offered a generalized framework for estimating the potential hazards of different classes of experiments.[1]

Notwithstanding the fact that the Asilomar scientists and the newly established NIH advisory committee had not discovered a unifying principle for grading the potential risks of the vast number of possible experiments, there was an effort to produce a document that had a rational basis. An edifice of the dos and don'ts of cloning was created from several independent principles, an undefined number of conjectures, and a body of information from virology and bacteriology. Unlike the theoretical constructs that scientists often work with, the end result of the risk assessment could not be confirmed or disconfirmed. There was no empirical investigation that one could undertake to show that one class of experiments was potentially more hazardous than another. But even without the rigorous theoretical basis for the specifics of the NIH *Guidelines*, there was a logic that guided its broader outlines. Nevertheless, that logic had its severe critics, a testimony, perhaps, to the unreliability of prediction and the lack of confidence in theory in those areas on which the risk analysis had depended.

There were two independent processes involved in setting up the research guidelines. The first consisted in grading the potential risks for a substantial number of possible genetic experiments. The second involved mapping these risks to the various experimental conditions. These conditions included the type of host organism permitted as the recipient of the recombinant molecule, laboratory procedures and practices, and the physical construction of the laboratory. Thus, even if agreement could be reached on how experiments involving artificial genetic recombinations should be graded according to risk, there was still room for considerable debate on satisfactory laboratory conditions.

An expression of general principles behind the guidelines was made by Maxine Singer in her presentation in 1976 before the Director's Advisory Committee. She cited four principles:

1. In the light of current information, certain experiments may be judged to present sufficiently serious potential hazards so that they should not be attempted at this time.

2. The group of experiments that pose either lesser or no potential hazards can be performed provided
i. the information to be obtained or the practical benefits anticipated cannot be obtained by conventional methods;
ii. appropriate safeguards for containment are incorporated into the design.

3. The more potentially hazardous the experiment, the more stringent should be the safeguards against escape of the agents.

4. There should be an annual review of the guidelines.[2]

The proscription of certain experiments was an issue vigorously debated at Asilomar. Some scientists argued that the real danger was to freedom of inquiry and that the impact of such a restriction would be far more injurious to science than any potential biohazard would be to the public. Taking the more cautious approach, NIH followed the voices of those insisting that certain experimentation be *verboten*. The contents of this *verboten* area of experimentation had evolved over time, and, like much of the guidelines, represented a negotiated settlement.

Four categories describe the prohibited experiments: organisms, genes, dissemination, and scale. They illustrate how the guidelines were developed out of disparate principles: a patchwork of a priori logic, science, and intuition.

There were certain organisms stipulated by the guidelines (or interpreted by RAC) that were not to be used to create new recombinant molecules, regardless of what part of the genome was considered for use. The list came from two sources. The first was from the Center for Disease Control (CDC), which published a classification of etiologic agents. Organisms designated by CDC as types 3, 4, and 5 pathogens were included on the list of forbidden experiments. A second grouping of organisms came from the National Cancer Institute. Its list of moderate-risk oncogenic viruses was also included. But once rDNA experiments involving agents that infect other organisms are restricted, the next logical question is how to handle the agents they infect. NIH followed the reasoning and extended the prohibitions to cells known to be infected by the viruses. But the cells that, some believed, may harbor latent oncogenic viruses were placed in the permissible category.

Some prohibitions were outlined for specific categories of DNA, such as the genes that code for toxins, plant pathogens, or drug resistance. Genes that synthesize botulinum or diphtheria toxins or venoms from insects and snakes were cited as examples. The concept of risk, thus, was directed not only at bacteria and viruses, that is, whole organisms, but also at DNA sequences.

The domain of restricted experiments included a proviso for not deliberately releasing recombinant organisms into the environment. The use of recombinant methods in agriculture or for oil spills was being seriously studied. With the rise of industrial interest in gene splicing, some scientists recognized that society was not far from testing a new recombinant organism on some ecological system. Their concern was that this would be done without the full understanding of the impact of the organisms on the ecology. This restriction on dissemination had little or no implications for the laboratory scientist.

Another restriction in the area of dissemination was directed at the transfer of drug-resistance traits. This was an issue that was actively discussed at Asilomar. There was growing evidence that the dispersal of genes that confer antibiotic resistance to organisms (R factors) into the environment was a public-health problem.

The final category that describes the limits placed on the research in the first NIH *Guidelines* is scale. Certain experiments involving more than ten liters of culture were proscribed unless exempted as an individual case. The exact wording is instructive:

> In addition, at this time large-scale experiments (e.g., more than 10 liters of culture) with recombinant DNAs known to make harmful products are not to be carried out. We differentiate between small- and large-scale experiments with such DNAs because the probability of escape from containment barriers normally increases with increasing scale. However, specific experiments in this category that are of direct societal benefit may be excepted from this rule if special biological containment precautions and equipment designed for large-scale operations are used, and provided that these experiments are expressly approved by the Recombinant DNA Molecule Program Advisory Committee of NIH.

Within a year, new language had been proposed that effectively tightened the restrictions on large-scale work. The prohibition was placed on experiments using more than ten liters of culture "with organisms containing novel recombinant DNAs, unless the recombinant DNAs are rigorously characterized and are shown to be free of harmful genes." The new language guaranteed that each large-scale experiment would be reviewed by RAC.

Let us turn to the important structural logic of the guidelines. An analogy will help. In establishing a thermometric scale, one begins by choosing two end points, for example, the boiling point and the freezing point of water, and then one develops a gradation of temperatures between the two extremes. In a similar fashion, the architects of the NIH *Guidelines* identified upper and lower bounds of risk. The upper bound was represented by the region of forbidden experiments. The lower bound consisted of genetic exchanges that commonly occur in nature among bacteria. Between these end points, experiments were graded according to their relative potential risk.

The hierarchy of risks for rDNA experiments was based upon a phylogenetic-ordering principle. According to this principle, the potential risks of such experiments were considered to be greater for DNA transplants from donor organisms closer to man. Several arguments were advanced to justify this approach. DNA products from higher eukaryotes exhibit greater similarity (homology) to human proteins. Furthermore, the genes likely to be more harmful, that is, become functional in human systems, were enzymes, hormones, or other proteins that were similar in molecular structure to those already in man. Thus, it was conjectured that the DNA from higher animals had a greater potential for interfering with human biochemistry than the DNA from lower animals.

A second reason offered for assuming greater hazard posed by gene splicing involving phylogenetically higher eukaryotes pertained to the infectivity of viruses. The closer the source of the DNA is to man, the reasoning went, the more likely that it has been infected with viruses that can propagate in human tissue.

Concurrently, many scientists were coming to the opinion that bacterial hosts were safer than eukaryotic hosts for implanting foreign eukaryotic DNA. They reasoned that bacteria lacked the regulation sequences required for the expression of a gene product for eukaryotic DNA.

For the purpose of risk assessment, a distinction was made between the DNA isolated from normal tissue and that isolated from embryonic tissue. The statement issued by NIH maintained that the embryonic or germ line DNA was less likely to be contaminated by pathogenic viruses than was the normal adult tissue.

Following, then, the principle of phylogenetic proximity and the reduced risk of embryonic DNA, the ordering of potential hazards was as given in figure 3.

Within this format, additional distinctions were made that provided a fine structure to the risk assessment. Experiments involving well-defined segments of DNA were rated less hazardous than shotgun experiments. In the latter case, entire chromosomes are broken

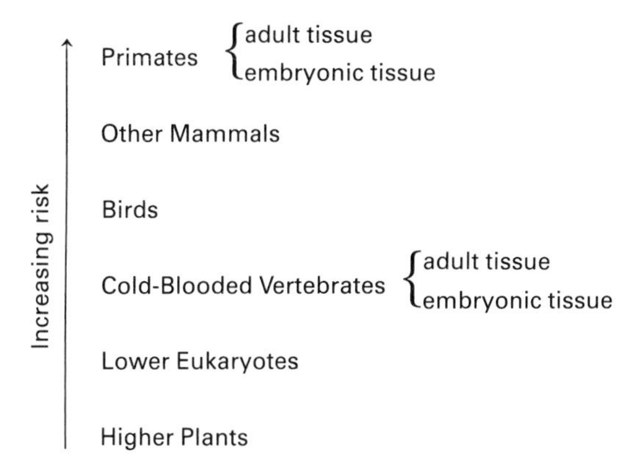

Figure 3 Order of potential hazards by the principle of phylogenetic proximity.

up with restriction enzymes and blindly shotgunned into host organisms. When the initial guidelines were released, these experiments were assumed to be of higher potential hazard because they involved unknown genes. The potential risks were considered lower for DNA that was well characterized, or pure, and whose product was known not to be harmful to humans, plants, or animals.

Biologists have studied to some degree the propagation in nature of organisms like *E. coli*. The result of these studies played a major role in the fine tuning of the *Guidelines*. Of special significance in this regard is the transmissibility of plasmids—the extrachromosomal circular elements of DNA that serve as vectors, or transmission vehicles, for gene-transfer experiments. Plasmids that exhibit a high degree of intercellular mobility are called conjugative; alternatively, the plasmids that transfer among cells at a very low frequency are termed nonconjugative. In order to limit the spontaneous exchange of genes once artificial genetic recombinations had been accomplished, the *Guidelines* specified that the hosts for such experiments, *E. coli* K12, contain nonconjugative plasmids. But while the nonconjugative plasmids do not initiate their own transfer at a high frequency, they can transfer with the help of other plasmids. This process is called mobilization because plasmid transfer for the nonconjugative types is promoted by the mobilization of conjugative plasmids.

We begin to see evidence at the molecular level of a complex system of relations. The properties of many genetic complexes are more than their internal relations, as in the case in which the ability of some plasmids to transfer depends upon the appearance of other plasmids. The limitation on plasmid exchange continued to serve as an important safeguard for those concerned about the inadvertent spread of certain gene sequences into the environment. On the other side of this issue were those organisms (prokaryotes) that were known to exchange genetic information with *E. coli*. For the purpose of cloning experiments, donor organisms that are natural exchangers with *E. coli* were treated as less potentially hazardous, all other things being equal, than donors that are not known to exchange with *E. coli*.

A second element factored into the strategy to reduce the risks of gene-splicing experiments was the use of defective or enfeebled host organisms. Ideally, one wants a bacterium whose genetic map is well understood and that will be robust during the time span of an experiment. In addition, such an organism should survive and propagate under the appropriate conditions, but should not survive outside of a restrictive and supportive laboratory environment. During the most active period of the rDNA debate, considerable attention was given to the host or recipient organism for DNA transplants. While *E. coli* K12 was chosen as the preferred host for such experiments, the critics developed sufficiently persuasive arguments to raise doubts in the minds of NIH officials about their long-term commitment to it as a host. The *Guidelines* stated, "*E. coli* K12 appears to be the system of choice at this time, although we have carefully considered arguments that many of the potential dangers are compounded by using an organism as intimately connected with man as is *E. coli*. Thus, while proceeding cautiously with *E. coli*, serious efforts should be made toward developing alternate host-vector systems."

The question of the safety of *E. coli* for experiments has several dimensions. These will be studied in greater detail, but a brief review will be useful at this point. In addition to the viability of *E. coli* K12 to live or colonize in the human gut, there were also concerns raised about the exchange of genes from K12 to wild strains of *E. coli* and the ability of K12 to be transformed into a human pathogen.

Another example of a transmission barrier that was incorporated into the *Guidelines* is the use of defective viruses as vectors for eukaryotic hosts. These viruses lack a key sequence that restricts their ability to reproduce without a helper virus. Upon infecting a cell, no new viral particles can be made with the defective strains. If an experiment requires that new virus particles be produced, a helper virus supplying the genes with the missing function can be used.

There is a simple logic associated with a process that establishes an ordering or weighting of experimental risks and then proceeds to match the individual experiments to a set of procedures and equipment designed to ensure containment. However, the *Guidelines* reflected a more complicated state of affairs. The experimental situation and the containment were not independent. To visualize this, let us generalize from the primary factors that describe an experiment and the conditions under which it is performed.

D(Di, . . . ,De) = donor(s) of the DNA;

V(Vi, . . . ,Vj) = vector(s) for implanting DNA;

H(Hi, . . . ,Hk) = host or recipient organisms of DNA recombinant molecule;

E(Ei, . . . ,El) = environment including the physical conditions of the laboratory and the procedures followed by the investigator.

An experiment is defined by elements from classes D, V, and H. The potential risk for each gene-splicing experiment is a function of the three factors, according to the underlying logic of the *Guidelines*:

r = F(D,V, H), where r = potential risk.

For any particular experiment, an r value is mapped onto a containment space that is a function of factors H, V, and E, where c = G(H,V, E). (c = containment space.) This simple schema shows the functional interrelation between the experiment, the relative-risk assignment, and the containment space:

F(D,V, H) = G(H,V, E).

The above equation illustrates the point that when the host or its vector is changed, one also changes the containment space, in addition to the experimental system.

For the purpose of constructing the *Guidelines*, factors V and H were defined by a single system, termed a host-vector (HV) system. Three categories of HV systems were stipulated: EK1, EK2, and EK3, referring to the use of *E. coli* K12 or a variant as a host, and specified plasmid or phage vectors in tandem with the host. Each donor of DNA, Di, was matched to an HV system and a physical containment environment, the combination of which provides the containment space:

Di→(H+V)i + Ei

The *Guidelines* defined four values for Ei, which were termed P1, P2, P3, and P4 physical containment. Each P value refers to a set of physical conditions and laboratory procedures, ranging from the application of some standard good microbiological practices (P1) to high-containment facilities designed for work with the most pathogenic agents known to man (P4). A simple two-by-two matrix in table 3 illustrates how a sampling of the many possible experiments was mapped onto the containment space. The column on the extreme right lists the forbidden experiments. Even for the purpose of risk-assessment experiments, the list of donors in this forbidden category

Table 3
Physical and biological containment for a sampling of rDNA experiments, NIH *Guidelines*, 1976[a]

	P1	P2	P3	P4	Prohibited experiments
EK1	Prokaryotes that naturally exchange genes with *E. coli*	Embryonic DNA from cold-blooded vertebrates	All or part of the DNA from plant viruses		Cloning of genes from pathogenic organisms of class 3, 4 or 5
		DNA from plants with no known pathogenic agents	Plasmids or phage from hosts that do not exchange with *E. coli*		Certain transfers of R factors
		DNA from bacteria of low pathogenicity	Mitochondrial DNA from primates		Genes for potent toxins
EK2: 10^{-8} survival factor		Prokaryotes that do not exchange genes with *E. coli*	Animal viruses free of harmful genes	All or part of an animal virus genome	
		Cold-blooded vertebrates	DNA of birds	Shotgun experiment with primate DNA	
		All or part of the DNA from plant viruses	Uncontaminated embryonic primate tissue		
EK3: systems confirmed by animal and human tests			Shotgun experiments with primate DNA		
			Animal viruses not known to be free of harmful genes		

a. Physical containment increasing from left to right; biological containment (variants of *E. coli* K12) increasing from top to bottom.

was out of bounds. Exemptions to that list were to be considered by RAC on a case-by-case basis.

Some classes of experiments were assigned two positions in the containment space. Table 3 shows that shotgun cloning with primate DNA was permitted under P3 + EK3 or P4 + EK2 conditions. An important assumption underlies this result: an operational equivalence was postulated between physical and biological containment. An alternative way of saying this is that one step up (down) in physical containment followed by one step down (up) in biological containment allegedly provides about the same overall containment for an experiment.

The concept of equivalence between biological and physical containment was not discussed in the NIH documents, nor was it a subject of debate among scientists, except perhaps within the small circle of RAC members who developed the *Guidelines*. The equivalence assumption had no empirical basis, but there was clearly a political basis for it. Some experiments, considered sufficiently worrisome, were designated for a P4 containment facility. But few such facilities existed in the United States, and none, for that matter, at academic institutions. A requirement of P4 containment was tantamount to proscribing the experiments for the vast majority of molecular biologists who had no access to a P4 facility. Even P3 facilities were not in widespread use, but they could be constructed in existing university laboratories at substantial but not prohibitive costs. By double-listing experiments in the containment space, scientists were making them as accessible as they could in the climate that existed at that time. While the NIH *Guidelines* were being developed, institutions like Harvard University and the University of Michigan were applying to NIH for capital improvement grants to build P3 containment laboratories on their campuses.

The *Guidelines* made explicit reference to a rule that was applied for determining the containment of recombinant molecules constructed from biological entities that were considered potentially hazardous. The rule stipulated that the containment for the experiment shall not be less than what is required for the most hazardous component used in constructing the recombinant organism. The three components are the host organism, the vector, and the donor DNA. The rule set a lower, but not an upper, bound on containment. The *Guidelines* also acknowledged that in most cases the containment for a recombinant experiment would be greater than the precautions used for an individual component when studied under conventional techniques.

Some scientists believed that the rule of the most dangerous component introduced some irrational constraints in the name of safety. Why, they asked, should a virus be treated under more guarded conditions when it is broken apart and attached to another DNA sequence than when it is used in its natural state, where it is allegedly most suited for infectivity? The issue here is whether the recombinant technique permits one to improve the virulence of certain strains of biological organisms. There were plausible arguments on both sides of the issue; hard empirical data was needed to resolve the differences.

The rule of the most dangerous component is another example of the dominance of a reductionist approach to the problems of risk and containment. The probability of a nonlinearity in the production of hazards was considered remote, as in the case where two relatively innocuous segments of DNA would produce something dangerous. The reductionist thesis proclaimed that "from nonpathogens, pathogenesis will not emerge."

Another facet to this problem arises in the discussion and debates over using viral DNA in experiments. The *Guidelines* were premised on certain beliefs about the hazards of viruses; some of these beliefs were still being debated among scientists. Can a portion of a virus be more hazardous than the entire virus? Is it possible to improve the host range of a virus? In this regard, could one create, through recombinant techniques, an animal virus that will infect humans whereas previously no such infections had been observed? Could the naked DNA of a virus be more infectious than the entire virus? These questions were addressed in subsequent years through special risk-assessment meetings and a few selected experiments.

The people who drafted the *Guidelines* and their associates at the NIH could hardly anticipate that lay persons would some day be struggling to understand them. A conspiratorial way of looking at the events is that the *Guidelines* were written so as to prevent those who were not specialists in the field from understanding the significance of the potential hazards and the containment system. The introductory section to the 1976 edition of the *Guidelines* made it abundantly clear that it was written for scientists on study sections and for other practicing scientists. It was designed for use within a system of self-regulation. Considerable responsibility was placed with the principal investigator, including interpreting the potential hazards and the containment levels for each experiment; establishing the degree of medical monitoring; and setting the training standards for researchers and technicians.

Even for those who were knowledgeable in the field of molecular biology, the *Guidelines* represented a cumbersome document that lacked simplicity, coherence, and internal consistency. Embedded within its general framework, which, as we have indicated, did attempt to provide some rational principles, were a host of individual exceptions, and special conditions. Most of those involved in its formation or in its critique were aware that it was a temporary measure. As more information became available, changes would undoubtedly be made.

In his introductory statement to the published *Guidelines*, the director of NIH, Donald Fredrickson, issued the following comment reflecting the pushes and pulls that he experienced: "In many instances the views presented to us were contradictory. At present the hazards may be guessed at, speculated about, or voted upon, but they cannot be known absolutely in the absence of firm experimental data—and unfortunately, the needed data were, more often than not, unavailable."[3] Fredrickson did not follow up this statement with a commitment to gather the data. It took several years of intense controversy in the public sector before NIH made some significant efforts to test the risks that had been speculated about. But, as we shall see in the following chapters, some of the dire consequences predicted for this new technology were not testable through laboratory experiments.

And as for the changes in the *Guidelines*, they came quickly. By 1977, efforts were under way to relax a number of the containment requirements. And no sooner had these changes been made in December 1978, than a new campaign started to exempt the vast majority of experiments from the *Guidelines* and from oversight by the local biosafety committees. NIH Director Fredrickson was openly supportive of the dismantling process. At the 15–16 February 1979 RAC meeting, he expressed optimism that the newly constituted committee would eventually eliminate the rules. Some day, he believed, the *Guidelines* would be under glass, an artifact of an unusual epoch. But he did not say how long that would take. However, nearly five years after the *Guidelines* were first issued, the April 1981 RAC meeting took up a proposal that would do away with their mandatory character, the first important step to their eventual elimination. The second step took place at the following meeting of RAC on September 1981, at which, by a 16–3 vote, the committee approved a motion to do away with special guidelines for rDNA research.

The Science and Policy Debates

14

After the Guidelines

As the NIH *Guidelines* were being prepared for a public airing, new avenues of dissent were developing around the country. The Director's Advisory Committee meeting in February 1976 served as a branching point in the debate. For the scientists who helped assemble the *Guidelines* and who now constituted the mainstream, or moderate wing, in the spectrum of DNA opinion, the institutional and professional "governors" that serve to modulate dissent within the ranks were no longer effective. Leading scientific figures who might be able to exert some influence over dissident colleagues were impotent against professional environmentalists and the hard-core scientist-critics who advocated a more open and shared system of decision making.

The DNA controversy began all over again at universities, state legislatures, and local municipalities. Issues that had been debated for three years in committees of NIH were now being debated by non-scientists as if the questions were being asked for the first time. Recombinant DNA "hotspots" flared up in a dozen sections of the country, usually over the construction of a new P3 physical containment facility.

Dissident scientists who considered the *Guidelines* inadequate to protect the public health joined forces with environmentalists, local citizenry, and university faculty. The new networks of scientists and lay persons made supporters of the NIH *Guidelines* uneasy; some were openly scornful of their colleagues' willingness to stray from disciplinary boundaries. At a meeting of the California Medical Association around November 1976, Stanley Cohen, one of the signers of the Berg letter, remarked, "What began as an act of responsibility by scientists, including a number of those involved in the development of the new techniques, has become the breeding ground for a horde of publicists—mostly poorly informed, some well meaning, some self-serving."[1]

In the four-year period following the issuance of the first set of guidelines, the DNA controversy took on an episodic character. Policy debates crested and subsided, giving way to new issues. The early debates centered on the safety of *E. coli* as a preferred host, the effectiveness of physical containment, the standard for establishing a host-vector system for moderate- and higher-risk experiments, and the procedures adopted by NIH to ensure public safety. Some scientific critics of NIH's policies for handling the DNA affair made a strategic decision to take the case to their university communities or to their local and state public officials. Hearings and the formation of study committees at the University of Michigan, Cambridge, Massachusetts, New York State, and San Diego prompted national media coverage, which began to draw attention from Congress. The new coalitions of scientists and environmentalists started out as informal networks and developed into a more organized and systematic force through the creation of the Coalition for Responsible Genetic Research in the fall of 1976 after the Cambridge controversy took hold.

Throughout 1977 and well into 1978, the subject of national legislation dominated the DNA debate, as over a dozen bills were introduced into Congress. While congressional subcommittees were holding hearings, scientists, concerned that their discipline was being singled out for regulation, followed two strategies to slow down or obstruct the legislative mood of Congress and to defuse the public's reaction to gene-splicing experiments. Some lobbied Congress with support from their host institutions and professional societies.

At the same time an offensive move was taken against the claims of DNA critics. Stanley Cohen of the Stanford Medical School released prepublication results of a study that, he claimed, showed that genetic exchange between distantly related organisms (prokaryotes and eukaryotes) occurs in nature. A conference held in the summer of 1977 brought together, for the first time, leading experts in the fields of infectious diseases and the ecology of *E. coli* to consider the safety of *E. coli* as a host for DNA recombinant molecules. Results of the meeting were widely publicized as vindicating the safety of *E. coli* K12 for gene-splicing work. An NIH-sponsored intramural study investigated whether an rDNA insert could make a bacterium pathogenic. The experiment involved the insertion of the DNA of mouse polyoma virus in *E. coli* K12. It was billed as a worst-case scenario since polyoma virus causes tumors in hamsters.

When the first session of the 95th Congress came to a close, the science lobby against legislation had begun to see signs of success. Forecasts of legislation appearing by the summer's end (1977) were proved false. Two leading bills competed with one another for support amid mounting opposition. The chief sponsors of legislation were Representative Paul Rogers (D-FL) and Senator Edward Kennedy (D-MA). The two bills reflected basic policy differences on the role of federal preemption and the governmental authority that should have regulatory and enforcement powers over rDNA technology.

As the public and its elected representatives were becoming increasingly aware of the promises and potential risks of genetic engineering through media attention and the publicity of environmental groups, many scientists were beginning to adopt a no-risk posture. During the Asilomar period, the spectrum of scientific opinion ran the gamut from "No guidelines were necessary" to "The research should be limited to a few national centers". When external pressures on the scientific community began to build, the polarization of views among scientists grew wider. The battle lines began to shift from "Set safe standards for research" to "Protect science from direct external regulation."

For the critics of the NIH *Guidelines* who persisted in bringing their cases to the public, the backlash from the science orthodoxy took several forms, from overt character assassination to a more subtle form of professional ostracism. Perhaps the most effective form of persuasion was the kind of paternalistic advice that comes from scientific colleagues. It draws attention to the fact that continued involvement as a critic could bring irreparable damage to one's career.

The connection between rDNA research and human cloning had drawn national attention in March 1978. The *New York Post* ran a front page headline, "Baby Born Without a Mother, He's the First Human Clone". The subject of the story was the controversial book *In His Image, The Cloning of Man* by science investigative reporter and free-lance writer David Rorvik. The book's author and its publisher, Lippincott, claimed that it was a factual account of a successful human cloning experiment. Names and places were withheld allegedly to protect the identity of the fourteen-month-old baby and his millionaire father, who wanted to leave a genetic legacy to the last detail. As far as the publicity was concerned, it did not matter that rDNA techniques were not involved in the human cloning technique or that most scientists dismissed the tale as a hoax. The book provided a forum for those scientists who had been concerned that the engineering of microorganisms was merely the first step in a broader program of human genetic engineering. Three scientists (Jonathan Beckwith,

Harvard Medical School; Ethan Signer, MIT; Liebe Cavalieri, Sloan Kettering), in collaboration with the People's Business Commission, a Washington-based advocacy group, petitioned the court to hasten a request they filed under the Freedom of Information Act. The FOIA request was for information on all federally sponsored human cloning experiments.[2] Results obtained from the FOIA request yielded no such information.

By the fall of 1978, the mood in Congress for legislation began to crest for a second time, as public interest groups like the Environmental Defense Fund, the Sierra Club, the Natural Resources Defense Council, and the Coalition for Responsible Genetic Research continued their lobbying efforts. But as conditions changed, the groups shifted their strategies. While initially citing faults with the 1976 *Guidelines,* in the face of new efforts by scientists to gain a substantial relaxation of containment conditions the environmental groups began to fight to retain the status quo until more information about the risks was made available. The rDNA battleground took place on two fronts: first, around the activities surrounding the passage of legislation; and second, around efforts under way to relax the *Guidelines.* Since a ban on most large-scale experiments was still in effect, what went on at the Recombinant DNA Advisory Committee meetings was largely irrelevant to the commercial interests. In an unprecedented situation, academic research was under federal guidelines, while research and development in the private sector went unregulated.

Other key targets for DNA critics were the procedural and administrative rules adopted by NIH. Formal protests were issued to HEW Secretary Califano that the Recombinant DNA Advisory Committee was too narrowly constituted. They argued that the "old-boy network" dominated the decision-making process and cited NIH's dual role as the promoter and regulator of the new research technology as an irresolvable conflict of interest for this agency.[3]

While DNA politics was still receiving its share of attention, the science was moving ahead rapidly.[4] Scientists were able to direct public attention to a bacterial clone that produced somatostatin, a substance that inhibits the pituitary production of growth hormone. Herbert Boyer reported the result at a news conference where he stated that "the man in the street can now get a return on his investment in science". In June 1978 a team of Harvard biologists led by Walter Gilbert reported that they had successfully induced bacterial expression of rat proinsulin, a precursor to insulin. The research was accomplished under the 1976 NIH *Guidelines.*[5] One of the first examples of the applied use of restriction enzymes for the study of human hereditary birth defects was discussed in the *New England Journal of Medicine* in July. Researchers from Harvard, Yale, and the Haceteppe University in Turkey produced an image of a single human gene that directs production of hemoglobin. The detection of the gene enabled doctors to screen fetal fluid for two types of anemia (alpha thalessemia and beta-delta thalassemia).

In the fall of 1978, the first bacteria-producing human insulin was reported by Genentech, a bioengineering firm located near San Francisco, and the City of Hope National Medical Center near Los Angeles. Immediately, Eli Lilly, the largest producer of insulin in the United States, announced an agreement it had with Genentech to take over industrial-scale production when the efficiency of the technique had been improved.[6] And in November *Science* reported that Paul Berg and his colleagues at Stanford University transplanted a rabbit gene that codes for hemoglobin into monkey cells. It was the first time a functional gene had been transferred from one mammalian species to another. Ironically, the transfer vehicle was the virus SV40, which, as an instrument for genetic engineering, had been one of the factors that precipitated the early debate over the safety of gene splicing.[7]

With limited resources devoted to rDNA technology, by the winter of 1978 environmental groups were finding it difficult to maintain the intensity of the lobbying effort. At the same time, environmentalists were drawing criticism from scientists who served on their advisory boards. The backlash reached the point where some prominent trustees of environmental groups, such as Friends of the Earth, the Environmental Defense Fund, and the Natural Resources Defense Council, had either resigned from these organizations or disassociated themselves from the positions on DNA research that had been adopted.[8] But the core staff responsible for the DNA lobby did not cave in under the pressure. Determined to see the issue through, they carried on their activities so long as they continued to receive support from their executive committees.[9] In response to the campaign waged by environmentalists, scientists countered with strong statements in support of freedom of inquiry and deluged Congress with articles and letters to forestall any legislative action.[10]

The year 1978 ended with a new set of guidelines from rDNA research and the failure of any DNA bill to pass out of committee for a congressional vote. But along with the relaxation of the rules for gene transplantation experiments, Secretary of HEW Joseph Califano restructured the Recombinant DNA Advisory Committee by expanding and broadening its membership. The outcome was that one third of the committee would be constituted by nonscientists. At the same time, by lowering the containment requirements for most experiments, the technology became available to many more investigators, who otherwise would not have had access to a P3 containment facility.[11] The broadening of public participation on RAC was not received kindly by many scientists who expressed fears that it would politicize the predominantly technical role of the committee. One veteran of the DNA affair, Maxine Singer, expressed the sentiment of many of her colleagues in her *Science* editorial "Spectacular science and ponderous process."[12]

No sooner had the new guidelines been in place than efforts were set in motion to make additional revisions that would exempt 85 percent of all rDNA experiments. Two scientists on RAC, Allan Campbell and Wallace Rowe, led this new initiative, which once again brought DNA critics to a defensive stance as they tried to hold the line on the current standards for conducting research. Supporters of the new revisions argued that scientists were now in general agreement that *E. coli* K12 is itself a safe organism to do most rDNA research and that there is no reasonable possibility of converting this strain into a pathogen. Publicity of the Falmouth Conference, the results of human feeding experiments, a "worst-case experiment" with polyoma DNA, and the outcome of an isolated experiment performed by Cohen and Chang were the principal sources of justification. Philip Abelson, editor of *Science*, wrote an editorial on the issue of legislation in which he stated that nature was performing many of the experiments that Congress was about to regulate. Some critics viewed this approach by Abelson, of extrapolating from a highly engineered laboratory setting to conditions in nature, as science by wish fulfillment.[13]

A major change in the DNA controversy took place in 1979 when discussion shifted away from the laboratory biohazards to the commercial development of technology. Several small firms, supported in part by venture capital from a few large corporations, were beginning to perfect the microbial expression of insulin, somatostatin, and interferon for large-scale production. With industrial scale-up seeming imminent and legislation more remote, NIH was left in an awkward position in relation to industry. Large-scale work with rDNA molecules (over ten liters) was prohibited under the 1978 guidelines; special exemptions could be granted by the director of NIH on a case-by-case basis after RAC review. But for nongovernmental institutions, NIH commanded no statutory authority or funding leverage to enforce compliance either for research or scale-up activities. The NIH director, after waiting out the legislation that never appeared, chose neither to neglect industry nor to lobby for legislation. Instead, he established a voluntary compliance program through which commercial institutions planning to undertake rDNA work could have their project protocols and industrial scale-up plans reviewed by RAC. Moreover, nonfederally funded institutions would also be given the opportunity to register their institutional biosafety committees (IBCs) with the Office of Recombinant DNA Activities (ORDA).

Some corporations, like Eli Lilly, favored this approach to legislation or pure laissez-faire conditions. Having the NIH backing for one's commercial activities helps in securing public confidence, which can affect liability insurance and the actions of local communities. Critics of the voluntary-compliance program argued that it gave the illusion to the public that NIH was certifying industrial processes. Despite its official and repeated affirmation that it was not a regulatory agency, the new voluntary compliance program for reviewing industrial projects made NIH appear as if it were taking on that very role. No one disputed the fact that NIH lacked both the resources for ongoing inspections of industrial plants and any enforcement authority over the private sector. Moral persuasion and adverse publicity were viewed by some as sufficient to keep industry in compliance. Under the voluntary-compliance program, RAC was asked to review large-scale proposals in closed session and with signed assurance by each member to protect information the company deemed proprietary. This new role for RAC raised questions about the ties certain members of the committee had with private industry. In addition, there were mutterings of resentment directed at those scientists who were principals in newly formed venture-capital firms. These scientists, who had been weaned on public funds, were not able to turn the fruits of their research into private gain. While some individuals divided their time between the academic and the commercial world, others (although far fewer) left their academic positions entirely. A period of readjustment had started within academic biology, and no one was sure where it would lead. But it was generally conceded that the industrial influence could reshape the disciplinary boundaries as it had in chemistry, electronics, physics, and computer science. There, too, scientists developed bridges to the commercial sector by which lucrative arrangements were established for faculty. Molecular biology was not setting any precedents in the relations between academe and industry; it was simply following an established pattern.

Since every recombinant mutant is a potential protein factory, the economic rights to these organisms became a primary concern to industrial and academic institutions. Organisms that could digest oil or that could express human insulin with high efficiency were the early subjects for patent requests. NIH had to consider whether its patent policies needed revision for processes or organisms developed by rDNA technology. The Federal Interagency Committee reviewed the patenting issue and concluded that no changes in its patenting policy were warranted because of rDNA technology.

The case for patenting microorganisms per se, in contrast to the uncontested right to patent novel processes that utilize those organisms, was brought before the Supreme Court in March 1980. Three months later, in a five-to-four decision, the Court upheld the decision of the Court of Customs and Patent Appeals. The appeals court ruled against the US Patent and Trademark Office, permitting patent rights to microorganisms per se (case of *Diamond vs. Chakrabarty*).

The rise of commercial gene-splicing activities brought out additional questions related to social ethics and professional responsibility. The great proportion of academic science is socialized with regard to funding sources. Should scientists be permitted to turn the knowledge they acquire through public grants directly into personal profit or the profit of commercial institutions? Is it appropriate for scientists to hold an academic position while they play a key role in a commercial enterprise? Some scientists justified a dual role as the means for expeditiously transforming scientific discovery into social use. But the manner in which scientists publicized advances in rDNA technology was criticized on grounds of professional ethics. The interest of small firms to excite Wall Street investors in biotechnologies was cited as having taken priority over careful, sober scientific assessment of their results.[14]

The second major revision of the NIH *Guidelines* was announced in January 1980. It cleared the way for most rDNA experiments to be done at the minimal (P1) containment requirement. NIH officials forecasted significant increases in gene-splicing work resulting from the newly promulgated rules. At the time NIH had funded 780 experiments. Another important breakthrough was also announced in that month. Charles Weissman at the University of Zurich and Walter Gilbert of Harvard, who had formed a company called Biogen, reported at a press conference that they had successfully produced bacteria that released interferon, a scarce antiviral substance found in humans and animals.

For several years molecular biologists had been concerned about the media's role in fueling the DNA controversy by overplaying conjectured biohazards. In 1980 the media began to turn their attention almost exclusively to the prospects of the new technology. *Newsweek*'s cover story read "DNA's new miracles" (17 March 1980); *Life Magazine*'s bold type read "The miraculous prospects of gene splicing" (May 1980). The *New York Times Magazine* spoke of gene splicing as accelerating the race toward better human health.[15]

As for any resurgence of legislative activity in that year, Senator Adlai Stevenson introduced the only new bill on rDNA technology (S.2234) which required institutions to register such activities with the Secretary of HEW. It did not draw much attention from the scientific community since its primary impact was on private institutions and firms that chose not to participate in the voluntary compliance program. While the media had been helping to shape a new public mood toward genetic research, aided by the extravagant publicity of scientists, the Recombinant DNA Advisory Committee was undergoing a change in its philosophy of regulation. A significant shift took place in how some RAC members, both scientists and lay persons viewed the burden of proof for setting physical and biological containment levels of experiments.[16] Instead of requesting evidence that an experiment or host-vector system was safe, some members on the committee now sought evidence that an experiment was hazardous. Failing to find such evidence was a signal to support minimal containment requirements, that is, exemption from the *Guidelines*. The activity of the committee was almost exclusively in the direction of dismantling the regulations. The major issue was, How long would the dismantling process take?

15

An Unwelcome Host

The Asilomar Conference successfully restricted the terms of the rDNA controversy to the question of biohazard. The social, ethical, and political questions had not been fully addressed—far from it. They were merely judged, by the scientists at any rate, not to be germane to the problem of how best to get on with the research.

Profound differences over values or politics do not disappear so easily, however. The inevitable result of forbidding such topics from the policy debate was that they were expressed in the form of scientific disagreements over the riskiness of the new technique.

There had been sufficient uncertainty about these risks for the principal scientists engaged in the research to pause and reflect. This impulse to act in a "socially responsible" fashion was itself a product of the social context of the times, in many ways a reaction of American intellectuals of the early 1970s to the events of the previous decade. By 1980 that uncertainty about risk and the caution that went with it had disappeared, replaced by "a consensus concerning the relative safety of certain types of recombinant DNA experiments."[1] The experiments were those involving the use of the bacterium *E. coli* as a host. Use of the organism in rDNA experiments had been of concern because of the intimate association it had with all warm-blooded animals, including humans. Indeed, in 1971 a promising experiment involving putting a lambda phage-SV40 recombinant into *E. coli* had been abandoned because of fears that SV40, which was an oncogenic virus in animals, might gain a new and catastrophic ecological niche in humans. Even the first set of NIH *Guidelines*, published in July 1976, was designed specifically with this *E. coli* "problem" in mind. A short three-and-a-half years later, the "problem" had disappared. What was the evidence that effected such a remarkable shift in perspective?

Escherichia coli, or *E. coli* for short, is a tiny rod-shaped organism less than a five thousandths of an inch long and a fifty thousandths of an inch wide. It is named after Theodore Escherich, one of the outstanding pediatricians of the turn of the century.[2] Escherich was trained at a time when the discoveries of Koch and Pasteur were revolutionizing medicine, and under their influence he pioneered the application of the new science of bacteriology to the care and treatment of infants and children. In 1884 he served as scientific assistant in a cholera epidemic in Naples and developed an interest in the intestinal flora. Shortly thereafter, in 1886, he published an important monograph on the relation of the intestinal bacteria to digestion in infants. In it he gave the classic description of the colon bacillus, then called *Bacterium coli commune*, but later renamed *Escherichia coli* to honor his contribution. (The organism had actually been described a year earlier by Hans Buchner but not studied in any detail.)

Although its presence in the intestine of warm-blooded animals is normal and universal, Escherich noted that under some circumstances *E. coli* could also cause disease. Most often this occurred when the organism invaded sites outside its normal niche in the intestine, causing bladder infections or skin abscesses. But Escherich also believed it was the cause of infantile diarrhea and other kinds of gastroenteritis, that is, it could be an intestinal pathogen as well. This observation was not confirmed until the 1940s, after decades of skepticism from the medical profession. It was the first of many surprises that this versatile organism had in store for modern science.

Bacterial cultures of intestinal contents usually produce *E. coli* as the predominant species. This had led the older bacteriologists to conclude that it was the principal organism of the gut. It turns out, however, that it represents less than 1 percent of the intestinal flora; the majority of organisms are from the anaerobic species, which are difficult to raise in culture. Moreover, despite the definiteness conferred on it by its formidable name, *E. coli* exists in hundreds of distinct variants or strains. One of these variants, designated *E. coli* strain K12, was first isolated in 1922 from the stools of a patient recovering from diphtheria.[3] It was maintained as a "typical" coliform organism in the stock culture collection of Stanford University from 1925 onward and used in laboratory exercises to teach students how to identify bacteria. At the time, K12 did not seem exceptional in any way. In the early 1940s, E. L. Tatum set out to produce mutant strains of bacteria.[4] He asked C. E. Clifton of Stanford to provide him with some *E. coli* cultures. What he got was *E. coli* K12. Tatum subsequently gave some of the K12 mutants that he produced to a young worker in his laboratory named Joshua Lederberg, who discovered that K12 actually carried on a rudimentary sex life, exchanging genetic information among its members in a mating process called conjugation. This gave scientists a powerful tool for analyzing the makeup of bacterial genes by studying the transfer of genetic information that occurs during conjugation. Most other *E. coli* strains are not usually fertile, so that Lederberg's discovery was a stroke of luck exploited by a talented scientist. This discovery of bacterial "fertility" later earned Lederberg a Nobel prize and earned the K12 strain a secure place on laboratory benches around the world.

In the late 1960s a center was established to preserve and disseminate standard genetic stock cultures of this important organism.[5] Activities of the center have shown that the original K12 strain has undergone a great variety of genetic changes since its first isolation. These genetic variations or mutations have been partially documented by tracing the K12 pedigree through records and notebooks, as progeny of the original culture were passed from laboratory to laboratory and maintained under many different conditions. Despite this reconstruction, the relationship between any of these mutant K12s and the original ancestral organism that inhabited the intestines of the patient in 1922 is no longer known with any precision. It is clear that K12 once could live in the human gut. By the mid-1970s it was being claimed that it could do so no longer because it had become too comfortable in the pure-culture conditions of the laboratory, with its enriched media and optimal temperatures. Yet, the difference between the original 1922 isolate and the current laboratory form stems only from a different complement of genetic information. Was it not possible that a K12 that served as host for a hybrid DNA molecule might take on new survival properties, or even become a pathogen? What other properties might emerge to confound our best-laid plans?

The environmental consciousness of post-Earth Day America was too skeptical of our mastery of complex ecological sytems to believe we knew all the answers to these questions. The molecular geneticists themselves were not immune to the general *Zeitgeist*, as the discussion of risks at Asilomar indicates. Their retreat from this skepticism was accomplished in stages and was aided, first, by the evident power of a technique that stimulated their professional (and sometimes commercial) appetites, and second, by the evolving polarization over assessment of risks between themselves and critics who had not bowed to the judgment that only scientists should control science.

At the time of the Asilomar Conference there was still enough doubt about what would happen if novel combinations of genes were placed in *E. coli* for scientists like Paul Berg to justify the caution he and the cosigners of his letter to *Science* had urged on their colleagues. The possible risks were many: Could the addition of new genetic information into *E. coli* turn the organism into a pathogen by any of a variety of means? Could *E. coli* pass on the genetic information to another bacterial species that would then itself become a pathogen or even to cells of the intestine, causing them to behave improperly? If there was a threat to health, was that threat confined only to laboratory workers or was there a risk of epidemic disease? There were also the threats to the nonhuman environment to consider as well: plants, animals, or perhaps even natural resources.

When, after Asilomar, the serious issue of writing guidelines was undertaken, these questions began to be looked at in a more systematic way. One of the first documents was a memo on "The ecology of *Escherichia coli*" from Stanley Falkow to the members of the newly formed Recombinant DNA Molecule Program Advisory Committee of NIH, commonly referred to as the RAC.[6]

In this unpublished paper, Falkow sketched the features of *E. coli*'s biology most pertinent to the rDNA problem. He discussed the organism as a human colonizer, a resident of the natural environment, and a pathogen. *E. coli*'s plasmids received special attention because of their importance to rDNA research. This informational memo serves as an excellent benchmark for what was and was not known about *E. coli* just after Asilomar.

E. coli has on its surface substances that can be used as immunological markers to distinguish various strains of the organism from each other, just as given names and surnames indicate family relationships in the human species. This kind of serological typing, as it is called, has been used by researchers to investigate the distribution of the organism in nature. Serotyping of fecal *E. coli* from individuals over time showed a continuous succession of different strains. The data base was small, however. One study showed that several strains coexist in the bowel at any one time but that some of the strains last only for a few days (transient strains), while others persist much longer, for several months or more (resident strains). The change from one resident strain to another is a gradual process and occurs without any symptoms for the individual. "The factors which influence the change of resident strain [are] not known," Falkow added.

"Attempts to replace the resident strain by ingestion of known *E. coli* types is rarely successful." On the other hand, changes in the resident *E. coli* flora must occur by ingesting a new strain. The source of these strains—whether other people, domestic animals, or plants—is not known.

Although *E. coli* represents less than 1 percent of the total bacterial population of the intestine, its numbers remain steady despite wide variations in diet. Referring to the possible deleterious effects of an *E. coli* organism that carried an rDNA molecule conferring the ability to digest cellulose, Falkow remarked that "[t]hese basic observations on the failure of diet to change the microflora may either mitigate or intensify such fears depending on one's attitude." This offhand remark could serve as a summary of the next five years of debate over the *E. coli* risk.

While diet does little to alter *E. coli* populations in the gut, other factors are more influential. Antibiotics used in both therapeutic and subtherapeutic doses have a marked effect on the "mix" of organisms found in intestines; it was thought by some, therefore, that they could profoundly affect the rate at which genetic information contained on bacterial plasmids was transferred from organism to organism. Since the possibility that a laboratory-acquired *E. coli* containing a foreign gene might pass that gene on to other organisms with unknown results was a primary concern, this was clearly an important point. It was one of the reasons that the Plasmid Working Group at Asilomar had recommended that people taking antibiotics not be permitted to do experiments using the rDNA technique.

E. coli is extremely widely disseminated in nature; it is found in soils in densely populated and sparsely populated regions, in insects such as beetles, grasshoppers, leaf hoppers, butterflies and moths, and in flies. Birds and fish can also spread *E. coli*, as do all warm-blooded wild and domestic animals. Falkow's assessment of the implications of this was sobering: "In the 'worst case' analysis of the dissemination of an *E. coli* carrying a 'hazardous' DNA recombinant molecule one would be faced with an enormous number of potential vectors for dissemination. The number of *E. coli* disseminated by these 'unusual' vectors is significant. Although only small numbers of *E. coli* have been isolated from insects, it has been estimated that a gull disseminates [50 to 100 million] *E. coli* per day (I cannot help but be reminded of a typhoid epidemic traced back to gulls feeding on sewage)."[7] *E. coli* can also live in fresh water, Falkow noted; in one case it survived for nine months in water contaminated with feces. Under certain favorable conditions it can not only survive, but multiply and prosper. "This is especially true of heavily polluted water containing organic compounds. In fact, the organisms may multiply to a level equivalent to that of a laboratory broth culture."

While most of Falkow's discussion dealt with the ecology of *E. coli*, he briefly described its role as a pathogen. Three pathogenic mechanisms had been reported at that time (a fourth has since been described). Two involved intestinal infections that frequently caused diarrhea in infants and sometimes adults. One form was similar to the food-borne infection caused by the *Salmonella* organism, the other to the disease bacillary dysentery (caused by various *Shigella* species). The third, also involving diarrhea, was produced by an *E. coli* that had acquired the ability to express a toxin. These "enterotoxigenic" strains were responsible for "traveler's diarrhea" and were endemic in many countries. The ability to produce toxin was controlled by a plasmid.

Falkow also touched on extraintestinal infections associated with *E. coli*. It is the most common cause of urinary tract infection and an important factor in hospital-acquired infections. The meager information available on the epidemiology of these infections suggested that substances variably present on the surface of the *E. coli* might play an important role in causing infection, especially of the bladder. In addition, these organisms were also more likely to carry certain specific plasmids as compared to other *E. coli*, say from the intestines. Thus, while *E. coli* is not ordinarily a highly virulent organism, "Certain [*E. coli* strains] have the inherent potential for invasiveness or may acquire sufficient accessory genetic information, usually plasmid-mediated, that is sufficient to tip the balance from a strain that is usually a commensal [a harmless resident of the host] to one which is capable of initiating overt disease. From the standpoint of recombinant DNA molecules, the documentation of the effects of plasmid-mediated determinants on pathogenicity must be viewed as one of the most cogent arguments for the potential biohazards associated with this research."[8]

Even if the genetic information contained on the plasmid did not affect *E. coli*, plasmids of *E. coli* K12 could pass from one *E. coli* to another and even from *E. coli* to other species during conjugation. Since this would further disseminate an rDNA molecule if an *E. coli* "escaped" from the laboratory bench, the plasmids chosen as cloning vehicles were nonself-transmissible plasmids; that is, unlike transmissible plasmids, they could not initiate conjugation and gene transfer on their own. They could, however, be "mobilized" (usually at a low rate) by a transmissible plasmid that existed in the same cell. Falkow cited studies to show that transmissible plasmids are very common in *E. coli* and "in the context of recombinant DNA molecules, the 'indigenous' plasmid flora of *E. coli* would represent (at least in theory) a ready body of vehicles to mobilize and recombine with the laboratory constructed molecules under the proper circumstances."

The potential for plasmid transfer obviously existed. But, asked Falkow, would this potential be realized under "natural conditions"? If K12 were the host, he was doubtful. Studies in his own and other laboratories indicated that K12 was a poor colonizer (i.e., did not become a resident strain) of the human alimentary tract. On the other hand, results indicated that not only could K12 survive for some days in the alimentary tract and even after excretion, "but . . . also can transmit and receive plasmids at a very low rate."

Some plasmids could also affect the survivability of K12 in the intestines. Two such examples were plasmids that conferred pathogenicity on some strains of *E. coli* (but not on K12). In calf intestine carrying neither, K12 survived for three days. In calf intestine carrying one of them, K12 could multiply and was excreted for eight days. And in calf intestine carrying both, K12 multiplied and was excreted for 14 days, after which the experiment was terminated. Falkow noted that this is a special case in that both plasmids evolved, at least in part, to confer a survival advantage to *E. coli* in calves' intestines. "Yet," he went on, "it may not be too far fetched to suggest that some DNA recombinant molecules could profoundly affect the ability of this *E. coli* strain [K12] to survive and multiply in the gastrointestinal tract."

Not only plasmids would affect K12's ability to colonize, but also antibiotics. Moreover, the difficulty encountered in implanting K12 in humans was by no means unique to this strain alone, since ingestion of other bacteria leads to the same results. If the subjects of these experiments had been taking antibiotics, however, K12 strains that contained plasmids conferring antibiotic resistance would have had improved survival and plasmid transfer. (It should be noted that the plasmids used as cloning vehicles in rDNA experiments carry antibiotic resistance genes for use as selective markers in cloning.) Falkow reported on a study in which three volunteers were fed *E. coli* containing antibiotic resistance plasmids. Without antibiotic therapy they lasted an average of three days, but when five days of antibiotic therapy were given at the outset, these strains lasted an average of fifty days.

Thus, the likelihood of plasmid transmission was hard to assess, although on the basis of available evidence and under conditions of no selection (i.e., in the absence of antibiotics to which the plasmid conferred resistance), Falkow doubted it would happen.

No such comparable data were available for the phage-cloning vehicles, and Falkow noted that "presumably this is one area in which it would be useful to have contract studies performed."

The overall picture presented by Falkow would seem to have justified the caution favored by many in the Asilomar period. In the memo he presented information to indicate that concerns about the effect of adding new genetic information were not implausible. Plasmids both affected survival of K12 in the intestine and conferred pathogenic potential on some strains (not K12 as far as was known). K12 both transmitted and received plasmids in vivo, some of which could conceivably mobilize an otherwise nontransmissible DNA hybrid in K12. And both colonization and plasmid transmissibility could be greatly enhanced if antibiotics were being used concurrently.

The question of K12's ability to colonize the human intestine was beginning to assume a central importance, although it was only one issue among many in the Falkow memo. If colonization could not occur, some believed, then neither could disease production or plasmid transmission. While no data in the memo were cited to support this, it clearly formed the basis for Falkow's judgment that plasmid transmission would be unlikely in vivo. In the Risk Assessment Workshop held at Falmouth, Massachusetts, the following year the survivability of *E. coli* K12 was the central question. Because the future course of the *E. coli* debate revolved around this issue, it is interesting to assess the colonizing ability of K12 inferred from the data in Falkow's memo. Specifically, the memo implied that a straightforward challenge of human volunteers that involved feeding them *E. coli* K12 was not a very appropriate test for the organism's colonizing ability. This conclusion, although not made explicit, can be obtained by reconstructing a logical argument from results given in Falkow's memo:

1. There is a demonstrated succession of resident *E. coli* strains in the normal human intestine. The factors that cause this change in strain are not known.
2. The new resident strain is implanted in the gut by ingestion.
3. Yet attempts to change resident strains by oral challenge (ingestion) in normal human volunteers are rarely successful.

4. Therefore, challenge experiments are a poor model for assessing changes in resident strains, and hence poor tests of colonizing ability.

Despite this, the evidence cited most frequently in the sequel for K12's lack of colonizing ability would be precisely feeding experiments.
At this point, however, most people were sufficiently uncertain about the *E. coli* danger to proceed with caution. This was reflected in the first guidelines' broad definition of an rDNA experiment: one that involved DNA molecules with segments from different sources replicating in a host organism. The addition of new genetic information to *E. coli* K12 was a possible biohazard and the NIH *Guidelines* took notice of this. Elaborate provisions for physical containment were drawn up, and a requirement that a specially enfeebled strain of K12 be used for many experiments testified to the concern. This latter stipulation was irksome to many researchers. The newly engineered "safe" strain was temperamental and difficult to work with. Some experiments would not work in it. Thus, initiatives to "reassess" the *E. coli* problem were welcomed by many rDNA researchers.

16

The Falmouth Affair

Shortly after the issuance of the first guidelines (July 1976), an informal meeting of NIH consultants was convened.[1] The group was called together after a member of RAC, Wallace Rowe, and his colleague Malcolm Martin, both of NIAID, realized that "an amazingly important segment of the biomedical community had not been heard from in the initial discussions on the subject."[2] Rowe and Martin organized a group of outside experts to give advice to RAC about how best to get input from infectious disease experts, epidemiologists, and enteric disease specialists. According to Rowe, as a result of that meeting, "We realized . . . that there was an awful lot of information—published primarily, but in part unpublished . . . which was amazingly relevant to the problems we were up against. We didn't know . . . that people have tried, by standard genetic crosses, to put genes into . . . K12 . . . to study what genes . . . would convert it to a pathogen. We found out, to our great surprise, that they could not create a pathogen out of K12. We said this is very important information, let's get broader input on it. We ought to have a workshop."[3] As a consequence, at September 1976's RAC meeting, a recommendation was made that a workshop be organized to bring together experts on the basic biology of *E. coli* for the purpose of risk assessment in rDNA research. Sherwood Gorbach of Tufts University, an expert on enteric diseases and one of the outside consultants, was asked to organize the meeting with the help of a steering committee.[4] Approximately fifty participants from the United States and abroad were invited to attend the workshop, which took place in Falmouth, Massachusetts, on 21–22 June 1977.

Rowe said later that a special attempt had been made to balance the Steering Committee scientifically and politically. But the conference was not an open meeting in the usual sense, and later was criticized for its lack of openness. MIT'S Jonathan King, for example, who presented a paper at Falmouth, later said that he had to go to the chairman and ask to be admitted. The conference, he said, "was not announced by the normal procedure for announcing scientific conferences, that is, in the scientific journals, Genetics Society of America, American Society of Microbiology. It was private. It was by invitation from the organizing committee. Many people were rather upset . . . to find out that a risk-assessment conference was taking place and they didn't even know about it until after the fact."[5]

The Falmouth Conference is of considerable importance because it was later to be cited repeatedly in justification of the belief that the fears expressed earlier about rDNA were "overstated." Chairman Gorbach's summary of the "Consensus agreement" was unequivocal: There was unanimous agreement, he said, that K12 could not become an epidemic pathogen by laboratory manipulations with DNA inserts. Moreover, the likelihood of transfer of a nontransmissible plasmid from K12 was also extremely improbable.[6] These conclusions, if valid, would effectively lay to rest the questions raised at Asilomar and in Falkow's memo just two years earlier.

The consensus just cited was contained in a letter from Gorbach to NIH Director Fredrickson dated three weeks after the conference. It is quoted verbatim in the *Environmental Impact Assessment* released by NIH in July of 1978 for the purpose of justifying the significant revisions of the first guidelines that NIH proposed at that time. Those revised guidelines exempted many experiments that involved cloning disparate DNA segments in *E. coli* K12.

> *Consensus Agreement*
>
> An important consensus was arrived at by the assembled group which I felt was of sufficient interest to be brought directly to your attention. The participants arrived at unanimous agreement that *E. coli* K12 cannot be converted into epidemic pathogen by laboratory manipulations with DNA inserts. On the basis of extensive studies already completed, it appears that *E. coli* K12 does not implant in the intestinal tract of man. There is no evidence that non-transmissible plasmids can be spread from *E. coli* K12 to other host bacteria within the gut. Finally, extensive studies in the laboratory to induce virulence in *E. coli* K12 by insertion of known plasmids and chromosomal segments coding for virulence factors, using standard genetic techniques, have proven unsuccessful in producing a fully pathogenic strain. As a result of these discussions, it was believed that the proposed hazards concerning *E. coli* K12 as an epidemic pathogen have been overstated. Such concerns are not compatible with the extensive scientific evidence that has already been accumulated, all of which provides assurance that *E. coli* K12 is inherently enfeebled and not capable of pathogenic transformation by DNA inserts.[7]

This is a very strong statement, made even stronger by such phrases as "extensive studies" (three times), "important consensus," and "not compatible with." Gorbach's "consensus statement" was not an official document of the conference, but a personal assessment of its importance. Gorbach's summary,[8] published a year later in the official conference proceedings in the May 1978 *Journal of Infectious Diseases*, is more circumspect. In it he defines the risks to be considered, and he places them into three categories: (1) Could a DNA insert into *E. coli* K12 convert it into a pathogenic strain similar to other pathogens already familiar to us; and could it then spread to others in the population? (This implied, said Gorbach, that the organism had gained the ability to colonize the host, cause disease, and survive in nature.) (2) Could a DNA insert be transferred from an *E. coli* K12 to another bacterium or to a somatic cell of the host? (3) Could a DNA insert code for some injurious product such as a toxin, a hormone, or some protein that could produce disease, say by an autoimmune mechanism?

Gorbach proposed to summarize the conference by discussing the potential hazards "within the general framework of microbial pathogenesis." The argument he actually exhibits pertains almost exclusively to point (1). A few studies bearing on point (2) are discussed, but Gorbach reports that it was the sense of the conference that further studies on transmissibility be carried out despite the comfort afforded by the extant negative studies (three are cited). Point (3) is not dealt with at all.

It should be noted that Gorbach is examining a very limited question. What is actually argued is that K12 cannot become a pathogen like any *known E. coli* pathogen. (Since the Falmouth Conference, a fourth mechanism of *E. coli* pathogenesis has been described in the *New England Journal of Medicine*. The final sentence of the accompanying editorial[9] noted that "the versatile *Esch. coli* may hold still more surprises, for which we must remain prepared.")

The reconstructed argument intended by Gorbach to show that *E. coli* K12 cannot become an epidemic pathogen goes as follows:

1. Epidemic *E. coli* (in general, not just K12)
a. must be able to colonize the intestines,
b. once colonizing the intestines, must possess certain virulence factors,
c. once causing disease, must be able to spread from one host to another.

2. But K12
a. cannot colonize the intestines, even after manipulation,
b. does not normally possess the virulence factors of pathogenic *E. coli*,
c. cannot be provided with enough, or the right kind, of these virulence factors, even deliberately; thus it is improbable that this could happen accidentally,
d. cannot spread easily from person to person.

---

3. Therefore, K12 cannot become an epidemic pathogen.

To unravel this argument we must examine each premise in some detail.

Premise (1). Epidemic *E. coli* must
a. be able to colonize the intestines,
b. possess certain virulence factors,
c. be able to spread from host to host.

Subpremise (a). Epidemic (pathogenic) *E. coli* must be able to colonize the intestines. (Here I have noted—in parentheses—that the issue argued is not epidemicity, but pathogenicity, a component of epidemicity.)

Gorbach presents several pieces of evidence for his general assertion about pathogenic *E. coli*. For *E. coli* that invade the intestinal mucosa, the ability to proliferate in that mucosa is part of the disease process.

For those *E. coli* that cause diarrhea by elaborating a toxin, however, colonization is a less obvious requirement. Studies in Gorbach's and other laboratories show that besides the ability to produce toxin, these disease-producing organisms also must be able to adhere to the mucosa of the small bowel. This ability, like the ability to produce toxin, is controlled by a plasmid. (There is some ambiguity here about the meaning of a word like "colonize," or "implant." Sometimes it is used in the sense of "multiply" or "proliferate." Sometimes it is meant to indicate that a particular strain becomes "resident" in the intestines, that is, is excreted for months, not just days, as would a transient strain. This ambiguity is pertinent because there is evidence that enterotoxigenic strains adhere to small bowel mucosa, proliferate there, and release a toxin that causes diarrhea, as Gorbach asserts, but that they also are eliminated completely from the small bowel in a few days.[10] On the other hand, much of the evidence about whether an organism can "colonize" or "implant" is given in terms of the "residence" meaning: If it is eliminated in four or five days from the stools, it is said not to implant. Gorbach's evidence that pathogenic *E. coli* must be able to "colonize" the gut is given in terms of the ability to adhere to bowel wall and proliferate there. For how long this must happen is not stated. The evidence given later, in premise (2), that K12 cannot implant is given in terms of the "residence" criterion.)

So much for the intestinal pathogens. But what of extraintestinal infections with *E. coli*, say in the urinary tract, skin, blood, or lungs? In Gorbach's argument only the urinary tract is considered. These infections are thought to arise by the "ascending" route, that is, by bacteria entering via the urethra and going up to the bladder and kidneys. Gorbach cites evidence that colonization of the vaginal area around the urethra in women is the initiating event in female urinary tract infection. The source of these organisms is almost certainly the nearby outlet of the intestinal tract, and in fact there is a high correlation between serotypes present in the urinary tract and those in the intestines. He sums up by asserting, but without citing any references, that the "evidence strongly suggests that urinary tract pathogens must be [at least] capable of colonizing the bowel of the host."[11] However, from the evidence given in the argument, only the presence of the organism in the bowel is necessary. Since many bowel organisms are only transient strains, there is again an ambiguity, or at least an imprecision. Perhaps the argument *could* be made more precise, but it is not.

In summary, the conclusion that all pathogenic *E. coli* must be able to colonize the gut is based on some rather unclear distinctions, and certainly is not completely supported by Gorbach's argument. The ambiguities in the nature of "colonization" aside, only urinary tract infections are considered for extraintestinal pathogens; and even there, a strong case is made only that the source of the *E. coli* is intestinal but not that the organism must be a colonizer of the intestines.

Subpremise (b). Once colonizing the intestines, a pathogen must possess certain "virulence factors" in order to cause disease.

One line of argument here has already been mentioned: Enterotoxigenic *E. coli* must possess two plasmid-controlled factors in order to cause disease. For *invasive* intestinal disease the argument is different. The incriminated strains are not a random selection of the existing strains in the gut; rather, a relatively small number of serotypes are represented disproportionately often. It would thus appear that they are more likely to cause disease; that is, they possess special properties vis-à-vis noninvasive strains that make them likely to do so. Gorbach then makes the same argument for urinary tract infections: While several strains may be present in the bowel at any one time, only one of them infects the urinary tract. The predominance of a minority of strains in disease states suggests that they must be "selected out" over other less virulent strains. One of the papers in the conference itself, however, raises doubts about this.[12] The predominance of a few strains in most urinary tract infections might just be a reflection of the relative abundance of those strains, not an expression of special properties. Gorbach does not mention this alternative.

Assuming that special properties of the organism are involved, the next question is, What are they, and how many of them are there? By cataloging common properties of organisms isolated from various sites in infected individuals, a few factors suspected of being associated with the ability to cause disease have been found. Others have been investigated by means of mating experiments to produce hybrids with some properties and not others that are then checked to see whether the result is virulent. As yet, however, no one knows exactly what properties are necessary or how many properties there are in any given case. (A terminological note: These properties are commonly called "virulence factors" but this is a misnomer. They should really be called "pathogenicity factors." Pathogenicity is the ability to cause disease. Virulence is the capacity to cause *severe* disease.)

Thus, the evidence supporting the proposition in subpremise (b) is, on the face of it, good, but of limited value since the nature and extent of the relevant factors are still largely unknown.

There is yet another kind of problem with the subpremise. The clear implication of the whole line of argument is that pathogenicity is a "property" of an organism. We know that this is an inadequate formulation of the concept, however. Pathogenicity is not a characteristic of a particular organism, but of the relation between an organism and its host. As McKeown has emphasized, microorganisms and their hosts have evolved together over millions of years.[13] Normally that relation is benign, not tipped in favor of one or the other. A parasite does not benefit by killing its host. But in individual cases that balance can be upset in favor of the organism, so that disease results. The balance is sometimes stable, sometimes unstable, but the concept of balance is still fundamental: two factors are involved, host and parasite. Measles is a generally mild disease in the United States, but often fatal in developing countries where nutrition is poor. An organism that is a harmless coresident in healthy individuals can become a virulent pathogen in someone chronically ill or with otherwise lowered "host defenses." The importance of host response was noted several times in the Falmouth Conference itself, where it was invoked to explain why enterotoxigenic *E. coli* sometimes produced a severe choleralike disease and sometimes a mild traveler's diarrhea.[14] At another point, a distinction was made between intestinal infections, for which characteristics of the organism are the important variables, and extraintestinal infections, for which host characteristics dominate.[15] In any event, any consideration of an organism's ability to cause disease must always include the host side of the relation, and a discussion of virulence factors concerned only with the organism is inherently one-sided.

We note in passing that one of the early concerns about rDNA technology was that new properties of the organism could be created that were outside the long process of balanced evolution. Thus, the proposition of subpremise (b) is restricted in two ways: it implicitly fails to recognize the host side of the equation; and it considers only the kinds of "virulence" that have appeared as a result of natural processes acting within the evolutionary framework.

Subpremise (c). To cause an epidemic, a disease-producing *E. coli* must be able to spread easily from person to person in the population.

Again there is a certain amount of loose terminology here. "Epidemic" is being used in the nontechnical sense of an illness spreading rapidly and widely in a population. There is always an additional connotation that the illness must be serious or plaguelike. But an epidemic, in the technical sense, is any unusual (beyond the expected) increase in disease incidence. Thus, a few cases of rabies is an epidemic, a few thousand cases of influenza might not be. Depending on the illness and the population at risk, we may wish to use one notion or the other. The Falmouth conferees did not come to grips with this.

The epidemic potential of *E. coli* was assessed by considering evidence of disease in the United States and by weighing the likelihood of its spread, given the known data about, among other things, its mode of transmission and infectivity. Eugene Gangarosa of the Center for Disease Control in Atlanta reviewed the epidemiology of diarrheagenic *E. coli* at the conference and concluded that the danger of epidemic spread was minimal in a country with a high level of sanitation.[16] He admitted that the evidence on the amount of enterotoxigenic *E. coli* diarrhea was contradictory; but he stated his opinion that this disease was not common in urban centers in this country. The principal reason he gave for his lack of concern was that relatively large numbers of organisms (inocula) are necessary to cause disease with *E. coli* toxigens. His estimates were that somewhere from a million to a billion organisms had to be ingested before there was much chance of suffering from diarrhea. Such strains are thus unlikely to spread by personal contact; they require some grossly contaminated vehicle, such as food or water. For the same reason, this disease is unlikely to be acquired in the laboratory, and indeed Gorbach cites a study of almost 4,000 laboratory-acquired infections of which only two involved *E. coli*.[17]

However, *E. coli* is a major extraintestinal pathogen as well, being the leading cause of community-acquired infections requiring hospitalization. It is also the leading cause of hospital-acquired infections. It is thus a serious "endemic" pathogen.[18] Trying to prevent its spread in the hospital setting, for example, is a major endeavor of hospital infection control committees that is never more than partially successful. It would seem, therefore, that the potential for spread is not only great, but in some circumstances, irresistible.

Thus, even subpremise (c) is subject to some qualifications.

Gorbach next tries to justify premise (2), which purports to show that *E. coli* K12 meets none of the requirements laid down in premise (1):

Premise (2). *E. coli* K12

a. cannot colonize the human intestine, even after manipulation,

b. does not normally possess the virulence factors of pathogenic *E. coli*,

c. cannot be provided with virulence factors deliberately, and hence is unlikely to acquire them accidentally,

d. cannot spread easily from person to person.

Subpremise (a). E. coli K12 cannot colonize the human intestine, even after manipulation.

The first point Gorbach makes is that K12 is different from wild-type *E. coli* in many ways: different substances on its surface; a rough coat instead of a smooth one; a chemical subunit in its membrane different from all other strains. Since the biological significance of these differences is not known, the point is purely rhetorical.

Second, there are "no instances in which [an] ingested strain of *E. coli* K12 has been implanted in the human intestine." By this he means that K12 cannot be recovered from the feces beyond four or five days in most every case. Smith reported at the conference on his attempts to improve on this by inserting a plasmid (Col V) into K12 that was known to promote survival of wild-type *E. coli*.[19] The results were no different. Similarly, Anderson had done feeding experiments on eight volunteers, and could recover no K12 past six days, with an average of three days.[20] Gorbach also reported on his own attempts to implant K12 in two patients whose ability to clear bacteria from the small bowel was severely impaired.[21] In each case K12 was eliminated after one day.

As discussed in connection with the Falkow memo of 1975, however, the relevance of these kinds of tests is open to question. It is not only K12 that fails to implant in a feeding experiment, but most other *E. coli* strains and other bacterial species as well. Sometimes prolonged implantation is obtained, but ordinarily it is not. Yet, under normal circumstances there is a regular change of resident strains of *E. coli* at intervals of a few months. Either host factors or some aspect of the microenvironment of the bowel is involved.

Such factors are easier to study in animals than in humans. Gorbach asserted that a number of investigators have tried "in vain to implant *E. coli* K12 in the intestinal tract of laboratory animals."[22] Included have been mice, rats, chickens, pigs, and calves. In two cases, however, their attempts have been rewarded. K12 can colonize the stomach of starved sheep. But because the digestive system of these animals is complex and unlike that of humans, Gorbach discounts this experiment. The other successful implantation was reported for the first time at the Falmouth Conference itself and requires a detailed account.

The work in question was by Rolf Freter of the University of Michigan.[23] He begins by acknowledging the difficulties: the microenvironment of the bowel is incredibly complex, and little is known about how the many factors important for *E. coli* survival interact. Given these facts, Freter said, "[O]ne must conclude . . . that nothing definite is known about the attributes that a bacterium must possess in order to be easily implanted into the intestinal tract. Consequently, there is no generally accepted test procedure by which one could determine or quantitate the ability of *E. coli* or any other microorganism to establish themselves in the intestine."[24] Freter also takes note of the literature that showed that feeding experiments on human volunteers never resulted in implantation of *E. coli* K12. Such evidence, he says, is not satisfactory: "To the casual observer, this kind of evidence may appear to be pertinent and conclusive. This, unfortunately, is not the case. Such experiments tell us little about the ability of bacteria to grow in the human intestine. It has been known for some time that the feeding of bacteria, especially those grown under the usual laboratory conditions, rarely results in implantation because the normal intestinal flora is antagonistic to the growth of invaders."[25]

Instead, Freter suggests an alternative. Published data indicate that if the ingested organisms are first adapted to conditions similar to the ones they are to enter, their chance of survival is improved. If this is true, an "*E. coli* strain may have less difficulty in spreading among people once it has succeeded in adapting itself to the intestinal environment of at least one individual."[26]

To test this hypothesis, Freter carried out the following experiment. He first adapted a natural isolate of *E. coli* (a wild-type strain) and a K12 strain to the intestinal environment of mice. He did this by implanting them in germ-free mice, animals raised from birth in a sterile environment. Implantation in these animals is almost always successful because there are no antagonistic or competing bacteria populations to interfere with colonization. After colonization was complete, the germ-free animals were caged with normal animals in a natural environment, and about two weeks later their gut flora became similar to that of their conventional counterparts. Both strains continued to colonize the animals for the entire duration of the experiment. In contrast, when these same strains were fed to normal mice, they were rapidly eliminated from the gut in most animals. (This was true for both the wild type and for the K12 strain.) Even more interesting was the fact that the conventional mice that were caged with the K12 colonized mice also acquired the K12 strain despite the fact that they had never been fed it. The strain had "spread" to the new population.

Freter also tried to perform the same experiment with the enfeebled K12 strain χ1776. It would not implant in germ-free mice, the only enteric organism that had failed to do so in his experience. Because of this, he ends his paper optimistically: "It . . . appears that problems concerning the possibility of implanting hybrid bacteria in the human gut can be satisfactorily overcome."[27] By this he seems to mean that problems arise with normal K12, but can be overcome by using K12 strain χ1776.

It is interesting to compare Freter's results with the summary of them given in Gorbach's letter to NIH Director Fredrickson:

> Dr. Rolf Freter reported on studies which showed that the intestine on germ-free mice could be colonized with *E. coli* K12. These experiments indicated that this organism, like other enteric bacteria, multiplied and colonized within this artificial intestinal environment which contained no competing bacteria. In other experiments involving the enfeebled EK2 derivative of *E. coli* K12 known as χ1776, Freter found that it was unable to become established in germ-free mice. This EK2 strain is the only enteric bacterium he has encountered which cannot colonize the intestine of germ-free animals.[28]

Gorbach's account in the published proceedings is more complete:

> A number of variables are known to influence the colonization of organisms in the intestinal tract. Implantation can be altered by antibiotic administration, starvation, the type of diet, reduction in gastric acid, and antimotility drugs. *It is clear that more implantation experiments need to be performed with attention to these variables, many of which may be found in laboratory workers exposed to these organisms.* It is important to note that once *E. coli* K12 has been established in the germ-free animal, the addition of wild-type strains to the flora fails to dislodge it. This finding may have implications for feeding experiments, particularly those involving subjects receiving antibiotics to which the specific K12 may be resistant.[29]

Thus, the whole question of K12's ability to colonize is left open pending further study. It must be reiterated that "colonization" is taken to mean long-term residence, not the ability to replicate or proliferate. None of the cited studies attempted to determine whether replication took place, although some of the evidence used to substantiate subpremise (a) of premise (1) employed "colonization" in this other sense of proliferation. Thus, it certainly could not be maintained that the findings of the Falmouth Conference established beyond doubt that K12 could not colonize the human intestine.

Subpremise (b). *E. coli* K12 does not normally possess the virulence factors of pathogenic *E. coli*.

The evidence that some characteristics are more often associated with pathogenic *E. coli* has already been discussed. There is no one characteristic that is present in all, or even a majority, of these pathogens. Rarely does K12 possess any of the features most often associated with the pathogens, however.

Subpremise (c). *E. coli* K12 cannot be deliberately provided with factors to increase its virulence; this is therefore unlikely to occur accidentally.

Several attempts have been made to convert K12 into a pathogen by conferring on it properties associated with pathogenicity in other *E. coli* strains. Minshew et al. reported on experiments in which plasmids that controlled two virulence factors were inserted into K12.[30] More than half of the pathogenic strains of wild *E. coli* with these virulence factors killed thirteen-day-old chick embryos, and 20 percent of normal fecal strains did so also. The augmented K12, however, was not able to kill these embryos.

A similar experiment involving plasmid-controlled virulence factors of enterotoxigenic *E. coli* resulted in diarrhea in one of eight new born calves.[31] Since these animals are very sensitive to *E. coli* enterotoxin, responding uniformly with severe diarrhea, this was taken to be a negative result. Autopsy of these animals failed to reveal any proliferation of the organism in the small or large bowel. The same paper also reported that the addition of plasmids to K12 could markedly enhance its lethality for chickens. When two-week-old chickens were given large intravenous doses of K12 strains that had a variety of plasmids, of thirty chickens only one died after injection by the most benign strain of K12 (containing a single plasmid), but strains with two or three virulence-associated plasmids killed almost all the chickens (26/30, 27/30). Thus, in this system, manipulation had greatly augmented K12's virulence. As in the case of Freter's article, Gorbach is silent about this result in his letter to Fredrickson (although he cites the negative result for calves in the same paper). He does mention it in the published summary. There he discounts the result by remarking that the dose of K12 needed to kill chickens, even with augmented virulence, was still large compared to the dose of wild-type *E. coli* that would produce the same result.

Also summarized at the Falmouth Conference were the attempts that had been going on for some years at Walter Reed Army Institute of Research to create a vaccine strain for *Shigella* infections by producing hybrids with K12 using conventional genetic crosses.[32] All attempts to colonize normal volunteers with these K12 hybrids, which carried complements of *Shigella* virulence factors, were unsuccessful—the bacteria rarely were excreted beyond five days. The conclusion of the investigators was that a multiplicity of genes determined *Shigella* virulence. Furthermore, since "invasive-type pathogenicity is determined by multiple genetic loci, [the authors] consider it unlikely that random insertion of foreign DNA into the *E. coli* K12 genome could supply all of the genetic information necessary to convert this organism into an invasive enteric pathogen."[33]

A review of the evidence for the proposition of subpremise (c) reveals that the "extensive studies" referred to by Gorbach in his letter to Fredrickson cited above refer mainly to these *Shigella*-K12 vaccine studies. Related studies had looked at few virulence factors in a small number of model systems. In one of these studies K12 virulence was markedly enhanced by genetic manipulation in the laboratory.[34]

Subpremise (d). *E. coli* K12 cannot spread easily from person to person.

The arguments against K12 becoming an "epidemic" pathogen are similar to those given for any other strain of *E. coli* in premise (1), subpremise (c). Moreover, K12 is very fragile in nature and has a perfect safety record in thirty years of use in genetics laboratories.

Conclusion (3). *E. coli* K12 cannot colonize the intestines, which it must do in order to cause disease. Nor can it be made to cause disease by conferring pathogenicity on it. And even if it could cause disease, it could not spread. Therefore, *E. coli* K12 cannot be converted to an epidemic pathogen.

Some general remarks can be made about the nature of the proposition that was to be shown: *E. coli* K12 as a host for rDNA molecules is safe. Some felt that the burden of proof should be on those who felt that it was dangerous. At issue here is the medical model versus the laboratory model. But, for the sake of clarity, the related issue of the medical model versus the legal model will be examined.

In law, it is always best to err on the side of innocence; a person is "innocent until proven guilty." In medicine, however, the reverse is true. One errs on the side of guilt; it is better to overdiagnose an innocent symptom, assuming *pro tempore* that it is serious, than to ignore it and lose the patient. Laboratory workers are like lawyers in this respect. A new hypothesis is never accepted ("guilty") until it has been proved valid by the scientific method. To prove the reverse, that the hypothesis was true until shown to be false, goes against the grain of most basic scientists. But because important issues of public and environmental safety and welfare were involved, a medical model was forced on them. As a result, they had to be satisfied with a lower standard of proof ("preponderance of the evidence") than they might have accepted had they been testing a new hypothesis ("beyond a reasonable scientific doubt"). This may explain why even some of the severest scientist-critics, such as MIT's Jonathan King, could be satisfied with an argument that had as many gaps in it as the one Gorbach gave at Falmouth. King even labeled the whole epidemic pathogen scenario a "straw man."

Other propositions were not addressed by the Gorbach argument, however. He considered the probability that *E. coli* K12 could become a pathogen similar to other *E. coli* pathogens very unlikely. But what about the possibility that it could become a novel pathogen? Thus, King suggested at Falmouth that K12 might produce in the human gut a protein or polypeptide that would evoke an allergic reaction directed against any host tissues that cross-reacted with the protein antigen, producing "autiommune disease."[35] It was at least a concrete suggestion that could conceivably be tested in the laboratory, and indeed an experiment to test it was proposed by one of the Falmouth participants. On the other hand, there were those critics who feared that the insertion of some DNA sequence would cause an entirely unknown type of "emergent" behavior by K12 that could have serious consequences. In such a case, since the precise hazard is unspecified, no empirical tests can be proposed. It may still be reasonable, if the fears of the consequences are of sufficient magnitude, to prohibit experiments on such grounds. But that would be a social decision made for reasons other than science, and no scientific reasoning could be brought to bear on it. It is true that scientists have tried to argue the case anyway, claiming that such an event is of vanishingly low probability. But the very nature of such assertions is that they pertain to events that cannot be predicted and measured.

For whatever reasons (and the logical tightness of the argument was not one of them), the question of converting K12 into a pathogen began to fade from the controversy. What continued to be of concern to many people at Falmouth and thereafter was the possibility that K12 used in cloning experiments could escape containment and then pass on its genetic baggage to the host's indigenous flora or even its somatic cells. This had also been of concern to Falkow, Curtiss, and others in the Asilomar per1od and was not resolved at Falmouth. It was discussed at some length, however.

Few experiments had been done on transfer characteristics of cloning vehicles from K12 to indigenous flora. All in vivo studies (three were reported at Falmouth) were negative—no transfer occurred—but in one laboratory study it was found that the indigenous organisms from three of eight volunteers carried transfer plasmids that could mobilize a nontransmissible plasmid out of *E. coli* K12. Anderson, who reported this work, concluded that transfer "would, therefore, be possible if a suitable conjugative plasmid entered a strain carrying a non autotransferring hybrid plasmid."[36]

Gorbach cited several reasons that in vivo transfer would be very unlikely. But uncertainty still existed:

> It was the consensus of the participants that the transmissibility studies, while given some comfort by their negative findings, are not sufficient in number or in scope to exclude the potential risk in this area. In addition, several discussants raised the possibility of transfer of genetic material to indigenous flora and to somatic cells of the host by unknown mechanisms. Within the complex milieu of the intestinal tract, it is possible that such transfer proceeds in ways not discernible in the test tube or in artificial laboratory conditions. More animal and human feeding experiments should be performed to confirm the previous observations and to provide further assurance that *in vivo* transfer of these plasmids would not occur.[37]

Rather than leave matters there, the Falmouth conferees held a "brainstorming session" to develop "Risk assessment protocols for recombinant DNA experimentation."[38] Six such risk assessment protocols were published with the proceedings of the Falmouth Conference. Together they touched on many of the human-health concerns that surrounded the widespread use of the rDNA technology.

Protocol I was designed to test whether *E. coli* K12, carrying DNA-cloning plasmids but without any foreign DNA spliced into those plasmids, could first, colonize the human gut, and second, transmit its plasmids to other bacteria in the indigenous flora. The studies were to be carried out in human volunteers so as to increase the likelihood of K12 colonization by (1) administering antibiotics to which the K12 was resistant, thus creating a selective advantage for the strain; (2) feeding the K12 to some volunteers after a twenty-four-hour fast to simulate the observed K12 colonization reported in starved sheep;

and (3) using volunteers who have been treated with drugs that slow down intestinal movement, thus perhaps promoting K12 implantation by causing a longer residence time.

The transfer studies were to be done using K12 with the cloning vehicle plasmids alone, and with K12 carrying additional special transfer plasmids known to induce mobilization at high frequency. While an rDNA experiment would never use such a system, it was felt that if "nonmobilizable" plasmids remained nonmobilizable under such an extreme test, they would "clearly provide a high level of safety."[39]

Protocol II was designed to test the transfer of plasmids in an animal model, specifically the germ-free animal model developed by Freter and reported on at the conference. The kinds of questions would be similar to those asked in protocol I, but the use of small animals was more flexible, and more variables and manipulations could be performed on larger numbers. K12 would again be the test organism together with typical cloning vehicle plasmids.

Protocols III, IV, and V would examine K12s that contained various kinds of inserted DNA. Protocol III was a general procedure to see whether K12 with some foreign DNA increased K12's virulence or its colonizing ability in mice. The intention was less safety testing than a method to see whether containment conditions were too strict: "This type of testing would be particularly useful in determining whether containment conditions for particular clones are appropriate or should be reduced. It could be developed into a standard protocol that could be carried out by a contractor."[40] The plan was to test the hybrid K12 strain for virulence by administering it to mice by various routes, such as orally and intravenously. Special model systems for testing extraintestinal pathogenicity were suggested. Intestinal colonization would be studied in a manner similar to protocol II. Protocol III was a general-purpose test in a mammalian system to see whether anything unexpected was occurring. Given the very large number of cloning experiments, however, it is difficult to see how it could be carried out on the very large subset of possible K12 strains with foreign genes, and no principle of selection or priority for testing was suggested.

Protocol IV was devised specifically to test the effects on the whole animal of carrying an *E. coli* K12 that produced a polypeptide hormone. One of the many benefits claimed for rDNA technology was its potential for using bacteria as miniature chemical factories to produce complex biologics such as vaccines or hormones such as growth hormone or insulin. What would happen, many people wondered, if an insulin-secreting *E. coli* became established in the gut of a laboratory worker? Because exquisitely sensitive assays have been developed to detect hormones, this protocol offered "unique means of testing general factors regarding risk."[41] Using germ-free mice to assure colonization, the assay methods would be used to detect hormone in intestinal contents and in the blood of the animals. Such studies would indicate whether the hormone could still be synthesized under conditions prevailing in the intestinal tract, and also whether the hormone could get through the intestinal wall into the general circulation.

Protocol V was also designed to test a recurring concern that one of the recipient clones of a shotgun experiment would have some unusual property that would present a threat to human health or the environment. Shotgun experiments involved the random fragmentation and hybridization with a cloning vehicle of the entire genome of an organism. Since most of the genetic sequences would be uncharacterized, it was feared that a harmful gene, say one that had remained latent in the original genome, perhaps in the form of an integrated oncogenic virus, might find its way into *E. coli* and create a hazardous situation outside its normal genetic environment. It might seem that the problem presented by this situation was similar to that of the previous protocols, but there is such a large number of different clones produced by a procedure of this sort that special pooling procedures are necessary to accomplish the testing. Once a method of pooling is selected (taking care to see that in the process the entire genome is represented), the procedures are similar to the previous protocols.

The final protocol, number VI, was "to identify the most important variables that determine whether a nucleic acid molecule can be transferred from an *E. coli* host into eukaryotic cells in vivo."[42] This experiment was actually suggested long before the Falmouth Conference, as early as 1975, by Sydney Brenner, who viewed it as a very sensitive test of the risk of the rDNA technique. On 5 December 1975 RAC formally asked NIAID's Wallace Rowe and Malcom Martin to perform the experiment. Because of the putative risk, very high containment facilities were required, so that by the time of the Falmouth Conference, eighteen months later, the experiment still had not begun. (A parallel study was at that time underway in Europe by an international consortium of microbiologists.) The proposed test involved the use of polyoma virus whose natural host is the mouse. By inserting the DNA of the virus, using rDNA techniques, into plasmid or phage-cloning vectors carried by *E. coli* K12, the ability to cause viral infection or tumors in the mouse could be measured. When infected by polyoma virus, the mouse produces an antibody that can be detected. Newborn rats and hamsters inoculated with large doses of the virus develop tumors. Using oral, intravenous, or other types of administration, the hazard of K12 carrying polyoma virus to mice and newborn rats would be assessed by protocol VI.

Thus, the two most important results of the Falmouth Conference were the assessment that K12 could not become an epidemic pathogen and the proposal for a series of risk-assessment protocols to permit a rational evaluation of the hazards and a scientific basis for establishing or revising the NIH *Guidelines*. The Falmouth Conference did not conclude that any *E. coli* K12 carrying a rDNA molecule was incapable of causing harm, only that it could not become an epidemic pathogen similar to other *E. coli* pathogens. Other risk scenarios were left open, pending execution and evaluation of the risk-assessment protocols.

17

Reassessing the Risks: The Guidelines *Revised*

One day after the Falmouth Conference ended, on 23 June 1977, the RAC presented a proposal to ORDA for revising the 1976 *Guidelines.* The revision process had begun much earlier and proceeded quite independently of the Falmouth Conference. A subcommittee of RAC was appointed in January, met several times, and presented its recommendations to the full committee in May 1977, a month before Falmouth.[1] Consensus was finally reached in June and a draft prepared over the summer and given to Director Fredrickson in September. The principal revisions exempted from the *Guidelines* all rDNA experiments that involved DNAs from any organisms that normally exchanged DNA with *E. coli* in nature. A list of such "nonnovel" DNA combinations was to be prepared, and henceforth would not be subject to the *Guidelines.* By the time a draft of the proposed revisions had been prepared, the Falmouth Conference had taken place and the "results" communicated by letter from Gorbach to Fredrickson. The conference also figured prominently in NIH's justification for the proposed relaxation of the *Guidelines,* although it spoke to a relatively narrow concern.[2]

The revision effectively redefined rDNA to exclude combinations of two DNAs each of which could naturally replicate in an organism. At a public meeting of the Director's Advisory Committee, held to consider the proposed revisions on 15–16 December 1977, John Littlefield, a member of the RAC subcommittee that proposed the revisions, explained it thus: "We believe that there has been ample opportunity in such a case for natural recombination between these two DNAs, and that they are not novel and not hazardous." In other words, if nature could already have done it, it is safe.

There was complaint at the DAC meeting that this revision was based upon "unpublished articles, informal reports from conferences, letters, and other documents which have not been submitted to traditional means of scientific review nor to public debate."[3] The reference to Falmouth is clear. The truth is that there was little in Falmouth that spoke to the question of hazard and genetic novelty, since the conference concentrated on whether K12 could be converted to a nonnovel pathogen, that is, a pathogen like other *E. coli* pathogens. But NIH was vulnerable to the irregularity in the process that the Sierra Club's Nancy Pfund hammered away at: "The fact that the NIH Advisory Committee gained access to new information and used it in developing the new *Guidelines* without securing outside comment while decisions were being made makes a mockery of public participation in this process. As a result, many of the changes seem arbitrary and baffling, even to those of us who, for one reason or another, follow this issue. If the NIH is committed to getting public input, it must equip the public with data its own Advisory Committee used to justify its decisions."[4]

The suspicion was growing that NIH was not interested in the widest range of opinion from the scientific community but only in opinions that accorded with the objective of deregulating rDNA research. Deregulation meant returning control to the scientists who were NIH's main constituency, while continued regulation established a precedent that neither NIH nor the scientific community appreciated. It was clear that any further efforts to deregulate could not give the appearance of taking place behind closed doors. When the final version of the revised *Guidelines* appeared, it included, besides the controversial exemptions, a provision for "public" members on RAC.[5] The scientists still dominated the committee—as much by the extreme technicality of the deliberations as by their numerical advantage—but the critics were coopted. And even as the new RAC met for the first time, a new set of revisions was in the making.

Within three weeks of the effective date of the first revision, a notice appeared in the *Federal Register* detailing a new set of proposals, this time containing an even more drastic relaxation of the original guidelines concerning experiments in *E. coli* K12. In the new proposal, virtually all experiments (except the prohibited ones) using *E. coli* K12 containing nontransmissible cloning vehicles would be exempted from the *Guidelines*. The proposal was recommended by the full RAC and published in the *Federal Register* in April 1979. A statement by Wallace Rowe of RAC, prepared for the May 1979 meeting, gave the reasoning behind this move.[6]

Rowe first offered his "perspective" on why the *Guidelines* were instituted in the first place. He held that a primary difficulty was that the problems presented by rDNA research found the scientific community unprepared in two ways. First, the broad range of expertise needed to assess the risks was too great for any of the people involved. In other words, if scientists had really understood the situation in 1974, they would have left well enough alone. This is a rather curious position, since even if Rowe is correct that no hazard existed—and many disagreed—the inability to assess that hazard by the scientists involved, for whatever reasons, made a pause in the research necessary. "Ignorance of the true nature of scientific knowledge" was no argument against the caution of investigators like Berg and Singer who first expressed concern.

But a clearer vision of how the scientific community was "unprepared" was given next:

> The second problem was the naivete of the scientific community, particularly in not foreseeing the inevitable, relentless movement from voluntary moratoria to external regulations. When the first guidelines were drawn up, in 1975, we on the committee were operating with a number of tacit assumptions in mind. Some of these were critical to the entire subsequent unfolding of events. 1) We assumed that by severely restricting the scope of recombinant DNA research to a few extremely well understood laboratory organisms as hosts, and permitting work to proceed under only extremely stringent conditions, both scientific and administrative, we would buy time and allow a more informed, rational, data-based set of guidelines to be developed. 2) We thought we were writing guidelines, meaning *guidance*, for local decision-making bodies; and 3) we assumed we had the freedom to amend, alter, and undo what we had done. On the latter two points we were completely wrong. We watched shocked and paralysed as a variety of forces moved the decision making away from the scientists, the RAC and the local biohazard committees and into the hands of lawyers and bureaucrats. For a period of two and a half years, virtually all decisions under the guidelines, by NIH staff, by ORDA, and by the IBCs have been made solely on the basis of what constitutes compliance and literal interpretation.[7]

Rowe then gave a brief accounting of the scientific data that led him to support the new revisions. First, K12 cannot escape from the laboratory. It is not spread by any route except ingestion, it dies on drying, and it is rapidly killed by heat or chemical agents. Rowe cited studies of lab personnel in England which indicate that in a "typical laboratory," personnel do not acquire K12 or its plasmids, "even the conjugative, readily transmissible plasmids." Second, K12 cannot become established in the ecosystem. "The Falmouth Report provided convincing demonstration that K12 itself cannot become established in nature." For the Falmouth conferees, "only transfer of the recombinant molecule to another host need be considered." Several reasons were then given to show that this transfer is unlikely. Third, a recombinant molecule carrying K12 will not be able to survive in an ecological niche because it would not have any selective advantage. If a piece of genetic material carries no advantage to its host, it will be quickly lost. How does Rowe know that it will have no selective advantage? He cannot imagine it: "It is most difficult to imagine how the cloned inserts could provide selective advantage to a naturally occurring bacterium, and I have never heard a single claim or suggestion as to how this could happen." Thus, in natural environments, competition would quickly eliminate any recombinant-containing K12. Finally, Rowe asserted, the hazard of pathogenicity or ecological disturbance is no longer remotely tenable. In his opinion, the Falmouth Conference and Rowe and Martin's own work with protocol VI to test the danger of K12's carrying a viral genome have settled the issue. Further studies showing that genes do not occupy contiguous stretches of DNA but are interrupted by other intervening sequences have also laid to rest any fears of untoward effects of shotgun experiments. One simply cannot produce harmful gene sequences in such a random fashion.

Rowe's overall conclusion was unequivocal: "I feel it is fair to say that the evidence is overwhelming that the area of research that we propose to exempt cannot pose a threat to man or the environment. The data continue to mount in support of this conclusion, while the assertions to the contrary are more and more based on the weakest of conjecture. We have taken on the task of assessing the risk of recombinant DNA research with *E. coli* K12; we have done the job thoroughly and well, and the answer is clear."

Little evidence is given for the first point. A single study from England is cited from an infectious disease laboratory with experienced and highly trained personnel. In the second point, the Falmouth report is cited for a result it did not establish. The third point gives Rowe's own opinion, which will be shortly called into question by an article in NIH's *Recombinant DNA Technical Bulletin*, which seems to show that χ1776 with a cloning vehicle plasmid survives better than one without. And the final point extends both the Falmouth report and Rowe's own controversial polyoma experiments far beyond their legitimate scope. The strength of Rowe's conclusion masks the thinness of the evidence.

But again it is the intrusiveness of the *Guidelines* into the world of science that is the real point. Rowe's comments here are remarkable for their bitterness:

> (1) The *Guidelines* are really not guidelines. They are Deadlines. The word 'deadline' is of Civil War origin and refers to a deceptively simple line of stakes around a prisoner-of-war camp. A prisoner who made one step out of the deadline was shot without further warning. These *Guidelines* have done exactly that; they put every recombinant DNA scientist in jeopardy of making trivial, paper violations, that in no way create a safety problem, but that can draw serious penalties on him and his institution. Such a jeopardy is not something that should be instituted lightly.
>
> (2) The *Guidelines* are wasteful of time and money; and they are inefficient, inflexible, and inhibitory. They may require expensive remodelling of laboratories or duplication of equipment. They thus prevent research, including risk assessment research.
>
> (3) They create a wasteful, expensive bureaucracy in the IBCs and in ORDA [Office of Recombinant DNA Activities], not to ensure safety or safeguard cherished values, but merely to ensure compliance.

Rowe summed up his position as follows: "Thus the *Guidelines* are not a trivial, neutral administrative system. If they are indeed necessary to prevent injury to the world outside of the laboratory, fine. If they are not needed, as I feel is true in the case of K12 research, then they are an intolerable, capricious, irrational, offensive intrusion. We have a great obligation to be sure that we do not lightly continue this set of restrictions one minute past the time they are no longer needed."

Rowe's was not a universal opinion, however. There were those who felt just as strongly that the second set of revisions had gone too far. They also cited evidence. One of these was Roy Curtiss, who had favored the first revisions, but who sent two strongly worded letters on the new proposals to NIH to protest the new relaxations of the *Guidelines*.[8] He gave six reasons for his disagreement: (1) Most of the data on the safety of K12 systems was developed for the enfeebled strains, not the usual laboratory strains whose exemption was proposed. Data on the enfeebled strains is irrelevant to the present proposals. (2) Data on the safety of the various *E. coli* K12 host-vector systems as were presented and discussed at the Falmouth meeting in 1977 were already evaluated by the membership of a previous RAC and were used to justify the lowering of containment required for many experiments as permitted by the revised NIH *Guidelines*. One can thus ask, What is the basis for the current committee's recommendation to use these same data to justify an additional lowering of containment? (3) Scientific data obtained under contract by NIH to evaluate the safety of K12 systems have indicated that transmission of rDNA from K12 might be a more likely event, not less likely, than was believed at Falmouth. (4) Data from Rowe and Martin's own protocol VI experiment with polyoma-lambda recombinants suggested that a new pathogenic agent could be created, even though it was of low infectivity. (5) "There must remain uncertainty since none of us is totally clairvoyant. Indeed, initial users of DDT, thalidomide and even the internal combustion engine were totally unable to predict the ultimate health hazards associated with their use." Moreover, pointed out Curtiss, untestable and unpredictable hazards aside, most of the risk assessment protocols suggested at Falmouth had yet to be conducted. Rather than using Falmouth to justify the new revisions, he suggested that NIH should acknowledge that Falmouth was being ignored: "It would thus appear that the current Recombinant DNA Advisory Committee no longer considers the advice of the Falmouth participants on the new data from the experiments they proposed to be relevant in assessing the risks of using the *E. coli* K12 host-vector systems." (6) "Adherence to the NIH *Guidelines* has not hindered our work." (Curtiss and his laboratory had been conducting a wide range of recombinant DNA experiments under P1, P2, and P3 conditions for some time.)

Curtiss's assessment of the situation was thus diametrically opposed to that of Rowe: "I therefore consider the exemption of all experiments using the *E. coli* K12 host-vector systems to be premature and not justified on the basis of objective review of available scientific evidence. I also believe that the [RAC's] recommendation to you was based more on the politics of science than on the data of science."[9]

As if to underscore Curtiss's doubts, in July 1979 four reports appeared in ORDA's own *Recombinant DNA Technical Bulletin*, as RAC was considering relaxing the *Guidelines* for the second time since their initiation. Workers at the University of Texas were investigating the recoverability of enfeebled *E. coli* K12 in the various unit processes in sewage treatment plants.[10] Those processes can be lumped into primary and secondary treatments; the former are essentially physical (screening, settling) and the latter biological (aerobic digestion of organic load). Many large cities only use primary treatment for their wastes (Boston is an example), although most have both. The investigators concluded that primary treatment removes 30 percent of one enfeebled K12 and 54 percent of another. Sixty-two percent of typical *E. coli* bacteriophage (bacterial virus) was also removed by primary treatment, but only 20 percent of the Charon 4A phage, a principal EK2 cloning vector. Even secondary treatment removed no more than 97 percent of one enfeebled host and only 80 percent of the other. Thus, a great deal of the enfeebled K12 that went into the model treatment plant passed through in viable condition. Much that did not was entrapped in the sludge that was produced by both primary and secondary unit processes. These sludges are often subjected to further anaerobic digestion and then disposed of in land fill or as a soil supplement. Nothing is known about survival of these sludge-occluded organisms. But of special concern was the observation that one of the enfeebled strains showed much greater residence than expected in the treatment process. The authors recommended that "the basis for this discrepancy be sought and that the possibility of replication of this organism in the treatment plant milieu be ascertained."

Theoretically, no K12 should ever reach the sewer system because the *Guidelines* require that all experimental cultures be chemically or physically disinfected before disposal. But the intense campaign to discount any danger from K12 has led many researchers to assume that no organisms could survive in nature; therefore they have regarded the disinfection requirement as a nuisance. This issue was raised, somewhat parenthetically, at a meeting of a consultants' group convened in August 1979 by NIH to advise it on whether Falmouth protocols I and II needed to be formally completed before allowing any relaxation of the *Guidelines*.[11] (The conclusion, ironically, was that K12 posed no danger to man or the environment.) In discussing whether the various requisites of good microbiological practice should be required, one of the participants (no names are given in the transcript) said he hoped that these were the kinds of things that everyone did anyway. But a colleague brought him up short:

First speaker: I have heard that a most respected scientist, a bacterial geneticist, says, "We couldn't think of decontaminating our standard *E. coli* preparations before discard. For a while we did decontaminate when we were doing some recombinants, but it is so much trouble to use we just don't want to do it." I was appalled; this was two months ago.

Another speaker: He said this in public?

Answer: No, but it is public knowledge that this is the way.

Another speaker: I think that has been one of the difficulties. . . .

people at Asilomar were coming with the idea that "Nobody is going to tell me I have to stop pouring my *E. coli* K12 down the sink." It was on that basis that the conference was convened.[12]

Not only survival in the sewer system was addressed by the new data. A second report in the same issue of the *Bulletin* showed that there was significant dispersal and survival of K12 in the course of normal laboratory operations.[13] All K12s, including the enfeebled strains, could survive for days on various surfaces in the laboratory.

Neither of these papers could give much comfort to many at Asilomar who believed that K12 could not survive in the environment. The organism in fact did quite nicely. But there was more unsettling information. Studies from Freter's lab on the survival of K12 in mice showed that all *E. coli* strains, including the enfeebled ones, could pass through the alimentary tract in a viable state.[14] Since this was not supposed to occur, this was quite a surprise. Freter also continued his previous work with germ-free mice, confirming that ordinary K12 that had become established in germ-free animals were not dislodged when the gut was later populated with conventional flora. He pointed out, therefore, that the ability to implant in the gut and the ability to maintain a stable population are two different characteristics and must be tested separately. The intact enfeebled strain could not implant even in germ-free mice, but in Freter's experiments mutations occurred in which χ1776 lost its bile sensitivity, and these mutants *could* colonize germ-free animals. Thus, an important safety barrier, bile sensitivity, was at times unstable.

The final article was the most surprising, however. Levy and Marshall of Tufts University School of Medicine tested the survival of K12 host-vector systems in human volunteers, and found that the survival of χ1776 was significantly enhanced by carrying one of the common cloning vehicle plasmids, pBR322.[15] No transfer of this nonmobilizable plasmid was detected, but the result was still worrisome. It raised anew all the old questions about the possibility that K12 with a DNA insert might take on totally new and unexpected survival characteristics.

As if these data were not enough, the November 1979 issue of the *Bulletin* carried a report by Cohen et al. from the University of Rhode Island showing that K12 strains are very good colonizers of normal (non-germ-free) mouse intestines, if the mice are given appropriate concurrent antibiotics.[16] K12 strains were not only good colonizers relative to normal fecal strains, but appeared to occupy an entirely separate niche in the mouse intestines, not competing at all with the indigenous flora.

All these published reports were occurring in the context of determined activity on the part of the rDNA research community and NIH to bring about the second revision of the *Guidelines*. The justification for these revisions was the need to extend the conclusions of the Falmouth Conference beyond their original scope while ignoring its concerns and disregarding the newly published data.

The "Director's decision document," published in the *Federal Register* on 30 November 1979, explains the reasons why the NIH exempted experiments involving propagation of recombinant DNA in *E. coli* K12 hosts.[17] Ten questions pertinent to the decision are asked (for example, "Is *E. coli* K12 pathogenic?") and answered in the document. Most data cited are either from Falmouth (eight citations) or excerpts from earlier *Guidelines* or studies cited in those *Guidelines* (six citations). Six new studies are cited as providing additional data. Included are studies showing that typical laboratory workers do not pick up K12 or their antibiotic resistance plasmids, two studies that suggest that K12 tends to lose implanted rDNA, and two studies that fail to show transfer of nonmobilizable plasmids. One of these latter studies was the Levy and Marshall work referred to above, which also showed improved survival for plasmid bearing χ1776. This result is not cited as relevant evidence, however. Nor are data on survival in sewage or on the laboratory bench mentioned, even though these were published in ORDA's own *Bulletin*. It is clear that there has been careful selection of evidence, heavily biased in favor of the intended outcome. It is interesting to note that the commentary on the Rowe-Martin experiment is quite noncommittal. (The experiment provides "valuable data on the question of the likelihood of an animal virus recombinant DNA insert being transferred out of an *E. coli* K12 host cell used in its propagation into a eukaryotic cell.") This is undoubtedly a reflection of the intense controversy these particular studies aroused. This lack of consensus made them less useful as "negative evidence" of hazard.

Susan Gottesman, an RAC member who voted against the exemptions, gave the most charitable explanation for RAC's action approving the revisions. Explaining in a letter to Fredrickson that her opposition was based on her feeling that the possibility of harm had not been fully assessed, she said that the split vote on the committee (ten in favor, four opposed) was a case of different people seeing the

Table 4
The prevailing view of the risks of using *E. coli* K12 as a host for rDNA experiments at various times

	Falkow memo (May 1975)	Falmouth Conference and first revision (June 1977)	Between first revision and second (summer 1979)	Second revision (December 1979)
K12 colonization	Doubtful	Impossible, except under very unusual circumstances (germ-free mice)	Possible only under unusual circumstances, but χ1776 survives GI tract	Possible with antibiotic pressure
K12 pathogenicity	Possible	Impossible or extremely unlikely	Impossible or extremely unlikely	Impossible or extremely unlikely
K12 as epidemic pathogen	Doubtful	Impossible	Impossible	Extremely unlikely (but survives in environment)
Transmission of rDNA plasmids-phage from K12	Possible but unlikely	Possible but unlikely	Possible (survival of Charon 4A phage in sewage)	Possible (survival of Charon 4A phage in sewage)
Survival of K12 in environment	Possible	Doubtful	Possible	Very possible

same data and reaching different conclusions. "The split vote of the RAC at least partially reflects [the] range of possible conclusions . . . and expresses the complexity of the problem more than anything else."[18] On the less charitable side is Roy Curtiss's judgment that it was a case of the politics of science, not the data of science. The whole history of the rDNA controversy suggests that Curtiss was the more accurate observer.

We began by asking what kind of evidence emerged in the three and a half years from the first *Guidelines* to the substantial relaxation of the second revision. The answer is not very obvious. Some new evidence was reassuring, some was not (see table 4). The conclusions of the Falmouth Conference, which are cited repeatedly as a basis for a variety of propositions, were extended far beyond their original intent. Even the highly restricted scope of those conclusions (that K12 could not become an epidemic pathogen) is not based on extensive data or rigorous argument; it was reached on the plausibility of the evidence.

But the evidence was not primary. The seemingly inexorable pressure to relax the *Guidelines* proceeded in a context of controversy and contention as scientists struggled to regain control over their working conditions. Their antagonists were those who saw outside control of a powerful technology as right and proper. In such a circumstance evidence was used as a weapon, not as data for a hypotheticodeductive argument that produces scientific judgments.

The stakes in the rDNA controversy were very high: the control of science and the control of an immensely powerful and potentially profitable technology. Scientists wanted to keep that control to themselves, and commercial interests were satisfied to give it to them. It was a tradition with which both were comfortable. But others believed that this kind of technology was too powerful, both for its positive and negative potentials, to leave to scientists. It is no wonder that the actual nature of the evidence should be secondary since control, not "safety," was to a large extent the main issue. "Safety" was only the strategic hilltop whose possession would help win the war. And as in many other wars, it was not who actually held the position that counted, but who was believed to hold it. And the battle was won by the scientists, for better or for worse.

18

A Worst-Case Experiment

The Asilomar scientists developed a number of scenarios for how hazardous products might result from rDNA experiments. They recognized there was little evidence to support or refute many of the hypothetical outcomes. Some biologists wanted to turn the speculative discussion about risks into an empirical one by testing one or more of the potential biohazards. But the call for risk experiments was not widespread; some believed that there was nothing meaningful to be measured or that there were too many unique cases for a few experiments to yield any general outcomes.

A few of the influential spokespersons at Asilomar began discussing the virtues of a worst-case experiment. The strategy behind such an experiment provides an interesting example of the rationality of risk analysis. The principal outcome of Asilomar was a stratification of risks according to potential hazard. While the gradation of risks was not without its severe critics, it made it possible to use a worst-case approach to evaluate the hazards of gene-splicing research.

The logic behind the worst-case experiment is as follows: If the upper bound of an ordered set of experiments is found to be safe, then the entire series of genetic manipulations deemed less potentially hazardous is also safe. The application of the logic is vulnerable on two counts. First, the classification of risk was built upon a series of assumptions lacking empirical verification. Second, because the types of experiments one could contemplate were so varied, no ordinal series could fully capture the comparative risks. For example, according to some critics of the *Guidelines*, one could not set relative potential risks between animal virus experiments and plant toxin experiments. But however one evaluates the scientific value of the worst-case experiment, its political value was undeniable.

The fourth meeting of the Recombinant DNA Molecule Advisory Committee was convened on 4 December 1975 at La Jolla, California. DeWitt Stetten, Jr., deputy director of science, and Leon Jacobs, associate director for collaborative research, NIH, presided at the meeting. One of those in attendance was Sydney Brenner, distinguished scientist at the Medical Research Council in Cambridge, England. During the meeting Brenner outlined an experiment that would test whether DNA transferred from higher organisms to lower organisms would be functional. The protocol consisted of inserting polyoma DNA into an *E. coli* K12 host, and using the host to populate new born germ-free mice.[1] After administration of the *E. coli* bearing the recombinant, the mice would be monitored by checking the serum for antibodies (indicating infection) or examining the tissue for other pathologies. RAC unanimously passed a resolution recommending that NIH sponsor such an experiment. In addition, RAC established a subcommittee to consider the design of other experiments that could provide useful information about biohazards.

Brenner's worst-case scenario closely resembles the experiment planned by Paul Berg that had initially activated the rDNA debate. Berg had planned to use lambda phage to transport SV40 DNA into bacteria. Brenner's model system was preferable in several respects. Polyoma virus is not known to infect humans. On the other hand, mice are easily infected by the virus. Beyond its usefulness as a risk experiment, it was of interest to virologists, who wanted to know what the results would show about the integration of mammalian viral DNA into a bacterial genome. The importance of the newly developed recombinant techniques would only be partially realized if it were not possible to implant pieces of a eukaryotic genome into a bacterium so that the foreign genes were able to express themselves. Bacteria (prokaryotes) are not infected with eukaryotic viruses. There are viruses indigenous to bacteria called bacteriophages, or just phages for short. The concern about implanting a eukaryotic tumor virus into a bacterium that might become a resident in the human intestinal tract was preeminent among the fears expressed in the scientific community.

A number of questions had to be addressed before the risks could be evaluated. Could the viral genome function in the bacterium: for example, would it activate the production of more virus particles? If the bacterial host that contains the recombinant (animal) tumor virus were implanted in the intestinal tract of an animal, would it induce infections? Would the viral recombinant induce tumors in susceptible animals? It was conceivable that the viral DNA insert in the bacterium would remain inert. Since there was more than one way to implant the viral DNA into the bacterium, part of a risk assessment would involve examining several routes of entry. Specifically, one could attach the viral DNA to a plasmid or a phage vector, which could then be integrated into the bacterium.

Another consideration in this risk experiment was whether the tumor virus DNA is implanted in its entirety or whether only parts of it are implanted (subgenomic parts). How would the subgenomic parts compare in infectivity or tumorigenicity to the entire genome? There had already been strong pronouncements by scientists on some of these questions. But a formal conference to evaluate the hazards of using viruses in recombinant techniques was two years away. It was held at Ascot, England, in January 1978.[2]

The NIH went forward on the RAC recommendation to undertake the risk experiment. The responsibility for planning and overseeing the work was given to NIH senior virologist Wallace Rowe and his colleague Malcolm Martin. Rowe was also a member of RAC. When a protocol for the polyoma experiment was developed, it was sent to the members of RAC as well as other scientists for their comments. At the 1–2 April 1976 meeting of RAC (three months before the new NIH *Guidelines* were issued), Rowe, unable to anticipate legal delays, reported to the committee that the experiments would be completed in eight to twelve months.[3]

It was also Rowe's belief that the polyoma experiment should be only one part of a larger risk-assessment program. He suggested other types of experiments that might determine whether certain DNA fragments were capable of conferring a selective advantage on bacterial hosts.[4]

The risk experiments with polyoma were in the class of prohibited experiments in the 1976 NIH *Guidelines*. RAC voted to exempt the experiment for the risk-assessment work. At RAC's mid-January meeting of 1977, Rowe informed the committee that the EK2 host strain (the Curtiss strain χ1776) was too sensitive to colonize mice. Rowe requested approval to carry out the experiment in a P4 + EK1 system of containment so that he could use the ordinary laboratory strain of *E. coli* K12. There was strong support from RAC for the use of the less debilitated strain. However, when the request was reported to NIH Director Donald Fredrickson, he decided not to make the exception to the newly promulgated *Guidelines*. The early months of 1977 saw considerable congressional activity over the regulation of gene-splicing experiments. NIH was under siege from many directions. A major exemption to the *Guidelines*, even for a risk experiment, could have fueled the controversy. The polyoma experiments were subsequently carried out with certified EK2 hosts.

In addition to the requirement that the sensitive Curtiss strain be used, there were other factors that contributed to a delay in this first major risk experiment. The NIH *Guidelines* placed strict requirements on P4 physical containment facilities. Certain tests were warranted to assure that the requirements were met. Initially, the facility designated to house the experiments was Building 41 at the Cancer Research Center located in Frederick, Maryland. It was engineered and designed to operate at a P4 level, but it had never been so utilized. The stated purpose of the facility was to serve as a research containment laboratory for human cancer viruses if they were ever isolated. Subsequently, a change in venue had been decided on. The experiments were scheduled to take place in Building 550 of the US Army's former chemical and germ warfare center at Fort Detrick, Maryland. The facility was converted to the country's first NIH-certified P4 laboratory at a cost of a quarter of a million dollars.

At the June 1977 meeting of RAC, Donald Fredrickson reported that a civil action was filed against NIH by a citizen and attorney at Frederick, Maryland, on behalf of his twenty-one-month-old son. The suit was directed at the polyoma risk experiments that Rowe and Martin, under NIH auspices, were preparing. The plaintiff in the suit requested that the court enjoin NIH from performing the experiments until an environmental impact statement was prepared in full conformity with the National Environmental Policy Act. The suit charged that the child suffered undue risk of "permanent and irreparable injury" if a polyoma DNA recombinant in *E. coli* escaped from the facility.[5] The legal battle was eventually ruled in favor of NIH, and the experiments were completed in late 1978, which, according to the investigators, represented nearly a two-year delay.

The first major revision of the 1976 *Guidelines* was issued by HEW Secretary Joseph A. Califano in December 1978. NIH proposed the revisions after a serious and sustained lobbying effort on the part of environmentalists and influential members of the scientific community. Califano, who was reputed to be a master at compromise, gave the critics of rDNA research some procedural gains as the *Guidelines* were significantly weakened. He expanded the rDNA Advisory Committee to twenty-five members from approximately half that size. And on 15 December 1978, he sent letters to his assistant secretary for health and the director of NIH requesting that a program of risk assessment be initiated without delay. He asked NIH to submit its first plan to RAC by 30 March 1979.

Late in 1978 early results of the polyoma experiment spread rapidly through informal networks in the scientific community. Investigators Rowe and Martin reported preliminary results to RAC prior to publication at its 15–16 February 1978 meeting. There were some scientists who believed that the outcome of this series of worst-case scenarios made it senseless to proceed any further with risk assessment. The NIH investigators were not willing to go that far in their interpretation of the results. They were clearly optimistic about its significance for a large class of potential hazards.[6] But another group of scientists disagreed with the interpretation of the results by Rowe and his colleagues. For them it was not simply a question of how far one could extrapolate the results. These scientists arrived at a contrary conclusion, namely, that the experiments proved there were novel risks associated with DNA recombinant methods.

To explain these differences, my focus will turn to the substance of the polyoma experiments, the outcome, and the controversy over their significance. The story illustrates several points worth emphasizing at the outset: Scientists exhibited considerable confidence in the data obtained by respected researchers, despite differences they may have had on political issues. Even for a well-defined experiment that was designed to test the risk of a technology, there was polarization over what the data showed; the roots of controversy can be traced to disagreements over which questions the experiments were designed to answer. The risk experiment using a worst-case scenario was embedded in a political context; different parties looked for what they could find that would vindicate their previously chosen positions.

The last point is reminiscent of conflicts between theories in which schools of thought compete to establish a dominant paradigm. I have termed these latter controversies "scientific disorders of the first kind," while the controversies over risk I have called "disorders of the second kind." In either case one anticipates a certain rate of conversion of scientists as new information becomes available. In comparison to the number of conversions that took place from 1976 through 1978, when a considerable number of scientists reversed their earlier positions and declared that they foresaw no hazards, there were relatively few conversions of major actors in the debate in either direction when the results of the risk-assessment experiment became known. Indeed, its principal value was political: to allay the public's fears that rDNA could produce a cancer epidemic. Let us now turn to the details of the worst-case experiment. (In fact, this was not the worst-case, but the best worst-case; the same experiment with wild-type *E. coli* as a host is generally conceded to be a worse case.)

The first major set of risk experiments in the gene-splicing debate was designed to study the ability of bacteria carrying a recombinant molecule containing polyoma DNA to cause polyoma virus infection in mice. It was part of a broader objective to investigate the extent to which DNA molecules can be transferred in a functional state from *E. coli* to mammalian cells under natural conditions.

Investigators cleaved the circular polyoma virus DNA, attached it to a vector, implanted the hybrid into *E. coli*, and observed whether there was virus activity when the modified *E. coli* was administered to mice or placed in contact with cultured mouse cells. Plasmid and lambda phage vectors were used for the experiment. Both systems, host plus vector, were certified as EK2 during the period of the experiment. (The plasmid host vector system consisted of pBR322 and *E. coli* χ1776, while the lambda phage system consisted of the phage vector λgtWES.B and *E. coli* K12 DP_{50} Sup F.)

Two preparations of the polyoma DNA were designed for the experiment. Single copies of the polyoma DNA attached to the phage or plasmid DNA are called monomers; double copies of the polyoma genome are called dimers. The dimer construction, whereby two copies of the polyoma DNA are joined head to tail, provides more possible mechanisms for excision of one complete copy of the polyoma DNA. From the outset of the experiment, investigators anticipated that the recombinant form with the dimer polyoma DNA was more likely to produce virus particles.

Four separate molecular constructions representing the plasmid and phage systems are given in figure 4. The point of the various constructions was to observe the conditions under which the polyoma DNA in a recombinant molecule could cause infection. Presumably, to achieve this the polyoma DNA would have to be excised from its linear state in the recombinant molecule so that it could form a circular genome. Constructions A and B in figure 4 show the polyoma monomer and dimer attached to a plasmid and inserted in *E. coli*. Similar constructions using bacteriophage lambda as the vector (C and D) were also used.

Since mice of all ages are susceptible to infection with polyoma

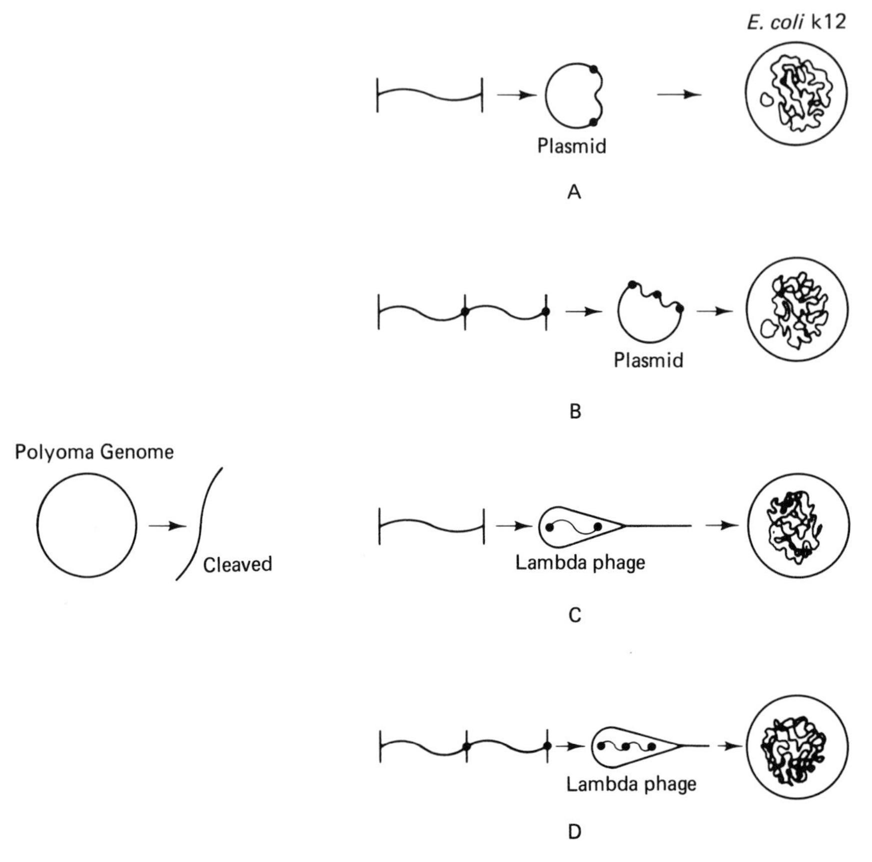

Figure 4 Plasmid and phage DNA recombinants for mouse polyoma infectivity experiments: (*A*) one copy polyoma DNA (monomer); (*B*) two copies polyoma DNA head to tail (dimer); (*C*) single copy polyoma DNA (monomer); (*D*) two copies polyoma DNA head to tail (dimer).

virus, the mouse systems were considered ideal for testing infectivity of DNA recombinants. The investigators tested for the appearance of infection by measuring antibody response to polyoma proteins. They collected baseline data on the infectivity of mice with ordinary (nonrecombinant) polyoma DNA and ran a control group of mice inoculated with various DNA preparations in saline solution.

The experimental results for this grouping of polyoma infectivity studies were published in *Science* on 2 March 1979 in two reports.[7] The published results played down the fact that these were risk-assessment experiments. For example, nowhere in the title were the words "risk" or "biohazard" mentioned. One report was entitled "Molecular cloning of polyoma virus DNA in *Escherichia coli*: Plasmid vector system." Similarly, the abstract made no reference to risk assessment: "A series of recombinant plasmids containing polyoma virus (PY) DNA were constructed, and their biological activity was evaluated in mice and in cultured mouse cells."[8] The first sentence of the text spoke of evaluating "postulated mechanisms of possible biohazard stemming from recombinant DNA research with *Escherichia coli*. Considering the significance given to their grouping of experiments in the risk assessment and its subsequent role in the revision of the NIH *Guidelines*, it is a curious fact that the results were positioned rather inconspicuously in *Science*.

One can reasonably assume that NIH did not want the risk-assessment studies to spark a new public controversy. But was there anything in the results that could have produced additional skepticism about the safety of DNA recombinants? Let us examine the results in qualitative terms, referring to the constructions in figure 4. (The summaries of the results for the polyoma and the lambda phage constructions are given in the matrix in table 5.) In experimental design A, a single copy of polyoma DNA was attached to a plasmid. Some of these plasmids were administered to mice directly (orally or by inoculation); some were inserted into *E. coli*, and the *E. coli* was then administered to mice orally or by inoculation; and some were brought into contact with cultured mouse cells. In none of these cases were infections (as measured by antibody responses) observed.

The investigators also tested whether the polyoma DNA became debilitated when attached to the plasmid. This proved not to be the case. When restriction enzymes were used to extract the polyoma DNA from the plasmid, the polyoma could still cause infection.

Table 5
Summary of polyoma infectivity risk-assessment experiments

	Appearance of infections in mice or cultured mouse cells				
	Vectors only		Vectors in *E. Coli* K12		Vectors only in cultured mouse cells
	Inoculated	Fed	Inoculated	Fed	
A. Single-copy (monomer) polyoma DNA on a plasmid	None	None	None	None	None
B. Double-copy (dimer) polyoma DNA on a plasmid	Dimer plasmids could not be isolated—postulated as unstable				
C. Single-copy (monomer) polyoma DNA on lambda phage	None	None	None	None	
D. Double-copy (dimer) polyoma DNA on lambda phage	Some	None	None	None	

plasmid + polyoma DNA ———————→ no infection
plasmid + polyoma DNA + cleaved→ infection

The conclusion drawn from the class A constructions is that polyoma DNA, when inserted into the plasmid, preserves all of its functions. Its inability to produce infection is attributed to its nonexcision from the recombinant. The investigators also attempted to isolate clones consisting of two head-to-tail inserts of polyoma DNA (dimer) in the plasmid. However, the dimer constructions of polyoma on the plasmid were not observed by the investigators. They attributed that to the instability of the construction. Experiments in set B yielded no results.

In the next two sets of experiments (C, D), either a single copy or a double copy of polyoma DNA was attached to the bacterial virus (bacteriophage), lambda phage. The phage serves as the vector for introducing the rDNA molecule containing polyoma in the *E. coli* host. In the cases where the lambda phage had a single copy of the polyoma DNA, no infections were observed in four separate instances (table 5C): when mice were fed or inoculated with the hybrid lambda phage or when phage-infected *E. coli* were fed to or inoculated in mice.

A positive result showed up when free lambda phage containing the dimer of the polyoma DNA was injected into mice. Also, when the total DNA of *E. coli* infected with lambda-polyoma hybrid was inoculated into mice, infections were observed.

In their published paper, just prior to introducing the positive results, the authors tell the reader that the positive outcome is theoretically plausible, which helps to mitigate the surprise element.[9] The authors also draw some conclusions about the degree of infectivity of viral recombinants in contrast to whole viruses or viral DNA. This is a critical area of concern for the safety issue. One of the questions that prompted the study was whether recombinant viral DNA will increase the risk of virus infectivity. The conclusion reached by the authors on the degree of infectivity for various forms of the virus is represented in figure 5.

In cases where the polyoma recombinants were infectious, they were generally less infectious by orders of magnitude than either the whole intact virus or the free viral DNA. This was interpreted by many scientists to indicate that the polyoma virus in recombinant form was safer to work with than by itself. The instances where poly-

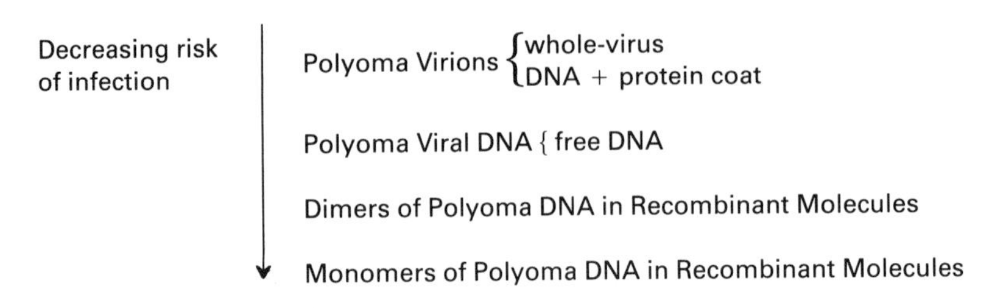

Figure 5 Relative infectivity for different forms of polyoma DNA.

oma dimers caused infection were interpreted by the investigators as (a) rare events since dimer formation is rare, and (b) of much lower infectivity than other polyoma forms.

The published results of the worst-case experiment in *Science* ended with a word of considerable optimism. The investigators explained that there were no unexpected outcomes and that the risks of certain classes of viral recombinants were considerably less than the risks associated with the virus itself.[10]

At a meeting of RAC in May 1979, MIT Professor of Biology Jonathan King made a presentation that provided a less reassuring interpretation of the polyoma risk-assessment experiment. King was among those who believed there was too little concern over the potential biohazards associated with gene-splicing technology. One plausible outcome of such experiments emphasized by King is that enteric organisms may be altered so that new pathways are developed to deliver disease. He published an article in *Nature* (2 November 1978) on that theme entitled "New disease in new niches."[11] The polyoma experiments were cited by King to illustrate the very biohazard he had anticipated in his *Nature* article. The argument he presented to RAC took the following form.

The positive results of the polyoma experiment show that polyoma-bacteriophage rDNA molecules are capable of infecting mice. Polyoma by itself does not grow in enteric bacteria. But by splicing it to bacteriophage, we have now created something that does not appear in nature. The polyoma virus has a new niche in bacteria, and therefore, we have effectively given polyoma a new route into humans. Contrary to the claims of others, these results are not anticipated for the dimer constructions; there is simply no way, he argued, to know what the outcome will be in this case.

The response to King's argument by investigator Wallace Rowe was that King was raising the wrong question with regard to the risk experiment. Rowe maintained that the experiment was designed to test whether a recombinant virus in a host bacterium would turn out to be as infective or more infective as the free virus. Since only one construction of polyoma and lambda proved infectious, and far less infectious than the virus itself, Rowe argued the results vindicated the use of rDNA technology for a large class of viruses. In no instance had *E. coli* carrying recombinant plasmids or phages with polyoma DNA caused infection (implying that virus particles were not produced). The investigators concluded that viruses could be handled more safely in the recombinant form.

In this episode, the locus of dispute was not the data, which were accepted by all parties. The controversy was over what questions the experiments were designed to answer. The NIH investigators dismissed King's criticism as irrelevant. The purpose of the tests, they claimed, was not to show that some polyoma-prokaryotic constructions would be infectious, but to examine how safe polyoma DNA would be when captured in a recombinant molecule and implanted in *E. coli*.

Were the positive results truly anticipated by the NIH scientists, or were they interpreted as such to defuse additional controversy? These are difficult questions to resolve. But it is useful to examine the original protocol of the experiment that was submitted to RAC and other scientists for review. I studied the document for any evidence that positive results for the dimer inserts were anticipated or that any positive outcomes were not to be considered significant. Two citations from the document that describe the goals of the experiment speak clearly to this issue:

> [A] single group of mice, colonized by an *E. coli* plasmid or phage system containing a polyoma component, can be tested repeatedly for whether productive virus infection has occurred at any time during the prolonged period of observation. . . .
> Progeny phage, as well as plasmid-containing *E. coli* will be fed to, and inoculated into mice, in the final step of the proposed experiment and the animals tested for evidence of virus infection. . . . In particular, since we are very concerned that a negative result will be misinterpreted as indicating that there is little biohazard in DNA recombinant work, we feel that a number of variables must be evaluated, and that the experiments should be designed to maximize the chances of a positive result. Also, if negative results are obtained with a given system, we feel that the experiments should be continued with modified

systems until we are reasonably convinced that a positive result is just not going to occur.[12]

At no point in this protocol was there mention of anticipated positive outcomes with oligomer constructions of polyoma. The previously cited passage suggests that *any* positive outcome would be a significant result.

Jonathan King's arguments did little to persuade other scientists on RAC of his view that the first part of the worst-case experiment showed that there were real biohazards. But King tried to deliver his message to a wider audience. First, he sent a letter, cosigned by MIT colleague Ethan Signer, to the *New York Times* (24 April 1979), outlining his interpretation of the polyoma experiments. The letter was published on May 3 (see note 18). King targeted another letter to a scientific audience and with two more signatures sent it to *Science*. But the journal never published it.

The first results of the polyoma risk experiments did not receive widespread coverage in the press. Considerably more attention was paid to the start-up of the experiments. The description of the renovated former germ-warfare laboratory was described in syndicated reports by Judith Randall (*Boston Herald American*, (30 March 1978), by the *New York Times* (18 March 1978), and in *Time* magazine (3 April 1978). A news conference was scheduled when the first results were published in *Science* on 2 March 1979. The experiments were frequently discussed in the context of the NIH initiative to ease the restrictions on rDNA research. On 21 September 1979 the *New York Times* editorial made an oblique reference to the polyoma experiments. "Even when scientists have deliberately tried to create recombinants that would cause cancer, the resulting organisms have turned out to be less infective than common cancer viruses." The editorial was written in the wake of the second major revision of the NIH *Guidelines*. In December 1978, when the first major relaxation in the *Guidelines* was issued, the *New York Times* reported that the decision was, in part, based on research studies done since 1976 (17 December 1978). *Science* covered the results of the polyoma experiment with a lead article entitled "Gene splicers simulate a disaster, find no risk" [203:1223 (23 March 1979)]. *Nature* reported a statement of Wallace Rowe to RAC that the risk experiments with polyoma virus that he and his colleagues carried out were further evidence that there is no cause for concern [279:360 (31 May 1979)]

Finally, there was an essay in the *Bulletin of the Atomic Scientists* by

W. A. Thomasson, a biochemist and free-lance writer, which discussed the polyoma results. It stated,

> When two copies of polyoma are inserted (as could happen inadvertently) infection does occur, but only at a rate one-billionth of that which follows direct administration of the virus. Thus, a novel route of infection was opened, but the probability of infection remained low. This type of experiment is therefore safer than handling the virus itself (indeed it has been seriously proposed that the smallpox genome be inserted into *E. coli* and all remaining stocks of smallpox virus destroyed) and a major lowering of the required containment levels seems clearly justified.[13]

The second set of risk experiments involving the polyoma virus was designed to test the ability of recombinants of that virus to induce tumors in hamsters. The results of the experiments were published in *Science* in September 1979 under the title "Molecular cloning of polyoma virus DNA in *Escherichia coli*: Oncogenicity testing in hamsters."[14] The investigators introduced the results by summarizing the polyoma infectivity studies. No mention was made of the positive outcome of those studies with the dimer-polyoma-phage constructions. Instead, the authors chose only to report the negative results when *E. coli* carrying the recombinants were fed or injected in mice.

Suckling hamsters were chosen for this part of the test because the animals are highly sensitive to tumor induction by polyoma virus and because complete virus particles are not required for tumorigenesis. Several constructions of plasmid-polyoma and phage-polyoma were used. As in the earlier experiments, some of the constructions consisted of single copies of polyoma DNA and others of double copies. All hamsters were inoculated with either *E. coli* carrying recombinant molecules or with purified rDNA. Animals were under observation for five months for the appearance of tumors.

A summary of the results of the tests is given in table 6, which indicates the number of hamsters that showed tumor production over the number tested with a particular form of the polyoma recombinant. The number of hamsters inoculated in all cases was too small for the results to provide statistical validity for the negative results.

No tumors were observed when the *E. coli* with the polyoma-plasmid DNA was inoculated in hamsters. Tumors were observed in hamsters that were inoculated with the plasmid-polyoma monomer rDNA. When the plasmid-polyoma recombinant was broken apart to free the polyoma DNA, a higher percentage of tumors appeared (9/17 versus 2/27). No tumors were observed when hamsters were inoculated with the lambda phage that carried a single copy of the polyoma DNA. However, of sixteen hamsters inoculated with the lambda phage consisting of a double copy of polyoma DNA, three had tumors. Comparable tumorigenicity was observed when hamsters were treated with nonrecombinant polyoma DNA.

After reporting the data, the authors emphasized the negative outcome of the results. First, they reported that the oncogenic activity of the intact rDNA and phage was comparable to or less than the pure polyoma DNA, which they pointed out is four to five orders of magnitude less active than the whole polyoma virus. Second, they indicated that the most relevant results for this part of the series of risk experiments was that polyoma recombinants in *E. coli* exhibited no tumors in hamsters. They interpreted the positive results with the purified DNA and the phage particles as simply showing extreme cases in which much greater amounts of polyoma DNA were injected into the hamsters. The use of *E. coli* as a host of the polyoma DNA, according to the investigators, "constitutes an effective barrier against the transfer of its DNA to eukaryotic cells."[15] Just as in the mouse infectivity studies, the authors concluded their paper with a high degree of confidence that the risks of cloning animal tumor viruses can be done safely: "The findings that no tumor viruses were induced with the viable plasmid-containing bacteria not only provides further evidence for the safety of cloning viral genomes in *E. coli* but also provides for the safety of cloning other postulated oncogenic gene segments."

In reviewing the results before the RAC meeting (21–23 May 1979) prior to their publication, Wallace Rowe stated that there were no surprises. The critical factor was the behavior of the polyoma when it is incorporated into *E. coli*. Rowe was taken to task for his interpretation of the data by Barbara Hatch Rosenberg, a research biologist at the Sloan Kettering Institute, and associate professor of biochemistry at the Cornell University Medical College. At the meeting, Rosenberg expressed some of her views and followed up with written statements that gave quite a different slant to the same data. First, she stated that it was not possible to do an accurate quantitative determination of tumorigenicity from the limited samples used. Second, she held that the data implied that the insertion of polyoma DNA into a vector does not significantly change the tumorigenicity of the viral DNA. Third, she questioned the interpretation of the results of inoculating animals with *E. coli* containing a polyoma rDNA molecule. Rosenberg argued that there are other routes of exposure for *E. coli* that would challenge the host other than by inoculation. Therefore, one could not use the negative inoculation results as an indicator that *E. coli* carrying a recombinant was not hazardous.[16]

There was little interest on the part of the investigators to respond to Rosenberg's remarks. Determined to get the message across to the scientific community, Rosenberg teamed up with microbiologist Lee D. Simon of the Waksman Institute of Microbiology, Rutgers University, and sent an article critical of the investigators' interpretation of the polyoma experiments to *Science*. The article was rejected for publication. The Rosenberg-Simon team sent a revised version of their analysis of the experiments to *Nature*. While the two scientific journals are about equal in status in the scientific community, there is a close relation between the *Science* editorial board and NIH. The receptiveness to criticism on this issue was greater at *Nature* than it was at *Science*. Since the polyoma results were being used to justify lowering the containment requirements for all types of rDNA research, interpretations of the results less favorable toward the conclusion of no risk could have slowed down the process of relaxing the *Guidelines*. In their *Nature* report, Rosenberg and Simon made the following points about the polyoma experiments: The conclusions of the investigators about the safety of cloning viral genomes in *E. coli* went far beyond the data. Some of their data confirm certain hazard scenarios that have previously been postulated. Additional experiments are necessary with the model polyoma system before the overall hazards of cloning viral DNA into *E. coli* can be assessed. The results of the experiments have been used without justification for relaxing the NIH *Guidelines* for gene-splicing research.[17]

The authors' scientific critique took the following form: The data (see table 6) show that polyoma DNA in recombinant form induced about as many tumors as the nonrecombinant form of the polyoma (19 percent tumors for the nonrecombinant circular [noncleaved] polyoma DNA versus 19 percent tumors for the dimer polyoma-phage construction). Even though statistical validity cannot be assured with the small samples, "for both single and tandem-insert (dimer) recombinants the tumorigenicity is of the same order of magnitude as that of free polyoma DNA." Therefore, inserting the DNA of tumor viruses in a recombinant molecule does not make it

Table 6
Tumor studies with polyoma recombinants[a]

Inoculum	Number of hamsters with tumors/number tested
E. coli χ1776 carrying polyoma DNA in recombinant plasmids	0/32

	Number of hamsters with tumors/number tested			
	DNA			
		Cleaved		
Inoculum	Uncleaved	Eco RI	Bam HI	Phage particles
Plasmid system	Recombinant DNA			
pPR18 + pPR21	1/11	6/9		
pPB5 + pPB6	1/16		3/8	
Phage system				
Monomer insert (λ-PY3 + λ-PY63)	0/20			0/8
Dimer insert (λ-PY30-5B)	3/16			2/12
	Nonrecombinant DNA			
Polyoma DNA	14/73 (19%)	29/64 (45%)	11/35 (31%)	

a. Plasmids pPB5 and pPB6 are recombinant plasmids consisting of pBR322 with the complete polyoma genome inserted at the Bam HI site in the two possible orientations; pPR18 and pPR21 are pBR322 with the polyoma genome inserted in both orientations at the Eco RI site. [Source *Science* 205 (14 September 1979)]

safer to use. Since the tumorigenicity of the DNA is not made safer in the recombinant, it is questionable whether the new packaging of the DNA might provide new routes of infection to other species.

Rosenberg and Simon concluded their review of the experiments with some comments about the negative results obtained when the polyoma recombinants in *E. coli* were adminstered to the hamsters. The *E. coli*, they argued, may not have been able to produce the minimal polyoma dose needed for inducing tumors. Thus, the experiments may have been biased in favor of negative results. Furthermore, the critique continued, other challenges to the hamsters could have improved colonization of the polyoma carrying *E. coli*, thereby amplifying the dose of polyoma and improving the likelihood of tumorigenesis.

It is surely difficult for the nonexpert to make sense out of the technical details of scientific controversy when questions of methodology and data analysis are at stake. In recapitulating the Rosenberg-Simon critique, it might help to place their comments in several categories: First, they cited *positive evidence* that the recombinant form of animal virus DNA is not safer than the pure DNA as far as inducing tumors is concerned. Second, they argued that the *evidence is consistent* with the view that new routes of disease could be established with gene-splicing methods. Third, they claimed the *negative evidence was too narrow* and possibly methodologically flawed.

The alternative interpretations of the polyoma experiments by Rosenberg, Simon, King, and several other scientists had no perceptible outcome on NIH policy toward rDNA experiments. As in the case with the earlier polyoma experiments, the media coverage viewed the results as supportive of the no-risk interpretation. The outcome of the experiments was also a key element of the justification for the second major revision of the NIH *Guidelines*.

Rowe and Martin's response to their critics was either to ignore them or to make perfunctory and bored replies. Thus, in a memo they wrote to Dr. Bernard Talbot, special assistant to the director of NIH, they commented ("as requested by Dr. Fredrickson") on biologist Stuart Newman's letter (dated 6 November 1979) in the following fashion:

> This letter does not raise any new issues or give any new perspectives on the polyoma experiment. . . . The underlining [sic] premise . . . is that K12 can supply a vehicle for the recombinant DNA to become established in the ecosystem, a point which has been thoroughly discussed, analyzed and rejected as a result of the Falmouth conference and the many other inputs on the biology of K12 host-vector systems. It must be remembered that the whole purpose of the polyoma experiments was to determine if, on the assumption that *E. coli* K12 carrying viral DNA did become established in the ecosystem, the DNA could move efficiently from the bacteria into the cells of the host; the results of the experiment were completely negative.

In a written statement to RAC at its May 1979 meeting, Wallace Rowe provided reasons for his recommendation that approximately 85% of current rDNA work be exempted from the *Guidelines*. Rowe, citing the polyoma experiments confidently, said, "I do not know of a single piece of new data that has indicated that K12 recombinant DNA research could generate a biohazard." Rowe was careful to mention that exempting most work with K12 was not tantamount to exempting the many other combinations of host, donors, and vectors.[18]

The principal locus of dispute between the investigators who executed the polyoma risk-assessment work and the scientists critical of their interpretation of the results is not the scientific discipline itself. The NIH team stressed repeatedly that they were looking at the factors that determine whether a DNA molecule can be transferred from an *E. coli* host to a mammalian cell.[19] They carried out other experiments, such as administering the DNA recombinants by themselves (outside of an *E. coli* host). But the positive results in those experiments were not deemed germane to their primary objective.

I have shown how others interpreting the same results were looking for novel routes of entry through which viruses could establish themselves in new niches. From their perspective, the result that viral DNA recombinants were less infective or less tumorigenic was not the only issue. Of equal importance was the creation of novel organisms with a host range equal to or greater than that of its component biological parts.

The most significant impact of the polyoma experiment was political in nature. The results were used effectively to gain approval for the relaxation of the NIH *Guidelines* and to stem federal legislation. Before the story is closed, however, there is a bit of irony to report. Three years before the results of the polyoma studies were published, a participant at an NIH-sponsored enteric bacteria meeting anticipated the outcome of a worst-case experiment in such explicit terms that, in retrospect, it almost seems the result was predestined. This extraordinary passage appeared in the detailed conference notes of Malcolm Martin, one of the principal investigators in the polyoma risk experiment. The source of the comment remains anonymous.

> It seems to me, there are two kinds of experiments that you can think about: One kind of experiment involves learning something about what's going on and those experiments require the standard sort of thing that we are interested in. The second . . . is a class of experiments whose aim is to convince people that some of their worst fears don't happen very easily. Now, one experiment in the latter case that I would argue is definitely worthwhile doing is to try every way you can, germ-free or otherwise, to make an epidemic of serious disease with cloned DNA. Just try to do it. I agree that it makes no sense. Just titer the survival of mice that have been injected with this material

and do that for a while and say, 'Look, there's no epidemic!' And the experiment gets published in the *New York Times*, not in the *Journal of Molecular Biology*. That's what ought to be done, that kind of thing. A kind of thing that will not live to be a famous experiment in the history of science; it's like all other experiments that end up in the front of the *New York Times*. It's something that captures the imagination of reporters. I would argue as strong as possible that, in the short term, that's what is needed.[20]

Of course, it did not happen exactly that way. The results were published in *Science*. Nevertheless, the deep cynicism expressed in the remark has a very unsettling effect on those who hold scientists in a position of trust. It is difficult to assess how widespread the cynicism was. But with public pressure for legislation mounting, it was inevitable that the frustrations of some scientists would supplant the quest for truth with a public-relations campaign.

19

Breaching Species Barriers

What had begun as a tool of inquiry quickly came to be seen as a powerful technology for modifying, reconstructing, or even creating organisms whose limits could only be guessed at. Speculations began to appear about the extent to which gene transplantation research would enable science to play a significant role in directing the evolution of species.

From the time the rDNA debate was in its embryonic stage until 1976, the term "breaching species barriers" was used widely in describing one of the unique outcomes of the new research tool. Strong advocates of the research did not shy away from the expression; on the contrary, the term had a Baconian ring to it. Science had the key to, and was prepared to penetrate, one of nature's deepest secrets. Witness this statement from an early brochure issued by the Cetus Corporation, one of the fledgling bioengineering firms, describing rDNA technology to potential investors.

> The significance of this power cannot be exaggerated. Perhaps the most important breakthrough has been the capability to transfer genes from one species to another. Classically, the definition of a "species" has been that an organism in such a group could *not* breed (i.e., exchange genetic material) with a member of another group. Nature, through evolution, has created barriers to the exchange of genes between species. Within a species, breeding for improvement is possible and has long been practiced. However, it is not possible to create an animal with the combined characteristics of a dog and a cat by mating them. These species barriers have so long been accepted as logical and almost absolute, that it is only within the past months that scientists have seriously contemplated the ramifications of breaking these species barriers.

After 1976, proponents of the research avoided using the term "species barriers" in public testimony, promotional literature, and essays describing the new technology. Moreover, some scientists who previously used this term later argued that there were no such things; they were fictions or hypothetical constructs. The assumption of the existence of such barriers, and of the ability to breach them, was the basis on which the new science could claim awesome powers. On the other hand, if it could be shown that the genetic exchanges that the scientists were achieving in the laboratories with rDNA techniques occurred also in the natural evolution of cells, there would be less force to the argument that science had developed a unique power over nature that must be carefully controlled by society. The debate on this issue signified a broadening of the potential hazards from laboratory health effects to environmental ecology and the control over evolution.

One of the most eminent scientific critics of rDNA research was biologist Robert Sinsheimer, formerly of the California Institute of Technology, more recently appointed chancellor of the University of California at Santa Cruz. Sinsheimer argued that the new gene-splicing techniques permit scientists to cross well-developed barriers between species, thus potentially creating dramatic changes in evolutionary patterns.

> Recombinant DNA technology—the fruit of 25 years of extensive research in molecular genetics—makes available to us the gene pool of the planet—all of the genes developed in the varied evolutionary lines throughout the history of life—to reorder and reassemble as we see fit. . . . Nature has, by often complex means, carefully prevented genetic interactions between species. Genes, old and new, can only reassort within species.[1]
>
> To introduce a sudden discontinuity in the human gene pool might well create a major mismatch between our social order and our individual capacities. Even a minor perturbation such as a marked change in the sex ratio from its present near equality could shake our social structure.[2]

Through his public lectures, Sinsheimer continued to draw attention to how rDNA can be used to reshape evolution. Biology, he emphasized, is following the path of physics and chemistry, which gave us the power to reshape physical nature on the planet with all its attendant catastrophes. His remarks combined an environmentalist perspective with theological overtones. He spoke of an "intricate web of life" and of novel organisms establishing themselves in new ecological niches.[3] He referred to science's role in tampering with creation. "We are becoming creators, inventors of novel forms that will live on long after their makers and will evolve according to their own fates. Before we displace the first Creator we should reflect whether we are qualified to do as well."[4]

Sinsheimer struggled with the tension between the positive rewards of scientific knowledge and the power of that knowledge to transform the world. He offered few definitive claims, but posed many rhetorical questions. One cannot find specific scenarios in his writing or his lectures for a biodisaster. But one does find several moral imperatives. He called upon his colleagues to take a longer perspective on the work they are doing, beyond their own lifetimes to subsequent generations. He rejected the fatalism of the technological imperative. That something can be done does not mean it shall be done. But Sinsheimer created the most turbulence among his scientific colleagues with his views on the limits to knowledge. He argued that there are certain types of knowledge that once attained would almost assuredly be applied. And where such knowledge could severely harm human society, Sinsheimer maintained, it should be proscribed.[5] It was a heretical position from an esteemed member of the biological community. And for his views, his colleagues, while treating him with respect in public, in private scoffed at as having undergone a religious conversion. Where should we draw the line? On rDNA research? Should we decide that there is some forbidden area of knowledge about the functioning or regulation of genes? Sinsheimer had called into question the most basic scientific value—a legacy from the Aristotelian heritage—knowledge is virtue. Not only were some types of experiments hazardous, he argued, but some *kinds of knowledge* may prove too dangerous for society to have in its possession. His views spawned scores of papers and a special edition of *Daedalus* devoted to the limits of inquiry.[6]

Ironically, while Sinsheimer was a scientist of exemplary credentials for discussing the rDNA technology, considered by many as the most formidable opponent, his views had little influence on the outcome of the public-policy debate. His remarks were not designed to be put into some operational mode, that is, this experiment should be banned but this other one is safe. Most of his criticisms were too open ended for an effort directed at developing workable guidelines for carrying on the research. The primary impact of Sinsheimer's remarks was to widen the boundaries of concern, which at this time had been centered around short-term laboratory biohazards. Two other scientists, Erwin Chargaff and George Wald, entered the debate with the same fundamental values. Both men were persistent in citing the evolutionary consequences of this new research program.

Erwin Chargaff, emeritus professor and emeritus chairman of the Department of Biochemistry of Columbia University's College of Physicians and Surgeons and recipient of the National Medal of Science (the United State's highest tribute to a scientist), was another critic of genetic engineering who took an evolutionary perspective. His published letter in *Science* entitled "On the dangers of genetic meddling" fomented considerable discussion and reaction from the scientific community and the general public.[7] While Chargaff warned his fellow scientists about the use of a "colon bacillus" as a host for genetic transplants, he cited the ethical issue as more significant than the public health problem: "[T]he principal question to be answered is whether we have the right to put an additional fearful load on generations that are not yet born. . . . Our time is cursed with the necessity for feeble men, masquerading as experts, to make enormously far-reaching decisions. Is there anything more far-reaching than the creation of new forms of life?" Chargaff dramatized the unique aspects of this new technology that sets it apart from other malevolent abuses of science. "You can stop splitting an atom; you can stop visiting the moon; you can stop using aerosols; you may even decide not to kill entire populations by the use of a few bombs. But you cannot recall a new form of life. . . . An irreversible attack on the biosphere is something so un-heard of, so unthinkable to previous generations, that I could only wish that mine had not been guilty of it."[8]

The role of species barriers also arises in Chargaff's assessment of gene-splicing technology. He referred to the eukaryote-prokaryote hybrids as freakish forms of life. Like Sinsheimer, he posed the moral rhetorical question, Do we have the right to counteract the evolutionary wisdom of millions of years of evolution to satisfy the ambition of scientists?

Chargaff placed rDNA technology in a class with other forms of scientific activity that possess a transforming quality. They reach beyond a passive state of inquiry to an active intervention in nature. In his *Science* letter, he called it "a destructive colonial warfare against nature."

In Chargaff's view, neither the methods of investigation nor the knowledge they bring can be neutral if, as a by-product of the inquiry, there is a rearrangement of life processes. And once transformed, it may be impossible to bring nature back to its initial state. He warned, "No genius will be able to undo what one cretin has perpetuated."[9]

George Wald, Higgens professor emeritus in biology at Harvard University, won the Nobel Prize in physiology for his work on the biochemistry of vision. He entered the rDNA debate when his own department had planned to renovate an old laboratory for use as a moderate-level (P3) rDNA containment facility. Wald and other Cambridge scientists expressed their concerns to local city officials about the potential hazards of having such a facility in a heavily populated urban area. Like Chargaff and Sinsheimer, Wald could not separate health issues from evolutionary issues. The notion of species barriers played an important role in his argument. That was how he distinguished this technology from other intrusions into the natural order. "All such earlier procedures worked within single or closely related species. The nub of the new technology is to move genes back and forth, not only across species lines, but across any boundaries that divide living organisms, particularly the most fundamental such boundary, that which divides prokaryotes (bacteria and blue-green algae) from eukaryotes (those cells with a distinct nucleus in higher cells and animals)."[10]

Wald introduced an argument that drew attention to evolutionary change. It takes million of years for a single mutation to establish itself as a species norm, he maintained. With rDNA, this process can be speeded up so that, in comparison with evolutionary time, the species changes are instantaneous. Wald did not criticize the breaching of species barriers per se, but only insofar as such activity will unwittingly result in some danger. A clear demarcation line is established by Wald between gene splicing that breaches species barriers and gene splicing that does not. The term itself focused attention on "science as intervenor."

The term "natural barrier," or "species barrier," encapsulates the claim that genetic intercourse between certain species does not take place. Robert Sinsheimer brought this issue to the attention of the scientific community when he posited the existence of natural species barriers separating the prokaryotic and eukaryotic worlds.[11] The positing of such barriers is the foundation of Sinsheimer's evolutionary perspective on the impact of rDNA Technology. But the issue whether there are genetic barriers separating species is a complicated affair. Partly, this is because most discussions have not clearly explicated the meaning of species barriers. Moreover, the term species has a more arbitrary meaning among prokaryotes than it does among eukaryotes. Inquiry into these matters must consider three questions: (1) What types of barriers divide species? (2) What forms of genetic exchange exist apart from what can be achieved through technology? (3) Which barriers, if any, can be uniquely breached by rDNA technology?

According to evolutionary biologist Ernst Mayr, a species is a protected gene pool.[12] For higher life forms, fertility barriers are the basis of species classification. Two normal animals of the opposite sex are of the same species if they can establish a progeny line. For these higher organisms "the reproductive isolation of a species is a protective device that guards against the breaking up of its well-integrated, co-adapted gene-system."[13]

Sexual reproduction is not the only way that genetic exchange can occur. Foreign genes can be transported to cells by both natural and artificial asexual means. Prokaryotes possess a variety of natural mechanisms for exchanging genetic information both between and within species. It is also possible to engineer interspecies genetic exchanges by laboratory techniques that do not involve the joining of genes in cell-free systems characteristic of rDNA methods. For example, cell fusion is a technique through which the nuclei of two cells are fused into a single nucleus. This technique has been used to produce live chimeric tissue culture cells from two species of mouse and between human and mouse cells.

Three types of naturally occurring genetic transfer that do not depend upon rDNA techniques are conjugation, transduction, and transformation.[14] Conjugation is the process through which chromosomal genetic material and plasmids (extrachromosomal circular pieces of DNA that naturally inhabit microbes) can be transferred from one cell to another via a transient intercellular bridge. Conjugation among prokaryotes occurs widely in nature; however, with one exception, plasmid exchange between eukaryotes and prokaryotes is unknown. The exception involves the Ti plasmid of *Agrobacterium tumefaciens*, which migrates to plant cells and is responsible for crown gall disease in plants.

Bacterial genes can also be carried from one bacterium to another through the vehicle of an infecting virus particle called a bacteriophage (a process called transduction) or without any vehicle, simply in the form of free DNA (or RNA) taken up from the surrounding medium by the recipient cell (a process called transformation).[15]

The extent to which these processes of genetic exchange permit genetic intercourse in nature between distantly related species is not known. In point of fact, rDNA techniques use transduction, but under specially engineered circumstances involving in vitro gene-cutting and -splicing operations. In an essay in which he discussed the potential risks of gene-splicing technology, Richard Novick summarizes what is known about genetic exchange promoted by plasmids and viruses: "[N]o plasmid or virus that can cross the eukaryote-prokaryote barrier has yet been encountered. . . . However, there are plasmids with extremely wide host range among the prokaryotes and viruses with extremely large host range among the eukaryotes—although the latter are usually relatively virulent (e.g., yellow fever) and have not been implicated in host gene transfer."[16] The previously cited case involving *Agrobacterium tumefaciens* is generally acknowledged to be the rare exception to the situation described by Novick.

These matters take up questions (1) and (2). Before taking up question (3), a distinction should be made between barriers of transfer and barriers of expression. It may be possible to achieve a transfer of genes without rDNA techniques in which expression does not result unless the latter methods are used. In this instance, it may be said that the barrier breached by rDNA methods is not the transfer of genetic material but its cellular expression.

Genetic recombinations that occur in the "wild" are, by definition, natural events. But that is only one standard way of defining natural barriers. If "artificial" laboratory inducements of genetic exchange between eukaryotes and prokaryotes can be achieved without rDNA techniques, is that evidence that species barriers do not exist? Does it follow that such barriers can be breached by methods other than gene-splicing techniques?

What shall serve as a standard of "species barriers": the laboratory or the natural environment? If we apply the standard of the natural environment, it can be argued that while, presently, we do not observe exchanges between eukaryotes and prokaryotes, such natural recombinations might be occurring at very low frequency. Alternatively, if the more liberal laboratory standard is used, then the claim that rDNA methods uniquely breach such barriers is justified only after other laboratory techniques have been tried and fail.

Another consideration in the discussion of species barriers is whether the focus of concern is the transfer of plasmid DNA or genes that become integrated into the chromosome of the cell. The transfer of the bacterial Ti plasmid to a eukaryotic cell is an example of a "natural" event. On the other hand, there are no examples of analogous "natural events" in which eukaryotic genes are integrated into bacterial chromosomes.

It will be useful in dealing with all these complexities to distinguish three forms of "natural genetic barriers," namely, ecological barriers, absolute barriers, and statistical barriers, to which are associated, respectively, the concepts of laboratory environment, natural environment, and frequency of occurrence. Two organisms are separated by an *ecological genetic barrier* if the exchange of DNA between them is not observed under conditions resembling those considered as natural for the species in question. Two organisms are separated by an *absolute genetic barrier* if the exchange of DNA between them is not observed under natural or artificially engineered environments limited to non-rDNA techniques. Two organisms are separated by a *statistical genetic barrier* if the exchange of DNA is observed with low but not necessarily zero frequency under natural or artificially engineered environments limited to non-rDNA techniques.

For the sake of simplicity I have not generated as many distinctions as there are realistic possibilities. Although the exchange of plasmid DNA among organisms of divergent species (particularly prokaryotes) is more commonplace than the exchange of chromosomal DNA, which is where the heart of each organism's genetic makeup lies, this is not included in the three forms of genetic barriers. In reality, there could be greater barriers to the *integration* of foreign DNA into the cell's chromosome than to the cell's *intake* of foreign DNA through the transfer of plasmids. Also, the three forms of genetic barriers do not distinguish between barriers of exchange and barriers of expression.

The three forms are pertinent to the issues whether species barriers exist, and if they do, whether rDNA technology is unique in breaching such barriers. The difference between ecological barriers and absolute barriers rests on the distinction between what occurs in nature and what can be induced under laboratory conditions. For example, human cells are not known to exchange DNA with bacteria. Several reasons are possible. The DNA of the two species may have no routes of exchange. Or there may be no regions of similarity (homology) between the nucleotide sequences of two species. Or a bacterial host may destroy DNA of human origin before it has a chance of attaching to the chromosome. But can the cells of these species exchange genetic information in an artificially designed biochemical environment? At least one study has shown that bacteria can take up the DNA of higher-order organisms (eukaryotes) under suitable laboratory conditions. The experiments were carried out by Cohen and Chang of the Stanford University Medical School and reported in the *Proceedings of the National Academy of Sciences*.[17]

It is important to look rather closely at Cohen and Chang's results since their experiments were used to refute the hypothesis of natural barriers, primarily among policymakers, when rDNA legislation looked imminent. In the introduction to their paper, the authors state, "Although methods for cleaving and joining DNA segments in cell-free systems have enabled construction of a wide variety of recombinant DNA molecules that have not been observed in nature, there is reason to suspect that restriction endonuclease-promoted genetic recombination may also occur intracellularly as a natural biological process." The paper reported that enzymes similar to those used in rDNA experiments are found in a wide variety of bacterial species; also, the enzymes that join genes (DNA ligases) are widespread in nature. The implication is that since the chemical materials that were used in their laboratory experiment are found in nature, what occurred in their experiment could happen in nature, although admittedly at lower frequency.

Their experiments consisted of exposing plasmids in *E. coli* to cleaved fragments of mouse DNA in the presence of specific enzymes. They obtained results showing that mouse DNA was connected to plasmids and that the genetic splicing that occurred intracellularly was indistinguishable from what is constructed in vitro by rDNA methods. While the experiment was directed at genes located on plasmids, the investigators felt it shed light on similar exchanges between eukaryotes and prokaryotes for chromosomal DNA: "[I]t seems reasonable to speculate from our findings that restriction endonucleases may play a major role in the natural evolution of plasmid, and perhaps chromosomal genomes."[18] What was achieved in the laboratory would ordinarily not be detected in nature, according to the authors' interpretation of the results. They had improved the likelihood of detection by choosing a eukaryotic DNA segment that was easily selectable. An analogy can be drawn with what toxicologists do when they test the carcinogenicity of a chemical by feeding mice large doses of it. Here, too, investigators hope to improve the probability that tumorigenicity will be observed in the small number of events they examine. But the Cohen-Chang experiment was questioned by some scientists as to its relevance to the issue of species barriers on two counts. First, it was engineered in a highly artificial laboratory setting, which, some held, had no resemblance to natural conditions. Second, it had no direct bearing on the exchange of chromosomal DNA and thus could not serve as a definitive refutation of the hypothesis of species barriers.

But even if it is assumed that the authors' speculation about intercellular recombinations with chromosomal DNA turns out to be correct, genetic intercourse of this type is consistent with the notion of ecological barriers. Referring to the Cohen-Chang experiment, plasmid biologist Richard Novick commented, "[T]he conditions under which this interpretation took place are so extremely artificial that there is essentially no chance of their occurring in nature; they do not, in my view, increase at all the likelihood that eukaryote-prokaryote genetic exchange is a natural phenomenon."[19]

What limited scientific impact the Cohen-Chang experiment had was greatly overshadowed by its use in the political arena. Preprints of the results were distributed to key congressional science advisers in the summer of 1977, during the period when legislative activity on DNA legislation had peaked. Release of the report was designed to persuade Congress that the hypothesis of natural barriers was false. Furthermore, it was argued that man, through his use of rDNA techniques, is not doing anything that nature cannot replicate given the opportunity—in other words, the laboratory was simply an extension of nature. Those scientists opposed to legislation could now say that there was "hard" scientific evidence proving that no species barriers exist between prokaryotes and eukaryotes. And for some policymakers, simplicity in such complex subject matters was openly accepted.

The use of prepublished data for influencing public policy was criticized by some scientists as a transgression of professional ethics. According to this argument, scientific results that play an important role in policy choices should be available to the general scientific community for comment and interpretation before policies are decided. By the time the results were published, there was scant attention in the scientific literature to the outcome of the experiment and its relevance to species barriers. If the results were important enough to shape public policy, then one can ask why were they not important enough to foster broad scientific discussion. The answer lies in the role the hypothesis of species barriers played in the progress of molecular genetics. In a word, it had no scientific relevance. But it did threaten to slow scientific activity; the idea of natural barriers projected the notion that certain transgressions were blocked by nature, possibly for some overriding, evolutionary reason. It was not surprising, then, that the concept of genetic barriers was criticized on other grounds.

Two veterans of the rDNA debate, Paul Berg of Stanford University and Maxine Singer of the NIH, responded to Sinsheimer in testimony before a Senate subcommittee. "No scientific fact or theory predicts or supports the existence of such a barrier, nor is there any indication that scaling the hypothetical barrier with recombinant DNA methodology would have the consequences Sinsheimer predicts. But neither is there convincing evidence to the contrary."[20]

As Sinsheimer's views began percolating in scientific circles, a major figure emerged who used evolutionary concepts to demonstrate that natural barriers between even the most distantly related species are a fabrication. Bernard Davis, an eminent bacteriologist at the Harvard Medical School, lectured and published essays in which it was concluded that the evidence is weighted more heavily against the existence of genetic barriers than for the contrary position. Davis's argument can be reconstructed in the following form:[21]

1. It has been shown empirically that bacteria can recombine with free DNA from prokaryote species both in the test tube and in the living animal.
2. Bacteria are continually exposed to free DNA from eukaryotes as a result of their proximity to dying eukaryotic cells.
3. There is no reason to believe that the mechanisms through which bacteria take up prokaryotic DNA would exclude eukaryotic DNA.

---

4. Therefore, bacteria must occasionally take up eukaryotic DNA in a natural setting.

There is nothing in Davis's argument implying that the exchange between eukaryotes and prokaryotes in nature must be observable. Indeed, Davis maintained that the uptake of eukaryotic DNA into bacteria is not an efficient process; therefore, if observed at all, it would be at very low frequencies. He also cited as a relevant datum against the hypothesis of natural barriers that prokaryotes are the evolutionary progenitors of eukaryotes. Therefore, he reasoned, in the millions of years of evolutionary history, it is reasonable to assume that some links between these cells have been sustained.

Novick interpreted Davis's position in a more extreme form as a target for his criticism: "During the several billion years of life on earth, all or most possible gene combinations have been tried by evolution through the process of mutation and genetic exchange, and any that are not now extant are counteradaptive. Therefore it is extremely unlikely that man could possibly create anything that is (i) novel, or (ii) adaptive for the novel organism. Therefore, there is no significant biohazard inherent in rDNA research."[22]

This position is tantamount to the view that evolution has essentially stopped. All genetic combinations have been tried and the adaptive ones have been selected out. In his critique of the argument, Novick used a thought experiment to show "that only a tiny fraction of all possible genetic combinations could possibly have been tried in the course of evolution."[23]

He imagines the earth to be a solid mass of bacteria containing equal numbers of unrelated species and that these bacteria have been multiplying every twenty minutes since the earth began (4 billion years ago). If one genetic exchange took place every time a cell divided (every twenty minutes), Novick calculated that 10^{65} new combinations would have been generated. (Since the earth is not a solid mass of bacteria, this number is an upper bound.) This quantity of novel combinations is minute in comparison to all possible genetic exchanges based on the conditions of the thought experiment.

Davis stopped short of arguing that every conceivable genetic recombination has already occurred. But he did assert that "nature has tried all classes of genetic clusters and selected those which will adapt."[24] Should a scientific investigator create an entirely new constellation of genes, according to Davis, it probably will not survive because it already failed at some point in evolutionary history. The argument does not spell out the criteria for a "novel class of genes." Furthermore, there is no way to determine whether certain groupings of genes have been tested out against natural selection. The argument is grounded on several a priori judgments that are given respectability through association with Darwinian concepts. If Davis is correct, why do we not see more eukaryotic genes in prokaryotes? His response: The genetic configuration of species has had a long period of time to adapt and improve. The incorporation of eukaryotic DNA into prokaryotes will result in less adaptive organisms whose adaptive properties will vary as the evolutionary distance of the donor species from the species host.

While it is true that Sinsheimer's arguments were conjectural, their purpose was to provoke a more comprehensive risk assessment. Davis, in countering Sinsheimer's speculations, appropriated some principles from the theory of natural selection, the suitability of which to genetically engineered microorganisms was questioned. The objective of his line of reasoning was to show that additional investigations into risk would be fruitless. Thus, Davis fought speculative reasoning with more speculation and theory. His critique of the hypothesis of the species barrier simply raises more unanswered questions. It fails to make a distinction between barriers to foreign DNA exchange through plasmid transfer and barriers to the transfer of DNA into the organism's chromosome. This is a distinction that some scientists considered significant to a resolution of the barrier controversy. Two other points are relevant. The first is the dubious extrapolation from what happens in the laboratory to what naturally occurs in nature. If there are no situations that qualify as offering absolute genetic barriers, there is still the ecological barrier to contend with. If, by rDNA techniques or other more conventional techniques, ecological novelty is created in the laboratory, there could still be cause for concern. The second point relates to the frequency of events. If eukaryotic cells, under natural conditions, are observed to exchange chromosomal DNA with bacteria, as Davis contended is highly probable, the frequency of these exchanges could prove to be very low, too low to have an important effect on the species population or the bacterial ecology. Novick summarized this point in a poignant way: "But just because an organism can be coaxed to take up foreign DNA in the lab, one has no right to assume that it does so regularly, if at all, in the wild; and further, even if *uptake* occurs, experimental evidence now available suggests that incorporation of foreign DNA into the cell's genome is a very special and unusual event."[25]

Consequently, even if absolute barriers and ecological barriers are found to have no significance, the concept of a statistical barrier may still be important. It suggests to us that rDNA technology, with its far-reaching capacity to transport genes and their regulation sequences, could introduce qualitatively unique pressures on the environment that arise from breaching a quantitative barrier.

20

Immunological Risk Factors

The two issues that dominated the discussion of rDNA risks were the transformation of *E. coli* K12 into a pathogen and the spread of a virulent form of a test organism into the environment. Except for two instances (chapter 8), the limited hazard of a noninfectious recombinant organism to the investigator or laboratory worker played almost no role in the debate until the Falmouth Conference in June 1977. In addition, the risks of large-scale work with rDNA molecules also was not in the mainstream of discussion since all work over ten liters was prohibited by the NIH *Guidelines*.

At the Falmouth Conference, Jonathan King introduced a risk scenario that was first conceptualized during the period he met with the members of the Cambridge Experimentation Review Board.[1] King's hypothetical hazard considers the possibility of a recombinant organism initiating an autoimmune response in a human host. Burke Zimmerman, while a staff scientist for the Environmental Defense Fund, also cited autoimmune disease as a possible hazard in his comments on the draft environmental impact statement.[2] The reconstructed scenario takes the following form:

1. An active polypeptide from a mammalian donor (nonhuman) is spliced into *E. coli* K12.
2. The *E. coli* establishes itself in the intestinal tract of a laboratory worker.
3. The recombinant bacteria produce the mammalian protein in the intestinal tract.
4. The mammalian protein is similar in structure to a human protein. (The antibodies to foreign antigens cross-react with human antigens.)
5. The human's immune system begins to attack the foreign protein; it does not discriminate between the foreign protein and its closely related indigenous protein.
6. The immune system then attacks its own tissue and commences a pathological process called autoimmune disease.

King did not claim he had evidence that this could occur. Moreover, he has no professional standing in the field of immunology. He merely cited this scenario as a possible state of affairs that should be examined by the relevant experts and through the appropriate risk-assessment experiment.

The autoimmune risk had not generated much concern among RAC or other scientists involved in biohazards. It was nearly a dead issue when in October 1978 it was resurrected by Jonathan Beckwith of Harvard Medical School. Until this time Beckwith had not been involved with the risks of rDNA research. His speech at the National Academy of Science Forum in March 1977 was directed at genetic engineering, particularly the misuse of genetics for solving social problems. Beckwith simply was not convinced that the risk scenarios about *E. coli* pathogens were sufficiently compelling for his active involvement. Indeed, Beckwith also represented a rare breed of scientist who openly questioned work with which his own laboratory was involved. Thus, when Beckwith wrote a letter to William Gartland, director of the Office of Recombinant DNA Activities (NIH), it was a departure from his earlier reservations about the biohazards aspects of the new research technology. And, as in the past, Beckwith's concerns were associated with work that was closely, rather than distantly, related to his own. His opening remark to Gartland stated, "I want to express a concern about a particular class of recombinant DNA experiments which may be hazardous and which may not have been considered in the drawing up of the NIH Guidelines. I have come to consider this problem as a result of experiments being done by a collaborator of ours who is using certain of our strains in recombinant DNA experiments."[3]

Beckwith described some recent work in the United States and France that involved the construction of hybrid proteins by fusing genes of human hormones with genes of bacterial proteins. The idea behind the construction was to export the human protein to the outer cell surface by making it a part of the bacterial surface protein. By having the human protein established on the cell surface rather than in the cytoplasm, it was anticipated that the gene product could be extracted from the cell more efficiently and cheaply. It was the placement of the hybrid protein on the surface of the bacterial cell that prompted Beckwith to raise the issue. Beckwith carefully phrased his letter, indicating that he had conferred with immunologists, infectious-disease experts, and scientists knowledgeable in rDNA. "In all cases, these individuals agreed that there was a real possibility that these modified peptide hormones could break tolerance—induce an autoimmune response in humans to their own hormones if they reached the appropriate sites in the body."[4] He suggested that animal experiments be carried out to test whether the bacterial strains that he had in mind could induce the breaking of tolerance.

The Beckwith and King risk scenarios were described by Wallace Rowe, while he was on RAC, in a letter that he sent to several immunologists for comment. Rowe made it clear in his correspondence of the importance of distinguishing "what is real and what is not real about risk scenarios of recombinant DNA research."[5] He categorized the concerns of King and Beckwith into two types: (i) the production of autoimmune disease as a result of altered antigenicity or recognition of host protein or (ii) the disturbance of hormone function as a result of an antibody response to the hormone. Rowe sent out the request to seven experts in the field of immunology. Baruj Benacerraf, Fabyan professor of comparative pathology and the chairman of the Department of Pathology at the Harvard Medical School, responded that both the King and Beckwith scenarios were gross exaggerations and built upon a set of hypothetical events. Benacerraf argued that immunological responses of the kind raised are difficult to obtain even in the best of laboratory conditions. Furthermore, he stated that autoimmune disease is a complex disease that was highly unlikely to result from a single antigen. The final portion of his letter responded to the remote possibility that the effects could occur. "I do not think that it is reasonable to expect that the creation of bacterial strains producing human proteins will automatically result in disease conditions. However, even if this proved to be the case, such an organism would not be a greater danger for mankind than *E. typhosa* or *P. pestis* or other virulent organisms grown in experimental or reference laboratories."[6]

Benacerraf continued that there is no such thing as a riskless endeavor. The relevant question, he said, is, Do the benefits outweigh the risks? And his conclusion for rDNA research was an unqualified affirmative.

A second respondent to the Rowe inquiry was Frank J. Dixon, the director of the Research Institute of Scripps Clinic in La Jolla, California. Dixon introduced his remarks by citing that King and Beckwith were novices in the area of autoimmune disease. Dixon maintained that there was no problem with gut antigens cross-reacting with normal human protein. It happens all the time with no immunological response; and where such a response occurs, there is little or no evidence that it is followed by pathology. Dixon picked apart King's published essay in the *Journal of Infectious Diseases*. King, he maintained, failed to make the important distinction between autoimmune immunological phenomena and a pathology called autoimmune disease. It is one thing to speak of an immunological effect, Dixon argued, but that is a long way from a pathological state.

Regarding Jonathan Beckwith's letter, Dixon was less certain of the results. He identified the positive outcome of such a scenario as remote. But Dixon did suggest a test that could lay to rest the scenario. One could administer *E. coli* with foreign protein to experimental animals and observe it for autoimmune or other pathological responses. Beckwith himself had suggested a similar experiment to resolve his concerns.

All the results of Rowe's inquiry were reported to RAC. The unanimity of those who responded to Rowe was that the King-Beckwith scenarios were farfetched. However, the risk-assessment working group of RAC did not dismiss these possibilities. In their report to the full committee, they suggested that consideration be given to arranging a workshop to address these issues. The National Institute of Allergy and Infectious Diseases (NIAID), having responsibility for the risk-assessment program of NIH for rDNA activities, responded to RAC's recommendations and convened a meeting in Pasadena, California, on 11–12 April 1980, to examine (i) direct adverse effects of hormone-producing strains of *E. coli* and (ii) the possible occurrence of autoantibodies or autoreactive cells due to the production of eukaryotic polypeptides (including hormones) by bacteria that colonize higher organisms. The workshop was chosen to coincide with the annual meeting of the Federation of American Societies for Experimental Biology.

Eighty-eight participants were invited to attend the Pasadena risk-assessment workshop, including some representatives of industry, private research institutions, NIOSH, EPA, and Congress. Participants were divided up into two workshops: (i) Peptide and Hormone and (ii) Autoreactivity Risks. The first was concerned with the production of biologically active peptides by *E. coli*; the second dealt with the potential immunological effects of those products.

The Peptide and Hormone Workshop raised questions of the following nature: If a recombinant bacterium produces a peptide, how much will it produce? What will be the fate of the peptide once it is released in a human host? If a bacterium is made to produce a human hormone, how will the quantity produced compare with what the body naturally produces?

In order to enhance the commercial value of genetically engineered bacteria, scientists have been working at increasing the efficiency of bacterial products. Ironically, as the efficiency is increased for commercial use, there is more reason to be concerned that the bacteria may take residence in a human host. Scientists at the workshop made a quantitative assessment of the effect of having insulin-producing *E. coli* in the gastrointestinal tract. The conclusions drawn were undisputed. The following is a summary of the argument.

Consider the number of *E. coli* in the human gastrointestinal tract (approximately 2×10^9, or 1 percent of the intestinal flora). At best, about 30 percent of a bacterium's protein-producing capability can be directed toward the production of a single protein. Assuming that all the *E. coli* in the gut are producing proinsulin (the precursor to insulin with a fraction of its biological activity, i.e., 1/20 to 1/50), about a half-unit of proinsulin will be produced (twenty-five micrograms). This represents only a fraction of the twenty-five to thirty units of insulin in a day that is produced by the normal pancreas. In the worst, but highly unlikely, case, which assumes that the entire gut is comprised of proinsulin-producing bacteria (aerobes and anaerobes), about 50 units of proinsulin will be produced, which, in terms of biologic activity, is still far less than the normal levels.

The outcome of these discussions suggested that there was no scenario that made insulin-producing bacteria a hazard to normal humans. The insulin calculation along with similar calculations for growth hormone were encouraging to the workshop participants; a small number felt that such calculations should be done for each worrisome product. But the conference report held that a majority felt a few calculations were sufficient to generalize about human hazards. A straw vote was taken and with only four dissenters they agreed to the following statement: "After analyzing the potential risks to man from the production of hormones such as proinsulin and growth hormone and other polypeptides by recombinant bacteria, we feel that the risks to the individual are minimal."[7]

However, the workshop participants did not wish to exclude other possibilities, which resulted in a somewhat paradoxical outcome in the minutes of the meeting. The consensus indicated little concern for individuals working with such substances, but called for an evaluation of other bacteria-producing peptides as they are developed. Workshop chairman Louis Sherwood summarized the workshop consensus in these terms: "Although proinsulin and growth hormones were not areas about which anyone was significantly concerned, there might be other peptides that might be cloned that have pharmacologic activity that will need to be evaluated. There needs to be continuous consideration of various peptides as one proceeds."[8]

In the Autoreactivity Workshop, the focus of discussion was whether *E. coli* carrying a foreign protein could induce an immunological response by the body and, second, whether the body would produce antibodies that could attack its own tissue (autoimmune disease). Participants cited many examples of autoreactive antibody responses to foreign substances, but viewed this either as rarely associated with disease or as having uncertain effect on the onset of disease.

There was no "compelling consensus," according to the chairman's report, on whether this risk scenario represents a public-health concern. There was also disagreement about the efficacy of new animal experiments that involved determining the effect of cloning in *E. coli* a gene that is likely to affect the immune system of a test animal. There were differences of opinion regarding the relevance of such data since animal systems may not react like human systems.

One of the key issues discussed was whether autoreactive antibodies induced by cross-reacting microorganisms played a significant role in producing disease. P. Y. Paterson argued that in the best-studied cases of autoimmune disease, the doubts are strong. During the discussion period, Jonathan King pressed the issue further, leading to this interchange.

J. King: Suppose next month a paper comes out which shows conclusively that . . . the antibody causes the heart damage. So that of all the papers coming out, the next paper does come out which shows the final connection, would you then conclude there is no reason for concern?

P. Y. Paterson: I'm going to answer that the way I've answered those kinds of questions for almost 30 years on the wards, for house officers and students, and in the experimental laboratories, where I've been. I don't have to consider the question because that's purely theoretical and I'm dealing with the facts before me as of today, right now at this moment, and going as far back in time as might bear on today. . . . Until it is a question that's real, there is no reason why I have to deal with it.[9]

The discussion period exhibited the tension between those scientists who wanted to see concrete experiments done to test some specific conjectures and those who felt nothing would be gained. Another point of controversy was the value of an effective medical surveillance program for industrial workers and investigators. One of those vigorously supporting a program of surveillance was John Robbins of the Food and Drug Administration: "In my opinion we will have an accident. . . . It's a good bet. In fact, I will be glad to take it [up] with anyone at any odds. But I don't think we can anticipate it . . . all we can do is to try to get the insight of those groups of people that will recognize it as soon as it is humanly possible to recognize it and react in the most responsible and clever way that we know how at the time."[10]

The workshop ended with a straw vote on two questions skillfully framed by RAC member Allan Campbell. The first question was whether experiments of the type discussed on autoantibody responses were desirable in general. The second question was whether such experiments were a necessary component of a risk-assessment program for NIH.

The first question passed handily. The second was voted down (7 in favor, 16 opposed, and 11 abstaining). The implication of the vote was that the majority found such experiments to be interesting but not especially important for risk assessment. In one additional straw vote, the body supported a motion to recommend to NIAID that they initiate discussions with the Center for Disease Control on an appropriate surveillance system.

The consensus of the workshop suggested that additional information was needed in three areas: the transfer of plasmids to anaerobic bacteria; the effect of polypeptides in both the normal and diseased bowel; and potential hazards of release of bacteria-producing peptides into the environment.

Several of those who attended the meeting reported to me that the lack of consensus among immunologists was rooted in the fact that a generally accepted model of autoimmune disease did not exist. Experts were not able to agree on the cause-effect relation for the best-studied cases. The emphasis on surveillance for some scientists was their way of signaling that the field of immunology was in its embryonic stages and that more human data were needed. But several scientists acknowledged that the problems of health surveillance were not simple. It was not always evident what one would look for, except perhaps for the obvious case in which specific phenotypic responses were anticipated, for example, when workers are exposed to bacteria that produce a human-growth hormone. Moreover, in cases where there is a long latency period between the cause and the effect and where an active peptide may only cause a chronic but low-grade disease, detection could be problematic.

When these issues were discussed at the June 1980 meeting of RAC, the complexity of the autoimmune scenario became evident. With scientists lacking confidence in the relevance to humans of animal research and with a failure on the part of the experts to reach a consensus on the value of health surveillance, an empirical resolution of the autoimmune scenario as a risk by-product of rDNA research was not a likely prospect for the near future. But NIH now felt obliged to make an official response to immunological hazards.

21

Commercial Prospects

In the early stages of the DNA controversy, scientists were balancing concerns about risks against intellectual benefits of research. But as the public gained a greater understanding of the potential risks, through increased media coverage and the sporadic flare-ups in local communities, scientists began focusing their attention on the social benefits. They cited the opportunity costs to society if this research were restrained. These costs would take the form of delayed discoveries that might result in cures for serious diseases or in developments of medically important products that the new technology was certain to spawn.[1] Some scientists and legal scholars continued to hold that the fundamental ethical posture of science should be to do no harm, which implies reducing the risks to a negligible factor regardless of the anticipated benefits. Others argued, however, that the public interest would be served by a favorable balance of benefits against risks.

The dialectic of these positions was played out at congressional hearings. Guided by their jurisdictional boundaries and historical missions, subcommittee chairmen were soon identified with one side of the equation or the other. This was a rare opportunity for society to anticipate the adverse consequences of a new technology prior to its full integration into our industrial sphere. The question remained whether society possessed the wisdom to distinguish between the positive and negative outcomes before irreversible damage occurred.

Ideas for industrial applications of rDNA techniques came simultaneously with research developments. Anticipating new commercial products, small venture-capital firms were established with the help of scientists who were leaders in the field of molecular biology. The earliest of these was the Cetus corporation, founded in 1971. In one of its promotional brochures, the company described the new commercial frontiers for genetic technology:

> What we have done in the past, while exciting and profitable, pales to what we can accomplish in the future. We have shown that we can grow, mutate and selectively screen large quantities of industrial microorganisms. But at present, all they produce is their own native

spectrum of products. Some of these are useful—we can eat or drink them, or use them in limited ways, to combat infection. By mutation we can alter them slightly, to increase yield, or create a modified antibiotic molecule. By mutating, we are hastening the small changes in DNA which are responsible for evolution. But no process of mutation or evolution will ever cause a microorganism to manufacture insulin, human growth hormone, or an antibody to save a patient dying of encephalitis. . . . Gene stiching can, and will make these things possible. Gene stiching finesses evolution; it breaks nature's species barriers. We propose to do no less than to stich, into the DNA of industrial microorganisms, the genes to render them capable of producing vast quantities of vitally needed human proteins.

By 1978 Cetus was joined by three other firms, Genentech, Biogen, and Genex, as leaders in the field of biotechnology. According to estimates given by the financial community, the combined value of these firms exceeded $500 million in 1980.[2] The four companies leading the field of genetic engineering were direct spin-offs from universities, and this transformation of the knowledge acquired from public-research grants to a technology for profit drew criticisms from some sectors of the scientific community.

Biotechnology began to attract investment from large industries. Drug companies like Hoffman-LaRoche, Pfizer, G. D. Searle, Upjohn, and Schering Plough began their own research efforts. Eli Lilly, the largest producer of insulin, entered into contract with Genentech for the development of an insulin-producing bacterium. Monsanto and Dow Chemical companies entered into contractual work with Biogen and Collaborative Research, respectively.

The prospects of a fuel-producing bacterium were responsible for Standard Oil Company's investment in Cetus corporation. On 10 June 1980 the Shell Oil Company announced it was issuing a $4 million grant to be divided equally between the American Cancer Society and the newly established Houston Foundation to finance tests of interferon for the treatment of human cancer.

The impact on academic science of the industrial influence in rDNA research was debated by David Baltimore and this author in *Nature* in January 1980.[3] In that interchange, I argued for the maintenance of a detached and free academic research community that does not serve the interests of industry. Baltimore, while expressing some sympathy for that position, argued that society has much to gain from a marriage of academic and industrial interests. However, he stipulated three provisos: scientists should be entirely open about their industrial affiliations; university laboratories should not be turned into "factories for the solution of industrial problems"; and students should be allowed to work on problems that arise from their intellectual interests rather than predefined industrial needs.

Within a period of three years, public concern over gene-splicing technology witnessed two peaks, as reflected in the attention given the issues by the mass media. During the spring of 1977, when federal legislation seemed imminent, *Newsweek* magazine ran a cover story entitled "The DNA furor: Tinkering with life." Three years later, in March 1980, *Time*'s cover story read "DNA's new miracles." The scientists had accomplished a significant sociological result—they had managed to put the issue of hazards to rest in the minds of the general public. Their attention was almost exclusively directed toward commercial breakthroughs, patenting, and sources of capital for research and development.

The range of anticipated benefits from this technology were vast, limited only by the imagination of the scientists. David Baltimore summarized this optimism to the media: "Anything that is basically a protein will be makable in unlimited quantities in the next fifteen years."[4] But making large quantities of proteins was only part of the future prospects. The rearrangement of genetic substance made it possible for the first time to rearrange species properties. Human society was now able, it appeared, to take a direct hand in the evolution of new life forms.

Probably the most widely publicized benefit for rDNA techniques is the bacterial production of medically important proteins for use in research or the treatment of disease. The statement on environmental impact issued by NIH in October 1977 offered several examples of the important products that one could reasonably expect from the technology: human insulin, human-growth hormone, clotting factors, and antibodies and antigens. Also frequently mentioned were the antiviral agent interferon, natural opiates such as beta endorphin, and vaccines. A leading figure in the field of biotechnology and former General Electric scientist Ananda Chakrabarty placed the following value on gene-splicing techniques in an essay outlining the areas of potential commercial use. "Their [referring to special proteins] fermentative production would make them available in large quantities at modest cost and thus would contribute significantly to the betterment of human health and welfare."[5]

However, the optimism over rDNA's medical value was not universally shared. The insulin case is particularly interesting in this respect. Once produced in bacteria, its payoff to society was widely acknowledged to be immediate. Estimates of the number of diabetics in the United States range from 5 to 6 million. In 1976 approximately one quarter of a million Americans received insulin. Most of the insulin taken by diabetics is extracted from the pancreas of cattle and swine. To engineer a bacterium with a human insulin-producing gene could offer a purer product and one that is an exact copy of human insulin. It promised to be of enormous benefit to the small percentage (estimated at about 5 percent) of people who react poorly to animal insulin. Moreover, it was possible, according to some experts, that the insulin demand would start to exceed its supply.

The microbial production of insulin was one of the case analyses in the National Academy of Sciences' 1977 Academy Forum. Irving Johnson, vice-president of research for Eli Lilly, concluded that there would be an overall decrease in the supply of insulin from conventional techniques, due mainly to diversion of animal pancreases away from insulin production, higher-purity products, and the 4–5 percent annual rate of increase of the diabetic population in the United States. To head off any possibility that future demand for pure insulin would exceed the supply, Johnson proposed that society take the opportunity to develop alternative sources—one of the most promising being the bacterial cloning of the insulin gene.

Harvard biologist Ruth Hubbard looked at the situation from another set of values. She held that the need for larger supplies of insulin has been widely overstated. Moreover, with most diabetics classified as of the adult-onset variety, proper diet and proper weight can frequently mitigate the need for insulin treatment. Hubbard summarized her position as follows: "But what we don't need right now is a new, potentially hazardous technology for producing insulin that will profit only the people who are producing it. And given the history of drug therapy in relation to other diseases, we know, that if we produce more insulin, more insulin will be used, whether diabetics need it or not."[6] Whether the medical value of insulin production through rDNA was what its supporters had claimed, the symbolic value of such an event was far reaching for the future of this new technology.

A drug competing with insulin in the race to be the first commercial product of the rDNA technology, but with much wider possibilities, is interferon. Discovered in 1957, interferon is a natural substance produced by the body's cells that interferes with virus production and appears to inhibit the growth of certain cancer cells. One of the conventional methods of producing the substance attests to its scarcity: It takes 65,000 pints of blood to produce 100 milligrams of the material. On 16 January 1980 scientists at Biogen announced to the press that they had engineered a strain of bacteria that expresses the drug.[7] Interferon was promoted in the media as a new wonder drug. Some even considered it as significant an antiviral agent as antibiotics are to bacteria. And with cancer as the second major cause of death in the United States, there was no lack of support for interferon's development and clinical use.[8] Another principal attraction of interferon in cancer treatment is that it naturally occurs in humans. It is also believed to be produced by the body's own defense system. Since the scientific community was showing an increasing interest in cancer as a pathology of the immune system, the race was on to begin extensive clinical trials with this drug once it was in sufficient supply. (Four months after Biogen's announcement, however, the *New York Times* reported new evidence that interferon was not the anticancer drug it had been made out to be.[9])

A third benefit of rDNA technology is the production of vaccines. The function of a vaccine is to stimulate the body's immunological response to an antigen (foreign body) without producing harm to the host. Vaccines are produced from inactive cells or defective viruses that are capable of inducing an appropriate antigenic response in the host. Viruses consist of a central core of a nucleic acid (DNA or RNA) surrounded by a protein coat. Scientists have learned that some of these coat proteins can induce an antigenic response by themselves. With rDNA technology, the DNA that codes for specific coat proteins can be isolated and amplified through bacterial cloning (or cloning in eukaryotes). By isolating the coat protein DNA from the entire virus, a vaccine can be developed for certain viral diseases that leaves no possibility for the entire virus to be produced.

The Animal Disease Center at Plum Island, New York, started plans in 1979 to use rDNA techniques to develop a vaccine for foot and mouth disease, a highly infectious virus that attacks cattle, swine, and sheep. Since the virus was classified as a class 5 etiologic agent by the Center for Disease Control, its cloning required an exemption from the Recombinant DNA Advisory Committee. Investigators at the center believed that they had identified a particular coat protein that would be suitable for such a vaccine. Safeguards were imposed by RAC to ensure that an intact virus could never be generated from the parts of the virus that left the island as cloned DNA segments. For example, when the DNA of the virus was sheared for cloning, certain active sequences considered essential to the production of the whole virus were kept on the island so that, even from a regrouping of broken segments, no intact virus could be reconstituted.

Since there have been no cases of foot and mouth virus in the United States for several decades, the immediate benefits of a vaccine would accrue to those countries that have not eliminated the disease. The United States would benefit indirectly, however, since it imports no beef from countries whose livestock is at risk.

The potential impact of rDNA technology on agriculture is great. Consider nitrogen fixation. Certain bacteria and blue-green algae have the ability to transform free nitrogen from the atmosphere into ammonia, which plants then utilize for their growth. The process is known as nitrogen fixation. The bacteria that carry out this natural process live at the root nodules of legumes such as soybeans and peanuts. Other plants, such as wheat, corn, and cereal grains, lack the root nodules at which such bacterial activity occurs. A discrete set of genes (nif genes) has been discovered that is associated with the nitrogen-fixation process.

For the plants that cannot utilize nitrogen-fixing bacteria, higher productivity has been achieved through the use of synthetic fertilizers. But the extensive use of chemical fertilizers has introduced serious environmental problems. First, its production depends upon a rapidly diminishing resource—oil. Second, the runoff of synthetic fertilizers into rivers, streams, and lakes adds unwanted nutrients that foster plant growth, deplete oxygen, and consequently cut off a life-support system for aquatic species.

Three types of suggestions have been offered for improving agricultural yield by engineering the genes for nitrogen fixation: Increase the efficiency of nitrogen fixation for the plant-bacterial systems that already possess this capacity. Develop new microorganisms with nitrogen-fixing properties. Transplant the nif genes directly into plant cells, thereby affording them the capacity for self-fertilization not dependent on a symbiotic relationship with a soil organism.

While many were citing such prospects for genetic manipulation of plants as the dawn of the second "green revolution," others were warning of technology's double edge. If the nitrogen-fixing capacity were brought to new bacteria, might they enhance the production of unwanted plants? Clifford Grobstein raises two potentially negative outcomes. Since the nitrogen-fixation process requires a heavy energy demand from the environment, its extension to new crops could conceivably reduce crop yield. In addition, nitrogen-fixing bacteria release ammonia into the soil. The excessive production of that chemical could have adverse environmental consequences.[10]

In the field of agriculture, nitrogen fixation was capturing most of the publicity devoted to rDNA technology. Meanwhile, a variety of other imaginative uses were being considered. By introducing cellulose-degrading genes into *E. coli*, which would then be introduced into humans, it was believed that people could obtain calories from vegetation that, although plentiful, was normally indigestible. Another idea under consideration was the development of protein products from bacteria that could serve as inexpensive dietary supplements.

Microbes are also being considered as protectors of the environment. For example, there are efforts to use bacteria for degrading environmentally undesirable by-products of industry into beneficial or innocuous substances. For example, in 1972 General Electric scientist A. M. Chakrabarty applied for a patent of a strain of the bacterium *Pseudomonas* that he had engineered without rDNA techniques to break down the main components of crude oil, which would make it useful in wiping up oil spills. Thought is also being given to the development of bacteria that can degrade persistent pesticides. In this way, it is believed that society can have its cake and eat it too; a spraying of a plant insecticide will be followed by a spraying of a bacterium to break down the insecticide.[11]

There is also considerable interest in the direct use of biological agents for regulating pests. Plant viruses are being studied as potential herbicides. Recalling the problems that the use of broad-spectrum chemical pesticides have introduced to the ecology, it is reasonable to ask, Will the plant virus kill off only the unwanted vegetation and leave unharmed the other biotic life?

Pheremones are hormones that regulate insect growth. Spraying these hormones during the critical stages of the insect's life cycle can inhibit maturation. Once a successful biological pest control is found, it will be left to rDNA technology to clone large quantities of the hormone. However, more efficient biocides may compete for the enormous insecticide market; an example is methoprene, a synthetic imitation of a natural insect hormone and the first commercial insect-growth regulator.

Microbial agents are being considered by the Environmental Protection Agency as one possible solution to the problem of toxic chemical wastes in land fills. Almost anything that involves a biological process is fair game for biotechnology. The mining industry has expressed an interest in the use of bacteria as part of a leaching operation to recover metals from lean ores.

The prospect of future dependence on Arab oil or nuclear power has directed the attention of energy researchers to a multitude of highly speculative alternatives, including the use of bacterial products as sources of energy. Chakrabarty speculated that rDNA technology could make the production of methane from cellulosic waste a profitable industry.[12]

Two additional applications of DNA technology in the energy field are the bacterial conversion of ethylene to ethylene glycol and the fermentation of biomass into ethyl alcohol. The hope is that the microbial converters will reduce the energy consumed, thereby enabling the processes to be achieved more cheaply than conventional methods.[13]

Reflecting a growing interest in the future role of biotechnology and energy, the Department of Energy established the Biological Energy Research program in 1979. This program, oriented toward longer-term fundamental research, had as one of its principal aims the use of microorganisms for converting biomass and other organic materials into fuels and chemicals. No one was excluding the prospect that bacteria might be created capable of synthesizing a hydrocarbon that would replace oil for fuel.

The commercial interest in biotechnology was greatly enhanced by the Supreme Court decision of 16 June 1980 in the case *Chakrabarty vs. Diamond* declaring the a microorganism can be patented *sui generis*, that is, independently of its role in a commercial process. The Court found no difficulty in classifying an engineered microbe as a product of manufacture, subject to the same patent laws as compositions of inert matter. By the time the Court issued its ruling, the US Patent and Trademark Office (PTO) had a backlog of at least 114 applications on patents for living organisms. The PTO reported to the *New York Times* that patent applications on all aspects of genetic engineering were mounting to about fifty per year.[14]

The Supreme Court's ruling came eight years after A. M. Chakrabarty filed a patent application on behalf of the General Electric Company for a strain of *Pseudomonas* that was capable of degrading crude oil.[15] The organism was promoted as an environmentally safe way to clean up oil spills. Announcement of the decision, however, fueled a new wave of controversy in the scientific community. Questions were raised about the private appropriation of genetic resources, the effect of patenting on the free flow of scientific information, the commercialization of basic research and the patenting of higher life forms.[16]

Most universities favored the decision, viewing it as an opportunity to cash in on basic research and at the same time reduce their dependence on government funding. The patenting decision additionally strengthened the financial community's feeling that the biotechnology industry was a sound investment opportunity. That conclusion was further reinforced by the results of a study released by the Office of Technology Assessment (OTA), a body of the Congress, entitled *Impacts of Applied Genetics*. One of the principal outcomes of the OTA report was a set of options to Congress for promoting commercial development of biotechnology.[17] Thus, the investment fever for genetic engineering was rising as the concern for the regulation of the new technology was diminishing.

22

Local Initiatives for Regulation

Between 1975 and 1979 nine cities and towns had considered proposals for an ordinance or a resolution on rDNA research. Legislation was also considered in seven state assemblies. Each debate that took place at the local level was a microcosm of the national controversy. But not just any city or town considered the issues. Certain conditions were satisfied before the controversy took hold at the municipal level. Generally, it started within a university or institutional setting where planning was underway to upgrade facilities for gene-splicing work. Here is a typical scenario.

Resident biologists or other university faculty, aware of the moderate-risk laboratories under construction, spearheaded a university debate. Newspapers carried the events, sometimes at the request of DNA critics, whereupon the stories drew attention from local officials. Finally, a group of scientists offered support and technical information to elected representatives. Without the perseverance of some group, however loosely assembled, that was able to persuade city and town officials of its credibility in this area, it is very unlikely that local governments would have tackled the issues.

In the cases where local communities did carry out their own investigations of the risks, the final recommendation on research procedures did not depart significantly from the NIH *Guidelines*. The biggest departures occurred in broad policy areas or in the philosophy of environmental protection.

The first nationally publicized debate over whether rDNA research should be permitted at an academic institution took place at the University of Michigan at Ann Arbor. The history of the debate goes back to 1974 when an informal group calling itself the Ad Hoc Committee for Microbiological Safety formed at the medical school. After the Berg letter appeared in July, this group sent a request to the vice-president for research calling for a study of the broad question of potentially hazardous microbiological research at the institution.[1]

In December 1974 seven faculty from three different colleges were constituted into the Committee on Microbiological Research Hazards with a charge to prepare a policy statement and establish guidelines for research activities. Some members of the committee attended the Asilomar Conference in February 1975. The report of the committee (submitted in April 1975) recommended that the university take full advantage of its scientific resources in molecular biology and support the development of rDNA technology. The committee estimated that within several years "a large number of basic scientists" at the University of Michigan will need to use materials and processes that were then considered by NIH to be too hazardous for conventional laboratories. To meet this need, the committee recommended that the university renovate three laboratories.

The committee report was reviewed by the Biomedical Research Council at the university. The council considered the report deficient for failing to treat the problem of biological research broadly enough. The council suggested a tripartite committee structure to look into the general area of biomedical research with potentially hazardous agents. The recommendation was approved with the following mandates for each committee: committee A was to examine the safety of physical facilities; committee B, to consider broad policy aspects of the research; and committee C, to have an ongoing responsibility to ensure that safety standards were met for individual laboratories.

Committee A, consisting of eight persons in the biological sciences and termed a technical committee, began its work in May 1975, thirteen months prior to the issuance of the first *Guidelines* by NIH and the origins of the controversy in Cambridge, Massachusetts. In contrast to committee A, none of committee B's eleven faculty were microbiologists. Its members represented diverse disciplines, including English, history, theology, and philosophy, law, and social work. Committee B, which began its work in September 1975, considered a range of policy options for rDNA research at the university, including its prohibition. A month after committee B began its deliberations, the university applied to the NIH for a laboratory renovation grant in the amount of $300,000. Although the administration claimed it would not spend the funds for renovation until after committee B's report was completed, it is most unusual for a university to turn down such funding.

The role of committee B was to examine the broader social and ethical issues in having rDNA research at the university. Since its membership did not include those who would be involved with the work, one would expect that the lure of the new research would not figure critically in this group's decisions. However, it appears that university officials did not leave the decisions of the committee members to chance. In June Goodfield's account of the events surrounding the committee selection, she offers evidence that shows the committee was stacked by the microbiologists with those who would be sympathetic to the research.[2]

The results of committee B's recommendation, issued in March 1976, anticipated all subsequent decisions by university and municipal advisory groups. It held that the research using rDNA technology could proceed with appropriate controls. Only a single member of the committee dissented. In his minority opinion, Shaw Livermore, professor of history, stated that the research should not be permitted at the university. He grounded his decision on his belief that this technology would give to society a "dramatically powerful means of changing the order of life." A strong influence on Livermore was biologist Robert Sinsheimer, whom he had quoted in his minority statement.

While the committee deliberations were still under way, opposition to the research at the university began to mount, but not from biologists *per se.* There were a few biologists who had strong reservations about the university's decision to proceed with the research under the NIH *Guidelines*, but, despite their differences, they chose not to become embroiled in a vigorous battle with the university. The group leading the opposition consisted of faculty from the Departments of Sociology, Physics, Mathematics, English, and Humanities. The first action of this group was to question the Board of Regents' decision to approve $300,000 for the construction of three P3 facilities in advance of a recommendation from committee B. They offered technical arguments to illustrate the types of issues that had not been addressed by committee B. To support its case, the ad hoc faculty group presented testimony and articles from biologists in other parts of the country. Letters from nationally prominent scientists were solicited for the occasion. In a letter to University President Robben Fleming, Robert Sinsheimer emphasized the uncertainties of creating novel microorganisms: "Few people that I know would argue that one could not, by these techniques, design some quite fearsome microorganisms, capable of initiating deadly epidemics. Are such likely to arise inadvertently? My view is that one cannot honestly calculate the likelihood."[3]

The decision-making process at the University of Michigan took over a year and included a variety of educational modalities. Frederick C. Neidhardt, chairman of the university's microbiology department, described some of these in his statement to the Board of Regents: "Discussions have taken place in meetings of school faculties, in departmental meetings, in a series of noon-time seminars and panel discussions sponsored by the University Values Program, in a 30-hour interdepartmental course (Cell and Molecular Biology 594) attended by 150–200 people, in numerous committee meetings, including the open meetings of Committee B, in a two-day Recombinant DNA Forum, in numerous feature articles, newstories, and letters to the Editor of the Ann Arbor News, the Michigan Daily and the University Record."[4]

The final decision on the status of the research rested with the Board of Regents. The resolution it issued in late May of 1976 came after it had arranged its own miniforum (May 12). To no one's great surprise, the board backed the report of committee B (by a vote of 6/2/1). Through its decision, the university ostensibly adopted the NIH *Guidelines* with only minor modifications.[5]

In contrast with other events of this nature, it is important to note that no biologists at the university were persistent outspoken dissenters of the report. Furthermore, some university administrators interpreted the entire episode as a waste of precious university resources, making it clear how they felt about the participatory process. Alvin Zander, vice president for research, spoke of the year's events as "elaborate and expensive redundancies."[6]

City officials in Ann Arbor kept a watch over university proceedings; indeed, minutes of important meetings were sent to the mayor. A few days prior to the Board of Regents' decision, the Ann Arbor City Council took up and tabled a motion that requested an environmental impact statement from NIH on the research. That was one way of gaining some time and educating the public about the differences in scientific opinion. There was virtually no participation by the city in the university decision. Speaking before a House subcommittee, Ann Arbor Mayor Albert H. Wheeler pointed out that the city government did not request, nor was it invited to share with the Board of Regents (a public body) in, the decision over rDNA research. Citing this as a problem in the process, Wheeler stated, "I believe that the value of such a partnership is that the approximately seventy thousand local residents, not directly affiliated with the university, have important rights, considerations and obligations that should be protected in an atmosphere of cooperation."[7]

But the mayor had done little more than explore the possibility of a local ordinance. The historical conditions for a university-city conflict did not exist. Municipal and university officials had good cooperative relations on most issues of mutual concern. The mayor of Ann Arbor had even received his doctorate in public health and microbiology from the University of Michigan. In this instance, a university was able to contain the issues within its own borders. But it was only a matter of time before some community would challenge a university's decision.

As the debate at the University of Michigan was coming to a close, the seeds of the next local eruption were being planted in Cambridge, Massachusetts, the home of Harvard and MIT. What had occurred in Ann Arbor did not precipitate the Cambridge controversy, which had its own independent gestation period. The events in Cambridge have been chronicled in several essays and newspaper feature articles.[8] More than anywhere else, the events in Cambridge dramatized the role of public involvement in the debate. But it was also easy for the media to romanticize the process through which the city arrived at its own decision. It was the body politic against the intelligentsia, the ordinary folk against the elites. There was far more at stake than rDNA research. A decision-making apparatus was on trial. Should the citizenry defer decisions on the risks and benefits of new technologies to scientific experts? And if not, is there another way? What is most significant about the Cambridge episode is not the outcome, but "the other way."

The context of the Cambridge and Ann Arbor events differed in some important respects. First, unlike the University of Michigan, there were biologists at Harvard and MIT who strongly opposed the use of rDNA technology and were willing to take their case to the community. Second, there was a core group of scientists in the Boston area who were involved in studying and writing about issues of genetics and social policy. These individuals constituted the loosely formed Genetics and Social Policy Group (GSPG) of Science for the People (SESPA). By the time the DNA debate was about to break in Cambridge, the GSPG had already taken on issues such as genetics and intelligence, the XYY male, and genetic screening.

When the rDNA controversy began to surface in the scientific journals, members of Boston's SESPA argued about whether the issue warranted the organization's involvement, and if so, what action should be taken. Eventually GSPG split on that question. One group continued to direct its work on science advocacy toward issues of ideology and science, especially the uses of genetics for social control. A spin-off group formed exclusively on the rDNA issues.

In addition to SESPA, there were spontaneous informal associations that coalesced to review drafts of the *Guidelines* and the environmental impact statement. Scientists from Harvard Medical School and Brandeis University formed the Boston Area Recombinant DNA Group, which drafted a widely circulated critique of the Woods Hole version of the NIH *Guidelines*.[9]

A third key factor in setting the context of the Cambridge events was the fragile relation between the major universities and the local residential communities. The universities are a major market for jobs in the area and are quick to point out the cultural contributions they provide the community. But community residents are cynical about university real-estate policies and their expansionist goals. Real estate purchased by universities is removed from municipal tax rolls, thereby placing additional economic hardship on the large numbers of lower-middle-income homeowners. Residents also compete with university students for apartments. By the mid-1970s, the vacancy rate for apartments in the area was in the 1% range—a widely accepted indicator of a rental housing crisis. The rent control ordinance passed by Cambridge affected all multiple-dwelling apartment units, including the substantial number owned by Harvard University.

In sum, the conditions for an anti-rDNA campaign existed in Cambridge in the form of widely perceived differences in interests between the universities and city residents. But it did take a few sparks to set off the explosive events. One of these came when a few Harvard biologists, among them Ruth Hubbard, raised some opposition to the renovation of a university biolab. It was being redesigned to meet the requirements of a moderate-risk (P3) physical-containment facility. Both viral and rDNA research were planned for the facility. Hubbard was more concerned about gene-splicing experiments than the work with infective viruses. She outlined her reasons for opposing the P3 facility in a letter to the chairman of Harvard's Biohazards Committee in April 1976:

> Much more worrisome to my mind than the work with viruses is the work with recombinant DNA. Viruses are more or less known quantites and the Biohazards Committee can restrict work to those it

> deems reasonably safe—say as safe as walking into a crowded lecture hall during flu season. But recombinant DNA is another story, because nobody can predict how genes will act in the vicinity of genes they have not been next to before. . . . I am frankly terrified at the idea of playing evolution, because it is an exceedingly dangerous game whose rules we will not know until it is too late.[10]

Harvard's faculty-wide Committee on Research Policy responded to the criticism by holding an open meeting. Harvard historian of science Everett Mendelsohn attributed the committee's open-meeting policy to the faculty's attitude toward classified defense research at universities: "A product of the late 1960's when students and faculty were questioning such issues as Defense Department funding of University research, the Committee on Research Policy had adopted a somewhat participatory mode of operation. Their expectation was that expert and lay persons alike should have the opportunity to be heard."[11]

Hubbard took advantage of the open meeting by alerting the local press and a city councillor. The events of the meeting were subsequently reported in a weekly newspaper, the *Boston Phoenix*. The lead on the article read "Biohazards at Harvard."[12]

There is some disagreement in various accounts of the events about how the issues moved from Harvard to the city council, especially the interaction of local scientists with local officials. My own review of the evidence leads unambiguously to the conclusion that the articulate and flamboyant mayor of Cambridge, Alfred Vellucci, read the *Phoenix* story and on his own initiative decided to bring the issue before the council. Once Vellucci's interest in this issue became known, Ruth Hubbard decided to provide the council members with some reading matter. Until this time, Hubbard's husband and Harvard colleague, George Wald, had taken a low profile on this issue. Much of his time had been devoted to the nuclear-arms build-up and radiation hazards. Wald was not inclined to start an intense involvement in a new controversy. But when he accompanied his wife to City Hall to drop off some reading matter and was invited in to talk to the mayor, he agreed to testify along with her at the council hearings on the dangers of the research. That started Wald's active involvement in the debate. In contrast to some accounts, Wald did not come to City Hall with the purpose of influencing Mayor Vellucci.[13] Vellucci wasted no time in informing the media about Wald's participation in the upcoming council hearings.

The hearings were held during two special evening sessions of the council, on June 23 and July 7. Visitors spilled over from the chambers into the halls and heard the events through special speakers that were installed for the overflow crowd. The mayor set forth the rules for the hearings and reminded the audience that "no one person or group has a monopoly on the interests at stake. Whether this research takes place here or elsewhere, whether it produces good or evil, all of us stand to be affected by the outcome. As such, the debate must take place in the public forum, with you the public, taking a major role."[14]

No one who was at the hearing will forget the mayor's imperative to the scientific speakers: "Refrain from using the alphabet. Most of us in this room, including myself, are lay people. We don't understand your alphabet, so you will spell it out for us so we will know exactly what you are talking about because we are here to listen." What a simple but powerful way of expressing the idea that scientists had the responsibility for communicating technical concepts in ordinary language.

NIH sent a special delegation to the hearings, including Emmett Barkley, director of the Office of Research Safety, and Maxine Singer, who carried a copy of the newly issued *Guidelines*.

The hearings did not reveal anything new, but for the first time the public had control of the forum. It was this dramatic displacement of authority that caused considerable unease among university researchers and administrators. The self-governance of science was concretely and symbolically threatened at this town meeting. The public approached this controversy with a skeptical attitude toward science and technology. It was easier, in some ways, to conjure up scare scenarios than to offer a defense of the research. Scientists who chose to defend the safety of their work made use of an interesting set of arguments and analogies. I have selected a group of these from the Cambridge City Council hearings. The fragments are paraphrases from the testimony. Reference to the author, transcript volume, and page number is given in square brackets at the end of each fragment.

It is misleading to refer to the bacteria which have been altered with new plasmids and bits of foreign DNA as altogether new organisms. [Mark Ptashne, V.I, 16] Over the past two years millions of bacterial cells carrying pieces of foreign DNA from higher order organisms have been constructed in the laboratories around the country and not a single dangerous organism has ever been produced. [Ptashne, V.I, 24] It is extraordinarily unlikely that the addition of a small piece of foreign DNA could impart to *E. coli* K12 strains the ability to survive in intestinal tracts, cause disease, and be transmissible to other animals, humans or plants. [Ptashne, V.I, 25]

The degree of risk involved in carefully regulated recombinant DNA experiments is less than that in maintaining a household pet; it is certainly less than that in maintaining a room of mice, hamsters, monkeys, and certain other animals, all of which are known to be carriers of certain human pathogens. [Ptashne, V.I, 27] If we were warranted in putting a stop to recombinant DNA experiments, then on the same logic we should put a stop to experiments involving animal viruses, animal cells, carcinogens and mutagens—this would signal the end of biomedical research. [Ptashne, V.I, 27] The risks are hypothetical . . . they are less than the typical kind of risks you meet everyday in walking across the street [Ptashne, V.I, 97]

The biologically contained organisms will provide the level of safety almost independently of how badly they are treated. [David Baltimore, V.I, 195] The danger of turning *E. coli* into some kind of a pathogenic monster by putting fruit fly genes or frog genes into it is very much less than the risk of eating a washed radish or a raw carrot or a piece of lettuce from your garden, or even drinking a glass of water from your well. Any of these will feed you thousands of *E. coli*. [A. M. Pappenheimer, V.I., 203] Even when we put genes of known pathogenicity into *E. coli* K12 we cannot convert it into a disease-causing agent. [Stanley Falkow, V.II, 159] The special requirement for containment facilities for doing rDNA work is like the third cable of an elevator or like putting fire sprinkler systems through buildings. You are safer if you put the systems in even if the building is fireproofed. [Walter Gilbert, V.II, 247]

The fragments illustrate the use by scientists of a logic of comparative risk to allay the public's fears, in a manner consistent with the mayor's request—to speak in the idiom of the people. However, the extreme polarization of scientists who testified about the risks was disturbing to most members of the council. The mayor proposed a two-year moratorium of all rDNA work in the city, regardless of where it fell in the NIH taxonomy of DNA exchange experiments. But the proposal was voted down. In a compromise move, the council decided to form a citizens committee to review the question of hazards to the city from gene-splicing experiments. That was the birth of the Cambridge Experimentation Review Board (CERB). This citizens committee consisted of seven Cambridge residents, one physician who practiced in the city, and the commissioner of health and hospitals who served *ex officio*.

Lear's account of the selection of the citizens for CERB commits the fallacy of *post hoc, ergo propter hoc*. He looks at the outcome of the choice and attributes intentionality on the part of the city manager. His account suggests that a grand design was used to fulfill geographical, ethnic, religious, racial, and sexual balance.[15] To be sure, the final outcome was a good mix; but the candidates chosen did not represent all the well-defined neighborhoods in the city. Some minority groups, such as the Portuguese and Hispanic sectors, were not represented, and one of the members was mistakenly identified as a Frenchman. The city manager was certainly conscious of diversity; he eschewed a committee of scientists since they were feuding among themselves. He wanted people whom he could trust and who could offer something unique to the process. Some of the board members were chosen because of professional expertise (medicine, structural engineering, nursing); others were selected because they were known and respected in the community.

As a participant on that committee, I have some personal reflections on how this extraordinary experiment in citizen assessment of social risk turned out. I shall begin with the learning process. Just as members of the committee represented very diverse backgrounds, the educational levels of the citizens also varied considerably. It was no surprise that the two medical people on the committee exerted a powerful influence because they possessed some knowledge of biochemistry and genetics. No planned or systematic attempt was made to educate the citizens or for some of the more knowledgeable members of CERB to share what they knew with others. It was the responsibility of each individual, through the extensive reading materials distributed, to reach an understanding of the issues. Members gravitated toward those problem areas with which they felt most comfortable.

I can recall the occasion at which I began to understand the rudimentary mechanisms of DNA expression. To share those ideas with other CERB members, from time to time I would request of the scientists who testified before the committee to explain the workings of molecular systems by using simplified models. My hope was that if we citizens achieved a certain level of technical understanding and were able to comprehend the sources of disagreement among the scientists, we could render a reasonable judgement from a public-interest perspective. One of the strategies I used was to play the role of cross-examiner to scientists on both sides of the issue. The same questions were repeatedly raised: In what ways was the rise of synthetic biology going to be different than the rise of synthetic chemistry? If genes are displaced from one system to another, can we expect emergent properties to result, or will the genes do what they usually do or nothing at all? Can *E. coli* K12 be transformed into a pathogen? What kind of exchange of genetic information can we expect to achieve from *E. coli* K12 to other hosts?

The meetings were held twice weekly starting at 4:30 P.M. and lasting until 7:00 P.M. One of the weekly sessions was devoted to hearing testimony from either a supporter or a critic of the NIH *Guidelines*.

I was eager to see that the most persuasive cases were presented on both sides. Harvard and MIT sent in delegations of scientists before the committee, accompanied by packaged resumés and statements of introduction prepared by their respective community-relations departments. Funds to invite scientists from outside the Boston area were not available.

In retrospect, it was too simplistic to divide the controversy into two camps. But that division was fostered by both the city council and the CERB hearings. Actually, there were significant variations in the critics' positions. But those who testified as proponents of the P3 facility and the NIH *Guidelines* presented a remarkably united front. If there were any differences in their views about the *Guidelines*, they were imperceptible or of little consequence. I made two contacts with critics, one prior to and one after an appearance. I asked Jonathan King to provide a concrete example of how gene-splicing experiments might cause an adverse effect. And when Jonathan Beckwith testified, his youthful demeanor convinced me that the committee would not be able to distinguish him from a graduate student. So I asked Beckwith, who has made significant contributions to the field of molecular genetics, to send a copy of his resumé for distribution to the committee.

Many people were curious about whether there was any external pressure placed on the committee. Prior to the decision of the CERB, there was no attempt of which I was aware to influence the outcome. Considerable lobbying by scientists and university officials took place after CERB's decision was issued, when the city ordinance was being drafted.

The event that best dramatized for me the degree to which CERB's deliberations were being watched was the agreement by NIH to hold an open-line telephone conversation with the committee. Director Donald Fredrickson convened a panel of four scientists at NIH headquarters in Bethesda, Maryland, to respond to questions put by the Cambridge citizens. The conversation, which lasted for over an hour, left a strong impression on Fredrickson, who subsequently referred to the incident in an article.[16] While no single event in the conversation was critical in shaping the outcome of the board's decision, the mere fact that the citizens were in direct contact with federal authorities added a degree of confidence.

Observers of the Cambridge events questioned the committee's potential for recommending an outright ban on the research. Some critics viewed Cambridge as a testing ground for grinding rDNA research to a halt. The conditions for such an outcome did not exist in Michigan. No one seriously expected a university to turn its back on all that research funding. But an independent citizens committee had no personal stake in supporting the research. Only one member of the committee expressed any religious opposition to rearranging genetic matter. That individual decided not to use her religious beliefs as the basis for a recommendation. The two physicians appeared to have decided quite early that a ban was not warranted. A few members, including myself, appeared to have kept the option of a ban open until CERB held its citizens court.

Early in our committee's investigations I introduced the concept of a citizen court to my colleagues. My goal was to draw attention to the unique role that a lay body can play in reaching a policy decision involving disputed scientific issues. The idea was acted on when, as part of the culmination of our inquiry, CERB held a courtlike proceeding with scientists as advocates arguing their case before the citizens. Initially I had suggested a grand two-day affair with scientists flown in from around the country as witnesses and with an opportunity for citizens from all parts of the city to attend the mock trial. Some members of the committee expressed displeasure at giving up so much time on a weekend. City officials did not delight over the idea of paying the transportation costs of out-of-town guests.

The compromise event took place one evening over a five-hour period in a small guest dining room located in the Cambridge Hospital. The facility could squeeze in thirty to forty people. There were two teams of scientists. Each team was permitted to have guest witnesses. The CERB members alternately questioned each team, with time allotted for rebuttals. There were periods when the debate between team members degenerated into *ad hominem* attacks until it was brought back on course. Each side had its Nobel laureate, David

Baltimore defending the NIH *Guidelines* and George Wald addressing its weaknesses. This was the single opportunity that members of the committee had to cross-examine scientists while listening to the responses of the opposing side. It was also an opportunity for the adversaries to bring in new experts to persuade the citizenry.

When the board was ready to begin its deliberations, I suggested the use of a process through which each person's concerns and suggestions would be given equal consideration. I worried that the authoritative position of the medical people would overshadow the suggestions of other members. A scribe recorded the suggestions of each board member, on large sheets of newsprint paper plastered around the room, and the complete list of recommendations was typed and distributed for consideration at the next meeting.

At this point it was clear that there would be no support for a research ban. When the committee met to decide on whether additional conditions would be required for the research to be done in the city, there was an eagerness to arrive at a consensus. If for no other reason, CERB members wanted to terminate the long and exhausting schedule of meetings. The committee spent over 100 hours listening to testimony and carrying out its deliberations. Once again I was reminded of a jury trial in which jurors reach unanimity through a process of wearing one another out. In this case, the committee members reached convergence with relative ease. The final recommendations included a city biohazards committee with oversight responsibilities for both NIH and private rDNA activities; a program to monitor the survival and escape of recombinant organisms; the requirement that all experiments undertaken at the level of P3 physical containment use an EK2 host-vector system; adequate screening to ensure the purity of the host strain and testing of organisms resulting from rDNA experiments for their resistance to commonly used therapeutic antibiotics.

The CERB gave no attention to the feasibility of implementation. It received conflicting opinion on whether rDNA organisms could be monitored and left that responsibility to the good faith of institutions. The outcome of the board's deliberations was a set of requirements superimposed on the NIH *Guidelines* and the addition of another layer of bureaucracy at the municipal level in the form of the Cambridge Biohazards Committee.

In its deliberations, the committee made no effort to address the ethical issues associated with the research. At the very outset it was agreed that our charge was restricted and that dealing with the public health issues was more than enough work. But the importance of those problems was emphasized in the introduction of our report to the city council:

Throughout our inquiry we recognized that the controversy over recombinant DNA research involves profound philosophical issues that extend beyond the scope of our charge. The social and ethical implications of genetic research must receive the broadest possible dialogue in our society. That dialogue should address the issues of whether all knowledge is worth pursuing. It should examine whether any particular route to knowledge threatens to transgress upon our precious human liberties. It should raise the issue of technology assessment in relation to long range hazards to our natural and social ecology.

The most widely cited declaration of the CERB report was on the issue of free inquiry:

While we should not fear to increase our knowledge of the world, to learn more of the miracle of life, we citizens must insist that in the pursuit of knowledge appropriate safeguards be observed by institutions undertaking this research. Knowledge, whether for its own sake or for its potential benefits to humankind, cannot serve as a justification for introducing risks to the public unless an informed citizenry is willing to accept those risks. Decisions regarding the appropriate course between the risks and benefits of a potentially dangerous scientific inquiry must not be adjudicated within the inner circles of the scientific establishment.[17]

The report issued by CERB was acted upon by the city council in February 1977 through the passage of the first DNA legislation in the United States. The Cambridge events touched off a series of local and state activities. By 1981 local ordinances were also passed in Princeton, New Jersey, Amherst, Boston, and Waltham, Massachusetts, and Berkeley, California; state legislation was enacted in New York and Maryland; resolutions were passed in the city councils of Emeryville and Del Mar, California; bills were introduced in the state assemblies of Wisconsin, Massachusetts, New Jersey, California and Illinois. Table 7 summarizes the key provisions of six pieces of rDNA legislation.[18]

The events in San Diego, California followed a path very similar to those in Cambridge, with the exception that no legislation resulted. In July 1976 the San Diego campus of the University of California (UCSD) began discussions with the mayor's office on a planned P3

Table 7
Comparison of Provisions in rDNA legislation for two states and four municipalities

Locality	Licensing or certification	Special office(r)	Special committee	Citizen participation on IBCs	Penalties	Liability	Health monitoring	Containment requirement identical to NIH	Additional containment requirements	Preemption	Enforcement provisions	Protection of proprietary information
Cambridge, MA			X	X	X		X		X			
Berkeley, CA	X						X		X		X	
Princeton, NJ		X	X	X		X	X	X				
Amherst, MA				X	X			X				
New York State	X	X			X		X	X		X	X	X
Maryland	X		X		X	X		X			X	

Source: Sheldon Krimsky, "A comparative view of state and municipal laws regulating the use of recombinant DNA molecules technology," *Recombinant DNA Technical Bulletin* 2(3):121–125 (November 1979).

facility. Anticipating major public interest, UCSD officials supported forums to educate the public. The mayor of San Diego instructed the chairman of its Quality of Life Board to conduct investigations. The San Diego DNA Study Committee was comprised of eight individuals, only two of whom were not MDs or PhDs. Like the CERB, this committee also acknowledged that the new technology raised ethical issues but chose to bracket those questions from its deliberations.

In its eighteen-page report, the committee cited nine areas of weakness with the NIH *Guidelines* and the current regulatory process. These included the failure of mandatory coverage of non-NIH-funded research, lack of standards to measure the effectiveness of local biohazards committees, the conflict in having NIH both the sponsor and regulator of the research, and its inability to inspect all sites. The DNA Study Committee recommended a city ordinance that would extend the *Guidelines* to private firms and called upon UCSD to keep accurate health records of its employees engaged in rDNA work. The report specified no changes in the technical demarcation of experiments, but like Cambridge and the University of Michigan it called upon the city to prohibit P4 experiments.

One member of the board who had no medical or academic affiliation offered the sole dissenting opinion. Shirley Wood cited the views of Burke Zimmerman and Robert Sinsheimer. While Ms. Wood was encouraged by the containment provisions, she was disturbed by the refusal of leading scientists to legislate their use. Her reference was to Paul Berg, who had consistently opposed legislating the *Guidelines* lest they be resistant to change. For Berg, the *Guidelines* were a living document—a first-order approach to the problems as initially framed that would require continuous refinement and modification as new information became available. Thus, with only one of eight of the study committee dissenting, the San Diego City Council endorsed the recommendations on 30 March 1977.

Another locale whose citizenry became involved in the aftermath of the Cambridge and San Diego events was Princeton, New Jersey. By December 1976 the ten-member Princeton University Biohazards Subcommittee had issued a sixty-seven-page report on rDNA research that drew attention to existing arguments among major scientific adversaries while providing its own commentary. The report even considered the scenario of the "Armageddon virus, which spreads lethally and endlessly." Citing this scenario as unlikely, the report stated that this kind of biohazard cannot be altogether dismissed.[19]

The Princeton University report departed from the NIH *Guidelines* in two significant respects: (1) It viewed biological containment to be a more effective barrier than physical containment, thereby rejecting the equivalence principle, according to which additional levels of physical containment can be traded off for lower levels of biological containment. (2) It rejected the distinction between nonembryonic and embryonic or germ-line tissue, the effect of which was that more experiments would be required under P3 conditions.

On 24 January 1977, nearly two months after the university report was issued, a committee was formed by the borough and township of Princeton to study the hazards on its own terms. The committee, a mix of scientists and lay people, produced its finding in May 1977 after holding its own investigations and hearings.

The final recommendations of the Princeton committee consisted of a patchwork of majority and minority opinions. Three members voted against building a P3 laboratory at the university. Among the reasons they gave: the broad ethical, social, and political implications of the research had not been thoroughly explored. By 9 March 1978, Princeton had enacted into law a municipal biohazards committee and a requirement that researchers be covered by liability insurance of at least $5 million for personal injury.

The committee's findings typify a number of trends in the local responses to the DNA controversy. First, each review process by a university or a local committee added something to the NIH *Guidelines* or to a prior review carried out by a university. But in no case did the majority view of these committees depart radically from the NIH *Guidelines*. The final results were always cosmetic touches on the prevailing rules. Second, scientists, physicians, and public-health officials serving on these committees were more likely than lay members to accept the views of the scientific orthodoxy. Nonscientists, however, seemed more inclined to be influenced by critics of the current NIH policies. They paid more attention to the potential hazards, even if they considered the probability of occurrence to be low, while giving comparably less attention to anticipated benefits. Third, certain shortcomings in the NIH process were universally accepted, such as their limited application to government-funded work. Fourth, none of the review processes tackled so-called ethical issues, although they were recognized by some groups as important.

By the latter part of 1978 it was becoming clear that the domino effect of local ordinances would not materialize. Attention had shifted to national legislation, and scientists turned from local involvement to the congressional arena to continue the struggle for control of their discipline.

23

Federal Legislation

During the summer of 1976, while news of the events in Cambridge were circling the globe, the momentum for national legislation to regulate rDNA activities in all sectors was growing. By the fall, the staffs of key House and Senate leaders in the 94th Congress were beginning to formulate bills. The most intense legislative effort took place in the first session of the 95th Congress. The session began 4 January and ended 22 December 1977.

In the late spring and early summer of 1977 most of the respected science-policy observers considered some form of legislation a near certainty. Even the scientists who felt that the passage of a federal law was a major setback to free scientific inquiry participated in the national debate by embracing the assumption that some bill would eventually emerge. Hence, they argued for the least restrictive legislation. By the early fall of that year, there was a turn of events. Congressional activity began to lose its vigor and focus as intense lobbying from scientists fostered doubts in the minds of key congressional supporters over different elements of the leading pieces of legislation.

Despite the fact that more than a dozen bills were introduced in both houses of Congress for regulating rDNA research, none was brought out of committee during the first session of the 95th Congress. The second session witnessed a resurgence of interest in rDNA legislation. A new compromise bill, which was drafted by House leaders, received support from several key scientific societies. It was favorably voted out of committee but never taken up by the House Rules Committee, which kept it from reaching a floor vote. What had begun as a colossal legislative cry to regulate gene-splicing technology turned into a minor whimper by the start of the 96th Congress.

Why, when so many keen observers had been convinced that a DNA bill was inevitable, had it failed to materialize? An answer to this question, requires tracing the legislative history of the rDNA episode from 1976 through 1980.

Several factors had prompted congressional legislative interest: accelerated patent decisions on discoveries in molecular genetics; the prospect of a patchwork of local legislation; media attention; lobbying by environmental groups; the efforts by a score of scientists to keep the issue alive; and finally, some key articles in the right publications.

Countervailing factors were intense lobbying on the part of scientific leaders and academic institutions; failure of prolegislation forces to maintain the pressure on Congress; the skillful use of scientific information to persuade congressmen that legislation was unnecessary; and the deployment of Lysenkoist arguments to cool the legislative fever. Because the rDNA issue touched upon several areas, including the risks to public health, benefits to society in such vital fields as health, energy, and agriculture, and the maintenance of the United States's lead in molecular biology, there were different committees in Congress that sought jurisdiction or oversight of a DNA bill. The action of one subcommittee to regulate was met with a reaction by another to protect science from direct government oversight. This helped to slow down the legislative process. Moreover, administrative support for legislation was half-hearted and lacked strong presidential backing. A combination of these factors, in conjunction with the competition that DNA legislation faced from other compelling bills of the 95th Congress, kept it from reaching a floor vote.

Senator Edward Kennedy had a head start on the rDNA controversy with the hearings he held in the spring of 1975. By the summer of 1976, when Cambridge held its own hearings on the proposed P3 facility at Harvard, the senator's office was receiving inquiries and letters criticizing NIH's handling of the regulatory aspects of genetic manipulation. On 4 June 1976 *Science* published two letters sequentially, covering two full pages. They criticized the path NIH was taking to ensure that gene-splicing research would be carried out safely. Erwin Chargaff questioned the wisdom of using *E. coli* as the preferred host and discredited the architects of the new guidelines as "advocates of this form of genetic experimentation."[1] Chargaff called upon Congress to act and cited four proposals to guide public policy: (1) prohibition of cloning experiments using hosts that are indigenous to man; (2) licensing such research by a public authority; (3) establishing all forms of genetic engineering as a federal monopoly; and (4) carrying out all research in high-containment laboratories, such as the Fort Detrick facilities.

Chargaff ended his letter with a warning to the scientific community that gene splicing may prove perilous to humankind and posing for consideration the assertion that science has a moral responsibility to future generations. Speaking about fusing eukaryotic genes in prokaryotes, Chargaff stated, "Have we the right to counteract, irreversibly, the evolutionary wisdom of millions of years, in order to satisfy the ambition and curiosity of a few scientists? This world is given to us on loan. We come and we go; and after a time we leave earth and air and water to others who come after us. My generation, or perhaps the one preceding mine, has been the first to engage, under the leadership of the exact sciences, in a destructive colonial warfare against nature. The future will curse us for it."[2]

The second letter was written by Francine Simring, representing the Committee for Genetics of Friends of the Earth. Simring urged that the risks and benefits of this technology be assessed by those who do not have a stake in its outcome before the intellectual and economic investments grow. Like Chargaff, she also called for federal legislation for gene splicing, citing parallels between the developments of nuclear energy and genetic engineering.

A month later, *Science* carried a rebuttal to the Chargaff and Simring letters by Maxine Singer and Paul Berg.[3] They summarized the long and deliberate effort made by scientists to establish safeguards and the response of the NIH in issuing guidelines. Replying to criticisms about NIH's role in promulgating guidelines, they wrote, "Certainly the principal biomedical research arm of the United States must be concerned with the health of laboratory workers and the public at large. Even if Congress or another governmental agency had intervened early and assumed responsibility in the area of recombinant DNA research, it is not conceivable that policy could properly be formulated without the involvement of NIH and informed members of the scientific community."

The first guidelines regulating NIH-funded rDNA experiments were published in the *Federal Register* 7 July 1976.[4] Less than two weeks later, 19 July, Senators Edward Kennedy, chairman of the Subcommittee on Health, and Jacob Javits, ranking minority member of the Committee on Labor and Public Welfare, sent a letter to President Gerald Ford expressing their concerns about the scope of applicability in NIH's rule-making effort. "[W]e are gravely concerned that these relatively stringent guidelines may not be implemented in all sectors of the domestic and international research communities and that the public will therefore be subjected to undue risks." The letter asked President Ford to consider ways to ensure compliance with the *Guidelines* by all sectors, whether by executive order or by legislation.

While the Cambridge moratorium on rDNA research issued by a vote of the city council during the summer of 1976 had been a major factor in accelerating congressional interest in regulation, many other individual actions such as letters, speeches and the beginnings of an intricate network of written communications also had an impact. Scientists sent out duplicate letters to their colleagues, environmental groups, and congressmen that had the effect of multiplying the significance of their position.

In the summer of 1976, the Elsevier-North Holland Publishing Company announced a new journal entitled *Gene* with Waclaw Szybalski as editor-in-chief. The appearance of the journal signified that molecular cloning and genetic engineering had prompted a disciplinary reorganization.

By the fall of 1976, NIH had issued a draft environmental impact statement (EIS) for the guidelines it had published in July. NIH Director Donald Fredrickson decided to publish the EIS in response to a number of commentators starting back at the February meeting of the Director's Advisory Committee.[5]

Some critics argued that NIH had turned the National Environmental Policy Act (NEPA) on its head by issuing a draft EIS after issuing the *Guidelines*. Fredrickson responded with the disclaimer that the public was best served by having some guidelines in effect as soon as possible. The other option was to continue the voluntary Asilomar recommendations until the publication of a final EIS. This could have meant a six-month delay. On procedural terms, NIH had violated the intent of NEPA. But Fredrickson was under considerable pressure from the scientific community to get things moving so that experiments could get under way. To have done otherwise might have driven some investigators to disregard the temporary standards.

Between the period when early drafts of the *Guidelines* were circulated and the date that the formal *Guidelines* were released, NIH had required its investigators, nonetheless, to follow the letter of the containment requirements. When rumors circulated or evidence appeared that someone was violating a provision, officials from NIH undertook an investigation; sometimes it proved uncomfortable for both parties. Perhaps it was the fear that adverse publicity would ruin this delicate system of peer governance that caused NIH to insist that its grantees follow the letter of the law.

In April 1976 DeWitt Stetten, Jr., chairman of RAC, wrote to Charles Thomas, Department of Biological Chemistry at the Harvard Medical School, requesting information about work that was going on in Thomas's lab. Stetten asked Thomas whether someone had been cloning genes in defective SV40 under approximately P2 conditions and whether Thomas had cloned the entire cauliflower mosaic virus in an EK1 host. Stetten also cited a published paper in which there was circumstantial evidence that one of Thomas's graduate students had disregarded the containment principles issued by RAC.[6] Thomas wrote back to Stetten expressing outrage against those who accused him.[7] He branded as hysterical those concerned about the risks. The fear of scientist spying on scientist was real. It was difficult for many investigators, who took for granted a free and permissive science, to accept a new idea of restraint and oversight—especially when, like Thomas, they did not believe there were any risks.

By the fall of 1976 there was an intensification of the activities of environmental groups. The Coalition for Responsible Genetic Research had formed in the aftermath of the Cambridge hearings. A mass letter-writing campaign was among the strategies developed to advertise the coalition and attract more members. The initial membership included George Wald and Lewis Mumford.

Spokespersons for Friends of the Earth (FOE) took NIH to task for not carrying out the mandate of the two-year-old Recombinant DNA Molecule Program Advisory Committee. FOE cited the issuance of guidelines prior to the establishment of programs of research to assess the risks as a principal flaw in the process.[8]

In November the Environmental Defense Fund and the Natural Resources Defense Council filed petitions for public hearings in an effort to widen the policy discussions. They pointed to gaps in the current policy, such as the lack of a surveillance system and an enforcement apparatus. Moreover, these groups emphasized that the *Guidelines* only applied to NIH-funded institutions.[9]

Dale Bumpers, an Arkansas Democrat, was among the earliest congressmen to become actively involved in proposing legislation. In late August Bumpers was influenced by an article that appeared in the *New York Times Magazine* entitled "New strains of life—or death," written by biologist Liebe F. Cavalieri of the Sloan-Kettering Institute.

Cavalieri outlined a number of risk scenarios associated with gene-splicing experiments. The most ominous of these was a cancer epidemic that he considered to be a possible outcome of the displacement of genes from cancer viruses into *E. coli*. By December Bumpers had communicated his interest in legislation to Kennedy, whose Senate Health Subcommittee had jurisdiction over any rDNA bill introduced in Congress.

Bumpers's message to Kennedy summarized an initial strategy for responding to the controversy. At that time Bumpers's first preference was for a complete moratorium on rDNA research for two or three years. But recognizing that such action could promote great enmity among scientists, he suggested the following alternative: a small number of national centers in which all rDNA research would be carried out, isolated from places of human activity; full public disclosure of all projects, including those carried out in the industrial sector; public evaluation of the safety and scientific value of the research; a goal to replace *E. coli* as the host organism of choice; no patent issuances for any recombinant DNA procedure.

Bumpers had in mind a model similar in concept to the national laboratories in which radiation research is carried out. Meanwhile, as Bumpers's staff was preparing legislation, an administrative agency of government issued its own policies on rDNA research.

On 13 January 1977 a notice appeared in the *Federal Register* from Betsy Ancher-Johnson, then assistant secretary of commerce for science and technology, which described a system of accelerated patent processing for rDNA projects. The statement indicated that patents would be speeded up for an industry that agreed to abide by the NIH *Guidelines*. One member of Bumpers's staff, George Jacobson, interpreted this action as accelerating an area of research at a time when many voices were requesting a slowdown.[10]

The decision of the Commerce Department was, in effect, setting policies into which Congress and NIH had no input. Senator Bumpers directed his concern to Assistant Secretary Ancher-Johnson and HEW Secretary Califano, requesting that the patent notice in the *Federal Register* be revoked. A month after Bumpers made his views known to Califano, the secretary replied that he and Secretary of Commerce Juanita Kreps agreed that the accelerated processing of patent applications should be temporarily suspended. The revocation of the original notice appeared in the *Federal Register* 9 March 1977.

It is not unusual for different agencies of government to become embroiled in jurisdictional conflicts. The function of the Commerce Department is to encourage the commercial development of scientific research. Since rDNA technology had been widely acclaimed for its potential contributions to the study of diseases, as well as a variety of commercial uses, the Commerce Department behaved in a manner consistent with that agency's objectives. The percentage of government-funded patents that had seen commercial development was not an impressive figure. Commerce saw this as an opportunity to improve society's general confidence in basic research.

Dale Bumpers introduced the first national bill (S. 621) to regulate rDNA technology 4 February 1977. It came just three days before the Cambridge City Council passed the first DNA ordinance in the country and sixteen days after Congressman Richard Ottinger (D-NY) introduced a resolution (H. Res. 131, 19 January 1977) in the House that called upon the secretary of HEW to use the powers invested in him through the Public Health Service Act to regulate gene-splicing activities.[11]

The essential features of the Bumpers bill (S. 621) are given in table 8. According to one of the principal architects of the bill, George Jacobson, its goals were to broaden the coverage of the *Guidelines* to non-NIH-funded institutions so as to take the regulatory role out of NIH's hands, since it was the agency financing much of the research, and to license all such activities.[12]

Representative Paul Rogers (D-FL) was the chairman of the House Subcommittee on Health and the Environment, which had jurisdiction over NIH. His bill, introduced 9 March (H.R. 4759), contained one provision that started a long and heated public-policy debate. The issue was whether local communities and states would be permitted to establish rules that were different than those issued by HEW. This was the first in a series of bills introduced by Rogers. It permitted state and local preemption so long as the requirements of these political divisions were at least as strict as those issued by the federal government. Through the bill Rogers was attempting to establish a balance between the concern over having uniform national guidelines and the need to respect the rights of states and municipalities to regulate their public health activities. The failure of preemption was a loophole that many scientists and their institutions wanted closed, and they proved willing to devote considerable resources to seeing that it was.

Table 8
Federal rDNA legislation—95th Congress, 1st Session: Summary of bills and resolutions

H. Res. 131	Resolution proposing that the secretary of HEW regulate rDNA activities through use of S. 361 of the PHS Act, which pertains to the prevention of the spread of communicable disease.
S. 621	Bill requiring the secretary of HEW to regulate rDNA research. A license for each project, issued by the secretary, would be mandatory. The penalty for a willful violation would be a fine of up to $10,000, imprisonment for up to one year, or both. Liability for harm would be imposed without regard to fault. No patents would be allowed unless the NIH *Guidelines* had been strictly adhered to and full disclosure of the process made.
H.R. 4232	Bill establishing a temporary, one-year study commission to make a report on rDNA issues.
S. 945	Bill requiring the secretary of HEW to regulate rDNA research. The NIH *Guidelines* would serve as the scientific standards. The penalty for a violation would be a fine of up to $10,000, imprisonment for up to one year, or both. The penalty for a willful violation would be a fine of not less than $10,000, imprisonment for not less than one year, or both. The bill would also establish a temporary, twenty-seven-month study commission to make a report on rDNA issues.
H.R. 4759	Bill requiring the secretary of HEW to regulate rDNA research. The NIH *Guidelines* would serve as the scientific standards. A license for each facility, issued by the secretary, would be mandatory. The penalty for a violation would be a fine of up to $1,000. There would be a limit of ten P4 centers. State/local governments would be able to impose their own requirements if they were at least as strict as the comparable federal standards.
H.R. 6158	The administration bill, introduced in the Senate as S. 1217. The bill would require the secretary of HEW to regulate rDNA activities. The NIH *Guidelines* would serve as the scientific standards. A license for each facility, issued by the secretary, would be mandatory. The penalty for a violation would be a fine of up to $5,000. The penalty for a willful violation would be a fine of up to $5,000, imprisonment for up to one year, or both. State/local governments would be able to impose their own requirements if they were at least as strict as the comparable federal standards.
H.R. 7418	Bill requiring the secretary of HEW to regulate rDNA activities. The NIH *Guidelines* would serve as interim standards. A license for each facility, issued by the secretary, would be mandatory. Local biohazards committees would be established to review each project and all license applications. A national advisory committee, with several operational mandates, would also be established. The penalty for a violation would be a fine of up to $50,000. The penalty for a willful

Table 8 (continued)

	violation would be a fine of up to $50,000, or imprisonment for not more than one year, or both. State/local governments would be able to impose their own requirements if the requirement is necessary to protect health or the environment and is required by compelling local condition.
H.R. 7897	(As reported by the Subcommittee on Health and the Environment) Bill requiring the secretary of HEW to regulate rDNA activities. The NIH *Guidelines* would serve as the interim standards. A license for each facility, issued by the secretary, would be mandatory. Local biohazards committees would be established to review each project and all license applications. The penalty for a violation would be a fine of up to $5,000. The penalty for a willful violation would be a fine of up to $50,000, imprisonment for not more than one year, or both. A national advisory committee, with some operational functions, would be established. State/local governments would be able to impose their own requirements if the requirements were necessary to protect the local health or the environment. A temporary, two-year study commission would be established to make a report on rDNA issues.
S. 1217	(As reported by the Committee on Human Resources) Bill establishing an eleven-member, free-standing commission to regulate rDNA activities. A license for each facility, issued by the commission, would be mandatory. Institutional biohazards committees would be established to approve each project at a licensed facility. The penalty for a violation would be a fine of up to $10,000. State/local governments would be able to impose their own requirements if the requirements were stricter than the comparable federal standards and were "material and relevant" to local health and environmental concerns. The commission would, within two years, study and report on rDNA issues.
S. 1217	(Amendment #754: an amendment in the nature of a substitute for S. 1217) Bill requiring the secretary of HEW to regulate rDNA activities. The NIH *Guidelines* would serve as the interim standards. A license for each facility, issued by the secretary, would be mandatory. Local biohazards committees would be established to review each project and all license applications. The penalty for a willful violation would be a fine of up to $2,000. A national advisory committee would be established. Other federal agencies would need the permission of the secretary before promulgating any regulation in this area. State/local governments would be able to impose their own requirements only if the Secretary granted such permission and if the requirement were stricter than the comparable federal standard, were necessary to protect health or the environment, and were required by compelling local conditions.

Source: US DHEW-NIH, *Recombinant DNA Research*, vol. 2, March 1978.

On 22 September 1976, with the advice of HEW Secretary David Mathews, President Gerald Ford issued a memorandum to heads of departments and agencies that outlined the formation of a new interagency committee, officially titled the Federal Interagency Committee on Recombinant DNA Research. Its function, in part, was "to assist in facilitating compliance with a uniform set of guidelines for the conduct of this research in the public and private sectors."[13] The bureaucratic wheels moved rather quickly on this matter. By 1 October federal agency heads were asked to nominate representatives. Some twenty agencies and departments were represented, including the Department of Defense and the Department of Transportation. The director of NIH was chosen as the chairman. At its first meeting, 4 November, the committee was advised that its role was to determine what actions, if any, were necessary to ensure uniform compliance with the NIH *Guidelines* by all public and private institutions.

A week after its first meeting, a petition to HEW, filed jointly by the Environmental Defense Fund and the Natural Resources Defense Council, requested the secretary to use the powers he has under a section of the Public Health Service Act to regulate rDNA research. According to section 361, the secretary is empowered to "make and enforce such regulations as in his judgement are necessary to prevent the introduction, transmission, or spread of communicable diseases." The transmittal letter was on EDF stationery and cosigned by Burke Zimmerman, who subsequently became a key staffperson for Congressman Paul Rogers and a major architect of DNA legislation.

By its third meeting, 25 February 1977, the Interagency Committee reached consensus on two points: first, that federal regulation of recombinant DNA research was desirable and second, that no single authority existed that could adequately regulate the research. A subcommittee, in which all regulatory agencies were represented, reviewed four major pieces of legislation that could conceivably be used to regulate rDNA activities before reaching its conclusion about the need for new legislation. These were The Occupational Safety and Health Act; The Toxic Substances Control Act; the Hazardous Materials Transportation Act; and the Public Health Service Act.

The subcommittee concluded that applying section 361 was not justified under the present situation. They argued that there would have to be a "reasonable basis to conclude" that the products of rDNA research may cause human disease and are communicable. Moreover, the subcommittee pointed out that the section failed to apply to environmental hazards. Failing to find any extant effective statutory umbrella, it recommended new legislation.

The report of the full committee was delivered to the secretary of HEW 15 March. It included key elements for the proposed legislation. It was adopted with the unanimous consent of the committee, save abstentions from the Council of Environmental Quality and the Department of Justice. In the introduction to its report, the committee gave the following justification for legislation: "[F]oreign DNA in a microorganism may alter it in unpredictable and undesirable ways. Should the altered microorganism escape from containment, it might infect human beings, animals or plants, causing disease or modifying the environment. Or the altered bacteria might have a competitive advantage, enhancing their survival in some niche within the ecosystem."[14]

Among the key components of legislation cited by the committee were primary responsibility for administering regulations, both in the private and public sector, should be given to HEW; licensing of facilities engaged in research or production of rDNA molecules; registration of all projects; and authority to inspect facilities.

The new secretary of HEW, Joseph A. Califano, acted on the recommendation with unusual haste. Two weeks after the submission of the Interagency Committee report, the administration bill was filed in the Senate by Edward Kennedy (1 April); a week later, Paul Rogers filed the bill in the House as H.R. 6158 (see tables 8 and 9).

Rogers and Kennedy were chairmen of the subcommittees in the House and the Senate having principal jurisdiction over prospective bills on rDNA research. Rogers chaired the Subcommittee on Health and the Environment of the Committee on Interstate and Foreign Commerce, while Kennedy chaired the Subcommittee on Health of the Committee on Labor and Public Welfare. For a novice of Washington politics, trying to make sense of the legislative process can be a perplexing affair. Rationality often takes a back seat to power and prestige. Bills are introduced for a variety of reasons, sometimes with a serious consideration that they will be passed, other times as a means of bringing an issue to the attention of key congressional leaders. If an issue is sufficiently important, that is, draws sufficient public attention, a House or Senate committee or subcommittee chairman may want to be the chief sponsor of the bill. Both Rogers and Kennedy felt that the rDNA controversy had passed that qualification and, as a result, it warranted their personal involvement. Thus, de-

Table 9
Key events in the legislative history of rDNA Research, 1977–1980

1977	
19 January	Richard Ottinger introduces H. Res. 131
4 February	Dale Bumpers introduces S. 621
7–9 March	Academy Forum on Recombinant DNA Research, National Academy of Sciences
8 March	Robert Metzenbaum introduces S. 945
9 March	Paul Rogers introduces H.R. 4759
1 April	Edward Kennedy introduces administration bill S. 1217
6 April	Paul Rogers introduces administration bill H.R. 6158
12 April	Curtiss letter to Donald Fredrickson on safety of *E. coli*
26 April	Resolution of the National Academy of Sciences on legislation
8 May	Nine principles on rDNA legislation issued by the American Society for Microbiology
24 May	Paul Rogers introduces H.R. 7418
20 June	Paul Rogers introduces H.R. 7897
20–21 June	Falmouth Conference on the use of *E. coli* as a host for rDNA experiments
15 July	Gordon Conference letter on legislation
22 July	Edward Kennedy introduces a substantially modified version of S. 1217
2 August	Gaylord Nelson amends Kennedy's version of S. 1217 in the nature of a substitute (S. 754)
6 September	Stanely Cohen's letter to Donald Fredrickson that eukaryote-to-prokaryote gene exchange occurs in nature
28 September	Edward Kennedy withdraws support from S. 1217 and substitutes a new bill as an interim measure
1 November	Harrison Schmitt introduces S. 2267, The National Science Policy Commission Act
1978	
19 January	Harley Staggers introduces a bill written by Harvard University, H.R. 10453
28 February	Harley Staggers and Paul Rogers introduce H.R. 11192
1980	
29 January	Adlai Stevenson introduces The Recombinant DNA Research and Development Notification Act of 1980, S. 2234

spite the alacrity of the administrative action on submission of an rDNA bill, and the early filing of the Bumpers bill, these were not destined to lead the legislative effort. The bills developed by Rogers and Kennedy emerged as the principal contenders for congressional action. Moreover, the Rogers and Kennedy bills provided two distinct models for handling research in, and the commercial use of, genetic technology.

Rogers became actively involved in writing legislation in late February 1977, several weeks after microbiologist and former scientific staff person for the Environmental Defense Fund, Burke Zimmerman, joined his staff. After Bumpers researched the issues in November and December 1976, he had written to Kennedy inquiring whether the senator had planned to take some initiative in this area. When Kennedy showed no signs that he was preparing to work on legislation, Bumpers introduced his bill. But by the early spring, Kennedy became centrally involved in drawing up his own legislation. Kennedy's bill (S. 1217) introduced the idea of an eleven-member national commission situated in the Department of Health, Education and Welfare, but whose members were appointed by the President. Modeled after the Atomic Energy Commission Act of 1946, the commission was invested with roles that went beyond those of NIH: to establish guidelines for gene-splicing activities; to enforce provisions of its rule making; and to monitor compliance.

Kennedy, a staunch supporter of public participation in science and technology policy issues, wanted a majority of the membership on the committee to be comprised of individuals who were not engaged professionally in biological research. The provision of a freestanding commission to oversee rDNA activities was one of two components of Kennedy's bill that infuriated the scientific community. The second element pertained to the issue of preemption. His bill stated that it was the intent of Congress to supersede any laws of other political subdivisions. But it contained a provision for the commission to grant exemptions under the condition that a local and state law was more stringent than the federal one and that it was "relevant and material" to health and environmental concerns.

By late June, Representative Paul Rogers had introduced three bills (H.R. 4759, H.R. 7418, H.R. 7897) representing variations of his basic legislative strategy (see tables 8 and 9). Like Kennedy, Rogers supported a licensing procedure for institutions involved in rDNA work. But from that point on, the two bills diverged. Rogers placed regulatory authority under HEW. The secretary would be aided by an advisory committee in contradistinction to Kennedy's freestanding commission. Rogers's staff also labored tediously on framing language that would achieve a broad consensus on the preemption question. At the time that H.R. 7897 appeared, the preemption language was weighted more heavily against local initiatives than the Kennedy bill. State and local regulations would be exempt from the general preemption clause if they were more stringent than the federal laws and were deemed "necessary to protect health or the environment." The decision to waive preemption was left with the secretary of HEW. The phrasing around the preemption clause consumed an inordinate amount of congressional staff time and was the principal target of the science lobby.

Anticipating some form of legislation, many scientists considered the Rogers bill significantly more favorable than the Kennedy legislation. There was little support for S. 1217 within government as well. HEW Secretary Joseph Califano strenuously opposed the idea of a special commission on the grounds that his agency already had more advisory committees than it could keep track of. Scientists rallied around Califano, arguing that such a commission would be a useless and expensive bureaucracy.

In response to fears of a weak preemption clause, the National Academy of Sciences issued a resolution in April that opposed nonuniform rules for gene splicing and expressed its concern that the establishment of a national regulatory commission to govern rDNA research would set a dangerous precedent for regulating other areas of science. Notable among the thirteen prominent scientists who signed the statement was a leading spokesperson for the conservative or go-slow approach to gene splicing, Robert Sinsheimer.

Kennedy also failed to get support from Secretary of Agriculture Bob Bergland and NIH Director Donald Fredrickson. In a letter to a key Kennedy staff member, Larry Horowitz, Fredrickson summarized his opposition to a commission: "I strongly believe that such a commission will cost more than the research it intends to regulate. It will not be capable of handling expeditiously the great burden of highly technical matter involved in the setting of standards, review of research protocols, and oversight of operations. The commission will quickly find it necessary to create a bureaucracy collateral to that already in place. There will be inevitable delays, duplication, and diffusion of responsibility, all of which could reach disastrous proportions."[15]

Despite mounting opposition, S. 1217 was voted favorably out of subcommittee by a vote of 13 to 1. But by late September, in the aftermath of an intense lobbying campaign on the part of US scientists, Kennedy announced that he was withdrawing support for his bill. Kennedy's change of heart was widely reputed to have resulted from the effective persuasion of scientific lobbyists as well as the senator's recognition that support for the bill among his colleagues had eroded. Senator Gaylord Nelson (D-WI), a member of both the Subcommittee on Health and its parent Committee on Labor and Public Welfare, introduced an amendment to S. 1217 in the nature of a substitute bill. Nelson's amendment, which transformed the Kennedy bill into a close variant of Rogers's legislation, was the target of intense lobbying by the American Society for Microbiology.

Another Senate leader who had turned sour on the Kennedy bill was Adelai Stevenson, chairman of the Subcommittee on Science, Technology and Space. Stevenson grew concerned over the impact the legislation might have on the freedom of scientific inquiry. *Science* reported on Stevenson's action to delay legislation:

> On 22 September, Stevenson delivered what the scientists have called a "masterful" and "statesmanlike" address on the floor of the Senate. He noted approvingly that scientists had called attention to the problem inherent in recombinant DNA research and, that the risk was not only hypothetical but seemed with "new" data, to be very, very small. Then, Stevenson struck the most responsive chord of all when he called on the Senate to put off any legislative action on the matter until next session lest, in haste, it enact a bill that would compromise "freedom of scientific inquiry."[16]

But even as congressional support waned, Kennedy did not abandon the idea of regulating rDNA activities. He introduced a substitute bill as an interim measure that called for a one-year extension of the NIH *Guidelines* to all rDNA activities, both federally supported and privately funded. His interim legislation also requested a special commission to evaluate the new evidence on the risks. But on the controversial issue of preemption, Kennedy's new bill was silent.

The House bill, meanwhile, was voted favorably out of subcommittee and taken up by the full committee in October. Once again, the preemption issue surfaced as the center of debate. Representative Barbara Mikulski (D-MD), responding to arguments by Friends of the Earth's Pam Lippe, offered an amendment that substituted the word "reasonable" for the word "necessary" in the key phrase "necessary to protect health and the environment." This would have had the effect of giving the secretary more latitude in permitting stricter local and state requirements. The weakened preemption language received the backing of environmental groups who generally favored public participation at the local and state levels in setting safety guidelines.

The Mikulski amendment was subsequently accepted by the full committee. Harvard University lawyers, angered by the change in preemption language, withdrew their support from the Rogers bill and appealed to a sympathetic Harley Staggers, committee chairman, to hold off on any further action until a new bill could be drafted. Persuaded by Harvard and other members of the scientific community who argued that legislation was no longer needed, Staggers was successful in getting the amended bill blocked by the full committee.[17] Thus, the first session of the 95th Congress ended with a stalemate on rDNA legislation. The passage of some form of legislation, considered a near certainty in the spring, had turned into a very dim prospect by the late fall.

The scientific lobby was the major reason for the change in congressional mood behind strong legislation. Scientists worked one on one with Senators and Representatives, created a major political force in the American Society for Microbiology, developed new coalitions among scientific societies, used scientific meetings as lobbying instruments, and applied the prestigious National Academy of Sciences as a persuasive force.

These lobbying tactics, which are a matter of style, were complemented by information directed against those who continued to warn society of the potential adverse impacts of rDNA technology. In some instances, as will be seen, the skillful use of a single piece of information proved to be more significant than a pile of letters to a senator.

Harlyn O. Halvorson, a microbiologist and head of the Rosenstiel Basic Medical Science Research Center at Brandeis University, was a member of the Public Affairs Committee (established in 1974) of the American Society for Microbiology (ASM) during the period that the Asilomar conference was being organized. Halvorson, who also held the presidency of the ASM in 1977, believed that this association should have a voice in developing policies that affect microbiologists since its membership of 25,000 made it the single largest society in the biological sciences.

ASM's involvement in the legislative debate was primarily through the initiative of Halvorson and other scientists and staff on its Public Affairs Committee who, once committed, stayed with the campaign for over three years. Halvorson was not a neophyte in political matters affecting science. Early in his career he had been an active opponent of investigations held at the University of Michigan by the House Un-American Activities Committee.[18] As the rDNA legislative activities progressed, the events were monitored by ASM's executive council. During February 1977, NIH Director Fredrickson invited Halvorson to review the early bills, and that was the point at which ASM's role changed from passive to active involvement.

At ASM's annual meeting in New Orleans in May 1977, a resolution consisting of nine principles was adopted by its Council Policy Committee. Through these principles ASM was able to maintain a consistent strategy on legislation in a period when many scientists shifted their positions. ASM's nine principles on legislation were[19]

1. All responsibility for regulating rDNA (both experimental work and commercial activities) should be under HEW.
2. An advisory committee to the secretary of HEW should consist of lay people with appropriate technical experts.
3. Institutions and not individuals should be licensed.
4. Maximum regulatory responsibility should be delegated to institutional biohazards committees.
5. Experiments requiring P1 physical containment should be exempt.
6. There should be no liability clauses; license removal is an effective deterrent.
7. Uniform federal standards should be favored.
8. Legislation should be flexible to permit modification.
9. Legislation should be sought only after "due and careful" deliberation.

Through a process reminiscent of political-constituency building, Halvorson helped ASM become a leading force in the shaping of legislation. His strategy included publicizing the nine principles through letters and editorials in journals, including *Science* and *Chemical & Engineering News*; sending mailgrams to key congressmen; and coordinating ASM's efforts with those of other organizations. Halvorson exploited the "freedom of scientific inquiry" issue in his effort to engage university groups, such as the Land Grant College Association. He consulted with lawyers in the American Civil Liberties Union and the American Bar Association. The rDNA episode helped to catalyze the formation of the Inter-Society Council for Biology and Medicine. The council, made up of the executive officers of the seven major societies in the fields of biology and medicine, including ASM, was active in supporting the nine principles for legislation.[20]

Halvorson led delegations of scientists into the offices of congressmen. He applied grass-roots political organizing methods to promote ASM objectives. The membership was divided into congressional districts that were matched with their representatives. Each member was provided with information about how to lobby for the ASM position.

The sophistication of his methods did not stop there. Halvorson consulted with experts on the Washington scene who could identify where the power and influence resided, which congressmen could be persuaded, and on what grounds. He received expert advice on how to run a campaign from individuals responsible for dismantling the House Un-American Activities Committee. He also sought support from Americans for Democratic Action, a liberal, pro-Kennedy organization.

To mount an effective campaign against the Kennedy bill, Halvorson selected out individual senators for more intensive lobbying. He urged Gaylord Nelson, reminding him of the benefits microbiology brought his own state of Wisconsin, to oppose the Kennedy bill. Nelson was persuaded and introduced his own legislation that met ASM objectives.[21]

One episode described by Halvorson to MIT's Oral History Program highlights the sophistication of ASM's lobbying machinery.[22] While at a visit to Duke University in March 1978, Halvorson learned of a new effort that threatened the passage of a compromise bill introduced by Harley Staggers and Paul Rogers as it was about to come before the full committee. He received permission to take over an office in the microbiology department at Duke with six phones and a book listing congressional staff. He then repeated, but this time on a grander scale, what he had accomplished on previous occasions. There were thirty-nine members on the Committee on Interstate and Foreign Commerce who had to be contacted and persuaded to vote down any amendments that would destroy the bill.

Each representative on the committee was identified by congressional district. Towns and the universities in those towns were then matched according to district. ASM members were called individually and asked to contact their representatives immediately. Halvorson described the strategy as building a system of pyramids, with each university as a base in each of the districts, and a single congressman at the apex. The entire process took less than twenty-four hours. Its effectiveness depended upon a loyal ASM membership.

The summer of 1977 had witnessed other developments in the science lobby. Since the city of Cambridge had passed its own ordinance early that year, Harvard officials were deeply concerned that federal legislation might not preempt the local act. Harvard joined together with Stanford, Princeton, and Washington universities and hired a professional lobbyist to work on a strong preemption bill. Harvard also developed an organization it called Friends of DNA to coordinate its lobbying efforts with those of other institutions. It sent delegations of administrative personnel and scientists to Washington frequently to speak with key congressmen and their staffs; it even offered its own model rDNA legislation.[23]

Professional meetings were ideal places to organize scientific opposition to legislation. In 1973 the letter from Gordon Conference participants had brought the rDNA affair out in the open. Four years later a new Gordon Conference letter received publicity. In a letter to *Science*, Harvard biologist Walter Gilbert publicized the conference resolution against legislation. The resolution, signed by 137 participants, 86 percent of those who had attended, stated, "Congressional, state and local authorities will set up additional regulatory machinery so unwieldly and unpredictable as to inhibit severely the further development of this field of research. We feel that much of the stimulus for this legislative activity derives from exaggerations of the hypothetical hazards of recombinant DNA research that go far beyond any reasonable assessment."[24]

Concurrently, environmental organizations were applying counterpressures to preserve the *Guidelines*, improve the public's input into the decision-making process, and remove preemption of local and state initiatives from the legislation. But molecular biologists, whose tolerance for the environmental lobby began to wear thin, were able to exert some influence over prominent scientists serving on the environmental organizations' advisory boards. Sloan-Kettering Cancer Center President and popular science writer Lewis Thomas resigned from the advisory council of Friends of the Earth in 1977 because of that organization's legal actions against NIH. Joshua Lederberg, then president of Rockefeller University, wrote to NIH to explain his disagreement with the position taken by staff people at the Natural Resources Defense Council (NRDC). Lederberg served as a trustee of NRDC. Stanford University population biologist Paul Ehrlich tried unsuccessfully to influence Friends of the Earth's executive committee and President David Brower to stop the campaign to legislate rDNA research and to end its legal action against NIH.[25]

At least as vital as the instrumentalities of influence developed by the scientific community were the ways it used information to defuse the public's concerns about biohazards. In a controversy of this type, where positions often become hardened with time, a conversion of a single individual scientist can be of enormous publicity value. This was the case with Roy Curtiss, who had the distinction of being cited by hawks and doves on DNA policy. Curtiss's August 1974 letter, which (by his own estimate) eventually reached over a thousand scientists around the world, had raised numerous potential hazards of gene-splicing experiments beyond that discussed in the original Berg statement and placed Curtiss in the camp of those who advocated a "proceed with caution" approach. Three years later, on 12 April 1977, Curtiss produced a twelve-page, single-spaced letter to Donald Fredrickson that detailed his new position on risks and expressed his concern about the dangers of overregulation. In what was to become his characteristically meticulous style, Curtiss reviewed the evidence that cloning into *E. coli* K12 could lead to adverse effects. He considered the bacteria's potential to colonize humans, be made into a pathogen, have its communicability altered, transmit genetic information to other organisms, and have conferred on it a selective advantage by introducing a foreign sequence. The letter stated that "the introduction of foreign DNA sequences into EK1 and EK2 host-vectors offers no danger whatsoever to any human being with the exception . . . that an extremely careless worker might under unique situations cause harm to him or herself."[26]

Curtiss had developed a reputation for being fair-minded and honest on the DNA issue. His conversion on the safety of *E. coli* K12 was treated with great favor by those who felt the cloning requirements in that organism were too severe. His letter made reference to his conversion process, which, he claimed, came about gradually from a scrupulous review of all existing knowledge:

The arrival at this conclusion has been somewhat painful and with

> reluctance, since it is contrary to my past "feelings" about the biohazards of recombinant DNA research. As a means to challenge the above-stated conclusion, I have taken some worst-case scenarios thought up my myself, by my colleagues and by others and subjected them to critical analysis by obtaining information from those scientists most knowledgeable about the genetic control, biosynthesis, mode of action, production, etc. of the foreign gene product(s) in question. In no instance have I found evidence that the necessary genetic information could be cloned in one step, would permit *E. coli* K12 to colonize the intestinal tract and lead to the production of the product(s) in the intestinal environment that would be harmful to the mammalian host. This is not to say, however, that an individual with considerable skill, knowledge (most of which is currently lacking) and luck could not construct in multiple steps a microorganism that would satisfy all these requirements.[27]

There were still the ecological consequences of releasing K12 with foreign genes into the environment. On this point, Curtiss was less forceful, but still confident that a modified organism would not inhabit a new ecological niche.

In concluding his statement, Curtiss argued that excessive regulation implies a diversion of funds into areas that will have no payoff for improving the health of the country. Moreover, he warned, if the scientific community does not believe in the regulations, they may develop contempt for them. Although Curtiss did not spell it out, the implications are that regulations lacking widespread support will be unenforceable.

The Curtiss statement, as impressive a work as it was, had limited effectiveness on policymakers, who most likely could not follow its technical vocabulary and penetrating discussion. In addition, there were no new data to satisfy those thinking along the lines of risk assessment.

The use of *E. coli* as the preferred host for cloning experiments had been the focus of considerable criticism. Chargaff, Sinsheimer, Goldstein, Wald, and members of Science for the People were among those who argued that it was dangerous to carry out gene splicing with an organism that was a symbiont to humans. On 7–9 March 1977, in a National Academy of Sciences Forum devoted to rDNA research, one of the ten workshops was entitled "Is it likely that *E. coli* can become a pathogen?" The session was cochaired by Richard Goldstein (Harvard Medical School) and Elena Nightingale (Institute of Medicine, NAS). The locus of disagreement was outlined in the workshop report:

Some of the participants wished to emphasize that *E. coli* K12 is being used in DNA recombinant research because after three decades of experience with this organism no known disease has emerged, because the years of experience have provided intimate knowledge of the genetics and biology of this type of bacterium, and because such knowledge enables understanding and control of the experimental situation. Others felt that these reasons for the continued use of *E. coli* K12 do not suffice because of the ubiquity of *Escherichia coli* and the readiness with which it exchanges information with so many other species of bacteria.[28]

For some scientists the only problem left with *E. coli* K12 was that of educating the public of its safety. Bernard Davis argued that the K12 strain was sufficiently different from other, disease-causing strains to warrant a new name and that the scientific community might consider such a change as a way of removing its guilt by association. But little seemed to be resolved in the matter of the safety of *E. coli* at the NAS Forum. The forum report was not particularly useful for influencing public policy mainly because it represented an airing of views in contrast to a condensation of opinion. Three months later, another conference at Falmouth, Massachusetts, held 20–21 June, which focused almost exclusively on *E. coli*, had a profound impact on the legislative tide.

The conference at Falmouth, sponsored by the Fogarty International Center, NIAID, and NIH, was chaired by Sherwood L. Gorbach, chief of the infectious disease unit at the Tufts University School of Medicine. The proceedings listed thirty-six participants. The conference brought together disciplines that until that time had never been coordinated in order to address the risks of rDNA research, including infectious diseases, epidemiology, immunology, and enteric bacteriology. The new faces were mixed in with rDNA veterans like Sydney Brenner, Allan Campbell, Wallace Rowe, and Richard Novick. The proceedings of the conference were published eleven months later in the *Journal of Infectious Diseases*.[29]

No sooner had the conference ended than efforts to characterize the consensus among the experts broke out in disagreements. Gorbach took the initiative by sending a five-page letter to Donald Fredrickson, providing his own understanding of the consensus of opinion. The letter was not circulated to the participants beforehand for comments. Yet, Gorbach's role as chairman of the meeting and his relative anonymity in the controversy up to that point helped to strengthen the importance of his remarks, especially to congressmen and their staff persons immersed in a legislative battle over an rDNA bill. The conclusions themselves were indistinguishable from those arrived at independently by Roy Curtiss (who also attended the conference) in his April correspondence.

Gorbach reported in his letter to Fredrickson that there was unanimous agreement on the issue that "*E. coli* K12 could not be inadvertently converted into an epidemic pathogen by inserted DNA molecules." Even if one deliberately tried to convert K12 into a pathogen, the letter continued, it would require twenty years of effort. On the second important question of the conference, namely, the ability of *E. coli* K12 to transfer DNA in vivo to other organisms, Gorbach reported that the consensus was it is extremely unlikely so long as nonmobilizable plasmids are used.[30]

The importance of this letter cannot be understated. It was reprinted in the *Federal Register*; it was cited on numerous occasions as one of the key pieces of evidence for the relaxation of the *Guidelines*; it was reprinted extensively in NIH reports, including the in-house newsletter *Nucleic Acid Recombinant Scientific Memoranda*; and it was cited as authoritative in newspapers and journal editorials.

Some Falmouth participants were upset with Gorbach's action, which came at a time when a major effort was under way to revise the containment conditions for many rDNA experiments. They did not fault him for what he had said, but rather for what he had left out or played down. Bruce Levin of the Zoology Department at the University of Massachusetts (Amherst) wrote Fredrickson that Gorbach's remarks only tell part of the story. There was, according to Levin, "absolutely no consensus reached which suggested that the probability of transfer of chimeric DNA by plasmids was sufficiently low to be disregarded".[31]

Richard Goldstein and Jonathan King, also participants at Falmouth, circulated a letter to scientists attending the summer Cold Spring Harbor Phage Meeting, which spoke of the "misleading and incomplete reports of the proceedings." They cited evidence raised at the meeting showing that K12 can persist for six days in a human intestinal tract. The transfer of DNA from K12 to wild strains was still a serious problem, according to their interpretation of the data and their assessment of the meeting.[32]

But these efforts to characterize differently the meeting's outcome simply did not counteract the remarkable impact that the Gorbach letter had on legislative initiatives.

Then, in early September, new experimental work by Cohen and Chang was skillfully used to thwart the efforts of those who were lobbying for strict regulations. Together with the Gorbach letter it packed a double punch. The latter said that *E. coli* K12 was fail-safe, while the former indicated that genetic manipulations in the laboratory were already going on in nature. Stanley Cohen had sent NIH Director Fredrickson a preprint of a paper, scheduled to be published in the *Proceedings of the National Academy of Sciences* (*PNAS*) a few months later, that indicated that he and Chang were able to promote the take-up of eukaryotic DNA by plasmids of prokaryotes in the presence of the appropriate restriction enzymes. Cohen explained the significance he placed on his work: "Our data provide compelling evidence to support the view that recombinant DNA molecules constructed *in vitro* using the Eco RI enzyme simply represent selected instances of a process that occurs by natural means."[33]

Cohen also provided the impression that this experiment, along with the Curtiss letter and the Falmouth Conference, were largely responsible for a major change in his position toward rDNA experiments. Once again, we have a key actor in the debate describing his conversion process:

> In the past I believed that the *in vitro* joining of different segments of DNA at restriction endonuclease cleavage sites resulted in the formation of genetic combinations that could not be made otherwise; for this reason it seemed important to call attention to possible biohazards that might be associated with certain kinds of novel gene combinations. However, along with virtually all of the other scientists who first raised these questions, I have since come to believe that our initial concerns were greatly overstated. Some of the important new information that has led to this changed perception has been summarized in Roy Curtiss' recent letter to you and in the Falmouth report.[34]

Like the Gorbach letter, Cohen's correspondence moved quickly through congressional circles, and before a general appraisal of his interpretation of the experiment was offered, it had persuaded many that the laboratory was an extension of nature.

There were efforts made to soften the effects of Cohen's lobbying. The Coalition for Responsible Genetic Research sent out mailgrams stating that the Cohen-Chang experiment was no less an engineered event than DNA recombination in the test tube. Burke Zimmerman, staff member to Paul Rogers, wrote that Cohen's actions represented a misuse of science that would not have been tolerated under normal circumstances, that is, circumstances in which scientific rationality and logic are not offset by fears of legislation.[35] By the time responses to Cohen's conclusions had reached congressional staff, many minds had already been changed. Nevertheless, there was still one last effort to enact legislation in the next term of Congress.

By December 1977 the original Rogers bill had experienced so many changes that even the latest version—H.R. 7897—was beginning to appear confusing and unworkable to those who drafted it. After Harvard withdrew its support for the bill, Rogers decided to take another stab at developing legislation when the second session of the 95th Congress convened. Meanwhile, Harvard lobbyists drafted a bill that Staggers introduced early in the new session. The Harvard-Staggers bill extended the NIH *Guidelines* to all sectors in which rDNA work was carried on. It failed to recognize that sections of the *Guidelines* were written almost exclusively for academic research institutions and were inappropriate for industrial-scale activities.

What Harvard could not achieve through H.R. 7897 in strong preemption language, it built into its own bill. It stipulated that the secretary may grant an exemption to federal preemption of a state or local act if (1) the act is necessary to protect health or the environment and (2) the act is justified by compelling local conditions.

Burke Zimmerman, chief architect of the Rogers legislation, considered the debate over preemption language as having no substance. He argued that the changes in language could hardly make a difference in how such decisions would be implemented by the secretary of HEW. Starting all over, Zimmerman tried to create a bill that would be supported by NIH, Harley Staggers, the White House, and HEW. Staggers agreed to cosponsor a bill with Rogers if it contained strong preemption language. The result was H.R. 11192, a two-year interim control bill that was filed 28 February. The bill extended certain sections of the NIH *Guidelines* to all rDNA activities. It took regulation and enforcement out of NIH's hands and placed it in HEW's. The bill also called for the establishment of a thirteen-member commission for the study of research and technology involving genetic manipulation. And a moderately strong preemption clause was written into the bill allowing the secretary of HEW to decide whether the application by a state or political subdivision for more stringent requirements "is necessary to protect health or the environment." Staggers shepherded the bill through the full committee as well as the Committee on Science and Technology. Both Staggers and Rogers took some initiative in getting the Rules Committee to move the bill to a floor vote. But that never took place. Burke Zimmerman surmised that the bill died from lack of interest. There was insufficient momentum on the part of prolegislation partisans. Also, Paul Rogers was a lame-duck congressman and many other bills were competing with his for congressional attention.

The bill did receive support from ASM and NIH Director Fredrickson among others. But because of its strong preemption language, some environmentalists did a flip-flop and opposed the legislation, taking their chances on the prospects for a stronger bill in a new session of Congress. But as the legislative debate continued, each new piece of legislation was a weaker version of what came before. On 29 January 1980 Senator Adlai Stevenson introduced The Recombinant DNA Research and Development Notification Act of 1980 (S. 2234), which requires all persons conducting rDNA research not funded by the NIH to file notification of their activities with the secretary of HEW. But the bill received little interest from environmental groups, which had begun a process of disengagement from the legislative effort. And for industry it offered nothing less than nuisance value at a time when the four leading biotechnology firms were estimated to be worth a half-billion dollars but were still without a single marketable product.

24

A Retrospective: Science and Society

Future generations of scientists and historians will undoubtedly refer to the rDNA episode as an example of the stresses that can develop between science and its social institutions. And when they look back at the events several questions will stand out among the many issues of this debate. First, why did scientists draw special attention to the potential biohazards of rDNA techniques? Today, many believe that distinguishing rDNA techniques from other areas of biological research was a mistake because the former is considered no worthier a candidate for concern than the latter. Second, why did the public respond in the way it did? There have been suggestions that the public's response can be credited to the media's "overindulgence" with doomsday scenarios and to dissident scientists acting from political motivations. Third, in what ways is the rDNA controversy similar to or distinct from other areas of technological innovation that pose uncertain risks? It has been argued that rDNA technology has received unprecedented attention inasmuch as no real hazards have been confirmed. In this respect, the argument continues, it is different from other technologies, to which society responded when the cause for concern was more than pure speculation. Fourth, what changes have the debates over rDNA technology brought in the relations between science and society?

In raising these questions it must be borne in mind that there is still uncertainty about NIH's future role as de facto regulator of rDNA research; the life expectancy of the NIH *Guidelines*; whether the institutional biosafety committees will become a permanent fixture at universities; where the commercial uses of genetic engineering will take us.

Inevitably, evaluations of the rDNA affair will generate controversy. Some leading scientists have already branded the decade as a period during which the public was obsessed with imaginary risks.[1] They question the wisdom of the Berg letter and the subsequent overtures for controlling research.[2] On the other hand, the more pragmatic-minded scientists view the public's intrusion into their affairs as a consequence of the coming of age of their profession. For the most part, the debate over university research has all but subsided. We have entered into a second phase of the DNA episode: commercial developments, including production sites and products. The prospects of continued public involvement in these areas are real. A report prepared for the US House Subcommittee on Science, Research and Technology states that "the extraordinary potency of gene splicing makes it all but impossible for the technique to disappear quietly from public attention, as scientists appear to hope. Commercial and practical aspects of recombinant DNA may possibly not escape regulation of some kind."[3] If this prognosis is correct, then we should anticipate a protracted period of developing relations between society and molecular biology. The first question however, takes us back to the origins of the DNA alert.

My investigation into the early history of the rDNA affair shows that several years prior to the development of restriction enzymes for genetic manipulation, scientists were beginning to express uneasiness with the way that viruses were handled in research. Those concerns were intensified with the adenovirus SV40 affair, which resulted in several attempts to regulate the laboratory use of hybrid viruses. The attention given to virus research was highlighted in the first Asilomar Conference in 1973, the same year that the National Institute of Allergy and Infectious Disease required investigators to sign a memorandum of understanding and agreement as a condition for obtaining certain viral strains.

The potential biohazards of conventional virus research did not receive wide public attention. Scientists did not request a moratorium. University research laboratories were not under scrutiny by local communities. But then came news to the scientific community that Stanley Cohen and Herbert Boyer had developed a plasmid system for transporting foreign genes into bacteria, and attention was drawn to the fact that gene splicing would soon become a working tool for molecular biologists, faster, perhaps, than most had anticipated. Scientists who learned of the Cohen-Boyer technique recognized that it would accelerate the construction of hybrid viruses and make it easier to produce concentrated amounts of virus material. The awareness of this technique was the basis for a sudden leap in anxiety about biosafety in the laboratories. Instead of looking comprehensively at the potential hazards of conventional virus research along with the new rDNA methods, scientists only publicized the risks of the new techniques.

Naturally, it is more difficult to reform procedures that are already widely used than to impose regulations on new procedures. But there was more to singling out gene splicing than that. The simplicity of the Cohen-Boyer technique suddenly opened up genetic engineering to a much broader spectrum of investigators. Among the veteran molecular biologists there was a concern that many of the neophytes might not be attentive to laboratory safety. Since rDNA techniques were considered capable of introducing qualitatively new hazards, it is understandable why this technology was singled out for vigorous oversight. The roots of the public's obsession with rDNA technology can be traced back to the early decision by leading scientists to establish it as the pariah of biological risk.

But why were the issues brought before the public? There were voices in the scientific community that warned of public overreaction to expressions of concern about biological risks. In the early 1970s a climate had developed in favor of caution and restraint in science and technology. The environmental movement was at its peak. There is no escaping the impact the Viet Nam War had on the campuses. University science was a target of the antiwar movement. Many scientists had become sensitized to the exploitation of science for military purposes. The field of bioethics was getting under way. Between 1969 and 1972 there was a flurry of papers on genetic engineering emphasizing the public's right to know and to decide the direction and use of science. Remarks like that of Salvador Luria had begun to gain favor: "Scientists must assume the responsibility to tell society, in a forceful and persistent manner, what science is discovering and what the technological consequences are likely to be."[4]

I believe the issues surrounding rDNA technology were brought to the public's attention as an affirmation by scientists, during a period when the public's confidence in science was severely strained, that science could indeed be responsible to society at large.

How do we gauge the public's reaction to this debate? Far from being a mass movement, it was by and large a response of well-educated and informed individuals. It is also important to distinguish the different sectors of the public that played a role in this controversy.

Local communities reacted to university-laboratory and some commercial ventures between 1975 and 1977 when vigorous debates among scientists were taking place. A new wave of community reaction surfaced in 1980 when firms like Biogen (Swiss), Collaborative Research (Waltham, Massachusetts), New England Nuclear (Massachusetts) and Genetics Institute (new US firm Boston, Massachusetts) discussed plans with community leaders to begin genetic-engineering activities. Although the scientific debate had subsided, neither the local media nor the citizens had forgotten the issues. There was little that a firm could do to keep the lid on once the media publicized a site plan request for commercial laboratory space and small-scale production facilities when the word "DNA" was mentioned.

Community residents were also skeptical about the voluntary compliance program that NIH administered for private firms. The bridges of confidence that universities had built to the communities did not automatically provide open access to commercial DNA activities. It took a different kind of public-relations campaign, which, in some cases, meant opening up new areas of negotiation that included assurances that firms would follow the NIH *Guidelines* and other recommended safeguards for large-scale operations.

Another sector of the public that became involved in the rDNA affair was a core group of environmentalists. The environmentalists added to their agenda of ecological concerns the unwitting release of potentially troublesome organisms into the environment. They called for legislated controls on the university and the private sector. The efforts of the public-interest activists were directed at lobbying Congress, participating in the open hearings held by NIH, attending RAC meetings, providing comments on changes in the NIH *Guidelines*, and fostering broader public understanding.

Scientists like Robert Sinsheimer emphasized the broad environmental impact of rDNA technology. But beyond that, environmentalists had a long-standing interest in genetics issues. In particular, there were campaigns directed against the narrowing of genetic variation on the planet, brought about, to some degree, by modern agricultural techniques, and a disregard of endangered species and their natural habitats. Friends of the Earth's Francine Simring was involved in the ozone-depletion affair and radiation hazards at the time the rDNA controversy broke. Molecular biologist Burke Zimmerman was working part-time for the Environmental Defense Fund in the aftermath of the Asilomar period. Science For the People had a Genetics Study Group that examined the political impact of genetic research and distributed policy positions at workshops and scientific meetings. The People's Business Commission (PBC), a Washington-based public-interest group headed by Jeremy Rifkin and Ted Howard, turned its attention to the industrial and political misuses of biological engineering. The PBC viewed the breakthroughs in gene splicing as eventually leading to human genetic engineering. Rifkin and his colleagues sent out public-interest alerts. Their articles and widely circulated book *Who Shall Play God?* helped to broaden the public's awareness of the DNA issues.

A third sector of the nonscientific public that played a critical role in the rDNA affair is the media. The breakthroughs in molecular genetics and the internal scientific debates hit the news like a bombshell. The turbulence within science served as an excellent stage for examining a classic conflict between progress and technological peril; between freedom of inquiry and the accountability of science to society. The proclaimed potency of the science provided the grist for literary metaphor, drama, and futuristic scenarios—all of which make for good copy. Once the public read that scientists were engaged in making "new life forms," the human imagination was set into motion.

Surely, media coverage frequently simplified the issues because discussions were written in nontechnical language and depended upon brevity. But it is critical to understand that the news media and popular-science writers generally depended upon scientists for their risk scenarios. When the source of those fears began to dissipate, the media shifted their attention to other issues.[5]

The media were an indispensable bridge between the scientists and other sectors of society. The diffusion of information about rDNA research was helped considerably by several books written between 1977 and 1979 that emphasized the revolutionary aspects of the new research. The phrases on the book jackets highlight how the issues were presented to the lay public: "man-made evolution;"[6] "new forms of unpredictable and possibly destructive organisms;"[7] "the most promising (and most threatening) scientific research ever undertaken;"[8] "power to dramatically alter existing life-forms and to create entirely new ones;"[9] "artificial creation of life."[10] Realistically, I would say that community reaction to such promotional devices was moderate. After all, there were many opportunities for communities to proscribe the research; yet none did. The apocalyptic visions of genetic engineering were tempered in community debates.

This discussion of the public's role would not be complete without some mention of the reaction by religious groups. Warnings about genetic engineering were issued by the Pope, individual church leaders from several denominations, and representatives from the World Council of Churches (WCC). There was no organized church opposition to the research or its applications. Since the WCC was the most active religious group on this issue its activities are of special interest.

In 1979 the WCC held a Conference on Faith, Science, and the Future. The proceedings were published in two volumes: the first consisted of addresses given at the plenary sessions; the second contained the recommendations and resolutions of the working sections. One of the papers delivered ended with the following warning: "Where are we going? Today, the cure for genetic disease; tomorrow the experimental improvement of the human character through the administration of psycho-active molecules? The scientist should stop playing God and listen to what God has to say in preserving his creation and its evolution into the distant future."[11]

In its working reports the WCC cited grave hazards in genetic engineering: the inadvertent production and escape of microorganisms pathogenic to humans or disruptive to the ecosystem; the production of pathogenic organisms for use in biological warfare; the transport by commercial enterprises of genetic engineering to developing countries lacking guidelines; ethically unacceptable modifications of humans and other vertebrates.[12]

The WCC made specific recommendations on human genetic-engineering experiments that included setting strict ethical boundaries for future developments. Their report also asked for broad public involvement in decisions concerning technological applications.

Several factors were responsible for public's reacting to rDNA research in ways that were distinct from its reaction to conventional biological hazards: the idea that "new life" was believed to be a consequence of rDNA techniques; the links between human genetic engineering and DNA research; the rich flow of metaphors conjuring up doomsday scenarios with such terms as chimera, breaching species barriers, Frankenstein monsters, cancer-producing bacteria; and the connection between laboratory science and the environmental movement. It is ironic that the entrepreneurial scientists were no doubt partly responsible for the public reaction. They were promoting the technique as unique, powerful, and revolutionary. They set it apart from more conventional methods. They promoted and participated in its vigorous industrial exploitation. Meanwhile public skepticism grew over whether the commercial developments in biology would escape the adverse environmental effects that similar developments in chemistry and physics had produced.

Some have argued that the public and the NIH overreacted to the potential hazards of rDNA technology. Among their reasons for this judgment are that no one has been known to become ill from such experiments and that there is no evidence to support the view that a product of rDNA research could be more hazardous than any of its component elements.[13] It is certainly the case that the furor precipitated by rDNA research was due to the spectre of disaster generated by hypothetical scenarios rather than hard evidence. In this respect, rDNA technology was handled differently from previous technologies, whose regulation was justified by evidence of real threats or imminent dangers to health or the environment. But we do not only regulate proved hazards to human health or the environment. Many chemicals are regulated without definitive evidence of their adverse effects, but merely on the basis of suggestive evidence from animal studies or short-term tests. The important point here is, Was there sufficient suggestive evidence that gene splicing could introduce additional hazards? Certainly, the record shows there was considerable agreement among leading scientists that, left unregulated, rDNA technology could be hazardous.

Scientists had spoken out and the public had reacted. If the rDNA affair proves to be a false alarm (and the answer may not come for many years) will it be the first false alarm? Hardly. There have been cases in which hazardous or even catastrophic consequences of technology have been mistakenly predicted by responsible individuals. When Enrico Fermi and his colleagues developed the first sustainable chain reaction underneath the stands of the Chicago stadium, some scientists postulated a nontrivial probability that the reaction would not be controllable. Prior to the lunar landing, biologist Joshua Lederberg provided a scenario of lunar organisms brought back by the astronauts and capable of infecting our planet.

Some biologists may well become more cautious in airing their concerns about science to the general public or the media. Responding to the issue of self-imposed silence on issues of future concern, David Baltimore remarked "If another issue appears on the horizon similar to the recombinant DNA controversy, I would hope that those who recognized the problem would not be afraid to speak out. I also hope that the scientific community will be more mature in its formulations and its responses so that the general public will be inclined to trust the activities of scientists rather than doubting their motives and their honesty."[14]

The final question in this retrospective examination concerns the changes that the rDNA affair has produced in the relations between science and society, in particular: changes in public consciousness; new institutions of social control over science and technology; and finally, the development of new relations between biology and the industrial sector.

Within a relatively short period of time, the biological sciences have passed from their age of innocence to their age of anxiety. The new rDNA technology was not the only reason for this transformation, but it has surely been the most important factor in raising the public's apprehension about biological research. The loss of innocence is irreversible. If it were simply a matter of bad press or the public's reaction to scientific dissidents, social attitudes could conceivably return to an earlier state. But there is no getting away from the fact that biology has undergone an objective transformation. From an activity that was primarily analytic, it has moved into a synthetic phase. The discovery and application of restriction enzymes has opened up possibilities for the controlled rearrangement of life's basic substance. These new powers will provoke enduring concerns about the impact of synthetic biology on our civilization.

As a direct consequence of the rDNA debate, biohazards have become a public affair in much the same way that the use of humans as experimental subjects has been for two decades. New institutions were created to respond to the potential misuse or abuse of genetic engineering. Since 1974 the Recombinant DNA Advisory Committee (RAC) had been making recommendations to the director of NIH on safety issues. This mix of biologists, ethicists, environmentalists, and public health experts on RAC has forced new avenues of communication between science and the public.

The creation of institutional biosafety committees (IBCs) is yet another example of the new forms of public participation in decisions affecting the laboratory and the commercial activities of biologists. The IBCs are clearly in a state of transition. The network of IBCs face two possible futures: either a phasing out or a broadening of their scope of responsibilities. The ambivalence over the permanency of the IBCs as presently constituted was expressed at a national meeting of committee chairpersons held in November 1980. A good number of the participants felt that the functions of an IBC could best be served by a single individual such as a biosafety officer or the principal investigator. In their view, there was no role for the lay public in laboratory safety.

But other IBC chairpersons, believing that their committees should not be restricted to rDNA activities, expressed a desire to see the mandate broadened to include the general area of biosafety. Since 20 percent of the IBC membership consists of community representatives, the fate of this NIH-mandated structure will determine the public's role in the oversight of laboratory science. Moreover, since NIH requries an IBC for commercial institutions that submit to its voluntary compliance program, community oversight of industrial genetic engineering is also at stake.

A second form of institutional change has developed at the local level. The laws passed by two states and eight local communities treat biological research and its commercial developments as a public-health issue. Nearly four years after the passage of its 1977 ordinance, the city of Cambridge once again convened a citizens committee, this time to review a proposed commercial DNA facility. The citizen group, working closely with the city's biohazards committee, recommended a new ordinance that includes a permit system for both small-scale and large-scale rDNA work. University research is not treated any differently than private-sector research and development in the new ordinance passed by the city in April 1981. The Cambridge citizen task force took a bold step in asking for external inspection and review of institutional facilities and monitoring practices where large-scale work was under way. This is a significant role for a municipality and reflects a new community involvement in overseeing the safe application of science and technology.

Some mistakenly believe that all the concern about rDNA research has subsided and that except for cities like Cambridge, involvement at the local level is no longer likely. The evidence is to the contrary. When faced with a similar situation, other cities and towns have begun to react like Cambridge. The city of Waltham, Massachusetts, passed an ordinance early in 1981 that placed all rDNA activities under the NIH *Guidelines*. More surprisingly, the Waltham law prohibits the use of human subjects in rDNA research. Boston, Massachusetts, passed a law in the summer of 1981 modeled on the Cambridge ordinance. In passing these local ordinances, communities are establishing their authority over academic research in molecular biology and its commercial developments. Whether the new initiatives of Massachusetts towns and cities in rDNA legislation is only a ripple effect from the Cambridge situation remains to be seen.

The third important area in which we can expect to see changes in the relations between science and society is in the university. The lure of commercial success is already being felt in academic circles. Once the products of genetic engineering become marketable and profitable, industrial organizations will undoubtedly want to establish spheres of influence in biology departments. And the departments themselves may undergo a metamorphosis as tensions develop between pure researchers and applied genetic engineers.

Harvard University, after first giving serious attention to establishing its own genetic-engineering company, withdrew the proposal when internal and external criticism began to mount. Conflict of interest was cited by some Harvard faculty as an important reason to keep the university out of direct commercial ventures. A *New York Times* editorial berated Harvard for its business overtures and offered an apt phrase to describe the plight of the commercial-academic—"market or perish."[15]

Stanford University President Donald Kennedy asked his own faculty to use caution and restraint in responding to the genetics companies that are courting the university. Kennedy cited the attraction of big money as a threat to traditional values in biological science, such as the sharing of information and the free exchange of cell lines.[16] Every new cell line is a potential patent application.

If molecular biology follows the route of other applied disciplines, we should not be surprised to find substantial funding from industry in the universities; a branching off of applied molecular engineering from basic research in molecular biology; a seeding of research-and-development firms around the academic centers of genetic-engineering research; an extensive network of consultant relations and university biologists serving as principals in venture-capital firms.

Although we have seen these things happen before, the public response to the research-and-development activities in genetic engineering is unlike anything we have experienced. Biologists and the Supreme Court may make little of the distinction between experimenting with inert chemicals and experimenting with living things, but the public sector believes quite the contrary. The manipulation of life forms touches off anxieties, even when the potential hazards are not well defined or confirmed. As the rDNA controversy passes through its first phase, an equilibrium between the public and the universities has been reached. The next test for gene splicing and society will occur when genetic-engineering firms begin sprouting up in communities around the country.

Appendix A

Chronology of Key Events in the Recombinant DNA Controversy

1971 June
Cold Spring Harbor Tumor Virus Workshop

1972 26–30 September
European Molecular Biology Organization (EMBO) Workshop, Leuenberg, Switzerland

1973 22–24 January
Conference on Biohazards in Biological Research (Asilomar I), Asilomar Conference Center, Pacific Grove, California

11–15 June
Gordon Conference on Nucleic Acids, New Hampton, New Hampshire

21 September
Letter published in *Science* from Maxine Singer and Dieter Söll, expressing sentiments of Gordon Conference participants on the potential hazards of new gene splicing techniques (181:1114). Singer and Söll were cochairpersons of the 1973 Gordon Conference

1974 26 July
Letter published in *Science* (183:303) by Paul Berg and ten other scientists on the potential biohazards of rDNA molecules. The signers comprised the Committee on Recombinant DNA Molecules, Assembly of Life Sciences, National Research Council, National Academy of Sciences

6 August
Letter from Roy Curtiss III to Berg committee on the biohazards and regulation of rDNA molecule research

7 October
The National Institutes of Health establishes the Recombinant DNA Molecule Program Advisory Committee

1975 January
Report of the Working Party on the Experimental Manipulation of the Genetic Composition of Microorganisms (United Kingdom) is published

24–27 February
International Conference on Recombinant DNA Molecule Research (Asilomar Conference), Asilomar Conference Center, Pacific Grove, California

28 February
First meeting of the Recombinant DNA Molecule Program Advisory Committee (RAC)

22 April
Senate Subcommittee on Health, Committee on Labor and Public Welfare holds hearings on genetic engineering (Edward Kennedy, chairman)

12–13 May
RAC meeting to frame guidelines for research with rDNA molecules

18–19 July
RAC meeting; subcommittee drafts provisional guidelines (Woods Hole Guidelines)

December
NIH workshop on the design and testing of safer cloning vehicles and hosts

4–5 December
RAC meeting; draft guidelines presented to NIH director

1976 9–10 February
Meeting of the NIH Director's Advisory Committee (DAC) open to public testimony on the guidelines for rDNA research

1–2 April
RAC meeting

9–12 June
Tenth Miles International Symposium on Recombinant DNA Molecules: Impact on Science and Society, MIT, Cambridge, Massachusetts. Meeting of the RAC working group on safer hosts and vectors

June
Office of Recombinant DNA Activities established at NIH

23 June
NIH Guidelines for Recombinant DNA Research released

28 June
Canadian government issues draft report on rDNA guidelines.

7 July
DHEW-NIH guidelines for rDNA research published in the *Federal Register* 41(131):27902–27943

August
Report of the Working Party on the Practice of Genetic Manipulation, United Kingdom, is published (Williams report)

August
Report issued by the Ad Hoc Committee on Recombinant DNA Molecules, International Council of Scientific Unions

19 August
NIH issues draft environmental impact statement on its rDNA guidelines

9 September
Draft EIS published in *Federal Register* 41(131):38425–38483

12 September
Meeting of RAC Working Group on Safer Hosts and Vectors

13–14 September
RAC meeting

22 September
Senate oversight hearings on the implementation of the NIH Guidelines. Joint hearing before the Subcommittee on Health, Committee on Labor and Public Welfare and Subcommittee on Administrative Practice and Procedure, Committee on the Judiciary

22 September
President Gerald Ford sends memorandum to all government department and agency heads requesting their cooperation in the formation of an interagency committee on rDNA research

October
Federal Interagency Committee established

4 November
First meeting of Federal Interagency Committee on Recombinant DNA Research

13 December
Meeting of a RAC Working Group on Safer Hosts and Vectors, Bethesda, Maryland

1977 15–16 January
RAC meeting

7–9 March
Academy Forum, National Academy of Sciences on Recombinant DNA Research, Washington, DC

15 March
Federal Interagency Committee issues report supporting rDNA legislation with suggested elements for such legislation

15–17 March
Hearings by House Subcommittee on Health and the Environment, Interstate and Foreign Commerce Committee on recombinant DNA bills

21 March
NIH Workshop on Safer Hosts and Plasmid Vectors, Bethesda, Maryland

29 March
Federal Interagency Committee issues recommendations on patent agreements developed under DHEW-NIH support

29–31 March
Hearings by House Subcommittee on Science, Research and Technology, Committee on Science and Technology on rDNA research and science policy. Hearings continued 27–28 April, 3–5, 25–26 May, 7–8 September

6 April
Senate hearings, Subcommittee on Health and Scientific Research, Committee on Human Resources, on rDNA bills

28–29 April
Meeting of the NIH Director's Advisory Committee on possible courses of action to regulate rDNA research

14–15 May
RAC meeting to consider revision of the NIH guidelines

20–21 June
Workshop on risk assessment of rDNA experimentation with *E. coli* (Falmouth Workshop) at Falmouth, Massachusetts

22–23 June
RAC meeting

14 July
Letter from Sherwood L. Gorbach, chairman of the Falmouth Workshop to Donald Fredrickson, on the consensus of the meeting

19 July
Workshop held on the design of tests for EK3 host vector systems by RAC

27 September
Proposed revisions of the NIH guidelines are published, *Federal Register* 42(187):49596–49609

October
DHEW-NIH publishes final environmental impact statement

31 October–1 November
RAC meeting

2 November
Senate hearings, Subcommittee on Science and Space, Committee on Commerce, Science & Transportation. Hearings continued 8, 10 November

November
Chang and Cohen experiment published in the *Proceedings of the National Academy of Sciences*; cited as evidence against the natural-barrier hypothesis

15–16 December
Meeting of the NIH Director's Advisory Committee

1978 26–28 January
US-EMBO Workshop to assess risks for rDNA experiments involving the genomes of animal, plant and insect viruses (Ascot Workshop), held in Ascot, England

20–21 March
Workshop on risk assessment of agricultural pathogens, sponsored by NSF, NIH, and the Department of Agriculture

6–7 April
RAC working group reviews the report of the Ascot Workshop

26 April
Meeting of RAC's Host-Plasmid Working Group

27–28 April
RAC meeting. Review of Ascot, Falmouth, and Agricultural Pathogens workshops

15–16 June
NIH Director's Advisory Committee meets to review proposed revised guidelines

28 July
DHEW-NIH publishes proposed revised guidelines for rDNA research with environmental impact assessment, *Federal Register* 43(146):33041–33178

2–3 August
RAC meeting

15 September
DHEW review committee established by Secretary Joseph A. Califano and headed by General Counsel F. Peter Libassi hears comments from the public on the proposed revisions to the guidelines

15 December
Secretary Joseph A. Califano announces (a) his approval of revised guidelines and that (b) he directed NIH to increase its risk studies for rDNA research, (c) RAC would have broadened public representation, (d) there would be a significant increase in public access to information, and (e) FDA-funded research would comply with the NIH guidelines

22 December
DHEW-NIH revised guidelines for rDNA research published, *Federal Register* 43(247):60080–600131; FDA issues notice of intent to propose rDNA regulations, *Federal Register* 43(247):60134–60135.

29 December
Fourteen new members named to RAC

1979 15–16 February
RAC meeting; first meeting with expanded public members

2 April
NIH publishes risk assessment program for rDNA research, *Federal Register* 44(64):19301–19314

1–4 April
International conference at Wye Agricultural College, Kent, England (Wye Conference) on the risks and benefits of rDNA research, sponsored by the Committee on Genetic Experimentation (COGENE) and the Royal Society of London

21–23 May
RAC meeting. Committee votes in favor of having mandatory compliance to the *Guidelines* for non-NIH funded institutions. Also votes to exempt from *Guidelines* rDNA molecules that are propagated and

maintained in cells in tissue culture that are derived entirely from nonviral components. Approved by director 20 July 1979, *Federal Register* 44(141):42914–42917

3 August
NIH publishes proposed voluntary compliance program, *Federal Register* 44(151):45868–45869

6–7 September
RAC meeting. Committee votes to exempt from *Guidelines* most cloning experiments in *E. coli* K12 when special vectors and nonconjugative plasmids are used

13 September
NIH publishes program to assess the risks of rDNA research as the final plan, *Federal Register* 44(179):53410–53413

30 November
Proposed revised guidelines published, *Federal Register* 44(232):69210–69234

6–7 December
RAC meeting

1980 29 January
Revised NIH Guidelines published, *Federal Register* 45(20):6724–6749. Includes policy on voluntary compliance by non-NIH-funded institutions.

6–7 March
RAC meeting

11 April
NIH publishes physical-containment recommendations for large-scale uses of organisms containing rDNA molecules, *Federal Register* 45(72):24968–24971. Large-scale recommendations are not a formal part of the *Guidelines*, but endorsed by the NIH Division of Safety

11–12 April
NIH risk-assessment workshop on potential immunological effects of rDNA products

5–6 June
RAC meeting

25–26 September
RAC meeting; first meeting of this committee with a nonscientist as chairman

21 November
Revised NIH *Guidelines* published superseding those issued 29 January 1980, *Federal Register* 45 (227):77384–77409

1981 8–9 January
RAC meeting

23–24 April
RAC meeting

10 June
NIH issues final plan for a program to assess the risks of rDNA research, *Federal Register* 46(111):30772–30778

1 July
Revised NIH *Guidelines* published superseding those issued 21 November 1980, *Federal Register* 46(126):34462–34487

9–10 September
RAC meeting; committee votes to make the *Guidelines* into a voluntary code of practice

Appendix B

Summary and Provisional Statements of the Asilomar Conference

Reprinted from *Proc. Nat. Acad. Sci. USA*, vol. 72, No. 6, pp. 1981–1984, June 1975.

Summary Statement of the Asilomar Conference on Recombinant DNA Molecules*

Paul Berg†, David Baltimore‡, Sydney Brenner§, Richard O. Roblin III¶, and Maxine F. Singer||

Organizing Committee for the International Conference on Recombinant DNA Molecules, Assembly of Life Sciences, National Research Council, National Academy of Sciences, Washington, D.C. 20418. † Chairman of the committee and Professor of Biochemistry, Department of Biochemistry, Stanford University Medical Center, Stanford, California; ‡ American Cancer Society Professor of Microbiology, Center for Cancer Research, Massachusetts Institute of Technology, Cambridge, Mass.; § Member, Scientific Staff of the Medical Research Council of the United Kingdom, Cambridge, England; ¶ Professor of Microbiology and Molecular Genetics, Harvard Medical School, and Assistant Bacteriologist, Infectious Disease Unit, Massachusetts General Hospital, Boston, Mass.; and || Head, Nucleic Acid Enzymology Section, Laboratory of Biochemistry, National Cancer Institute, National Institutes of Health, Bethesda, Maryland

I. Introduction and General Conclusions

This meeting was organized to review scientific progress in research on recombinant DNA molecules and to discuss appropriate ways to deal with the potential biohazards of this work. Impressive scientific achievements have already been made in this field and these techniques have a remarkable potential for furthering our understanding

*Summary statement of the report submitted to the Assembly of Life Sciences of the National Academy of Sciences and approved by its Executive Committee on 20 May 1975.

Requests for reprints should be addressed to: Division of Medical Sciences, Assembly of Life Sciences, National Academy of Sciences, 2101 Constitution Avenue, N.W., Washington, D.C. 20418.

of fundamental biochemical processes in pro- and eukaryotic cells. The use of recombinant DNA methodology promises to revolutionize the practice of molecular biology. Although there has as yet been no practical application of the new techniques, there is every reason to believe that they will have significant practical utility in the future.

Of particular concern to the participants at the meeting was the issue of whether the pause in certain aspects of research in this area, called for by the Committee on Recombinant DNA Molecules of the National Academy of Sciences, U.S.A., in the letter published in July, 1974** should end; and, if so, how the scientific work could be undertaken with minimal risks to workers in laboratories, to the public at large, and to the animal and plant species sharing our ecosystems.

The new techniques, which permit combination of genetic information from very different organisms, place us in an area of biology with many unknowns. Even in the present, more limited conduct of research in this field, the evaluation of potential biohazards has proved to be extremely difficult. It is this ignorance that has compelled us to conclude that it would be wise to exercise considerable caution in performing this research. Nevertheless, the participants at the Conference agreed that most of the work on construction of recombinant DNA molecules should proceed provided that appropriate safeguards, principally biological and physical barriers adequate to contain the newly created organisms, are employed. Moreover, the standards of protection should be greater at the beginning and modified as improvements in the methodology occur and assessments of the risks change. Furthermore, it was agreed that there are certain experiments in which the potential risks are of such a serious nature that they ought not to be done with presently available containment facilities. In the longer term, serious problems may arise in the large scale application of this methodology in industry, medicine, and agriculture. But it was also recognized that future research and experience may show that many of the potential biohazards are less serious and/or less probable than we now suspect.

II. Principles Guiding the Recommendations and Conclusions

Although our assessments of the risks involved with each of the various lines of research on recombinant DNA molecules may differ, few,

**Report of Committee on Recombinant DNA Molecules: "Potential Biohazards of Recombinant DNA Molecules," *Proc. Nat. Acad. Sci. USA* 71, 2593–2594, 1974.

if any, believe that this methodology is free from any risk. Reasonable principles for dealing with these potential risks are: (*i*) that containment be made an essential consideration in the experimental design and, (*ii*) that the effectiveness of the containment should match, as closely as possible, the estimated risk. Consequently, whatever scale of risks is agreed upon, there should be a commensurate scale of containment. Estimating the risks will be difficult and intuitive at first but this will improve as we acquire additional knowledge; at each stage we shall have to match the potential risk with an appropriate level of containment. Experiments requiring large scale operations would seem to be riskier than equivalent experiments done on a small scale and, therefore, require more stringent containment procedures. The use of cloning vehicles or vectors (plasmids, phages) and bacterial hosts with a restricted capacity to multiply outside of the laboratory would reduce the potential biohazard of a particular experiment. Thus, the ways in which potential biohazards and different levels of containment are matched may vary from time to time, particularly as the containment technology is improved. The means for assessing and balancing risks with appropriate levels of containment will need to be reexamined from time to time. Hopefully, through both formal and informal channels of information within and between the nations of the world, the way in which potential biohazards and levels of containment are matched would be consistent.

Containment of potentially biohazardous agents can be achieved in several ways. The most significant contribution to limiting the spread of the recombinant DNAs is the use of biological barriers. These barriers are of two types: (*i*) fastidious bacterial hosts unable to survive in natural environments, and (*ii*) nontransmissible and equally fastidious vectors (plasmids, bacteriophages, or other viruses) able to grow only in specified hosts. Physical containment, exemplified by the use of suitable hoods, or where applicable, limited access or negative pressure laboratories, provides an additional factor of safety. Particularly important is strict adherence to good microbiological practices which, to a large measure can limit the escape of organisms from the experimental situation, and thereby increase the safety of the operation. Consequently, education and training of all personnel involved in the experiments is essential to the effectiveness of all containment measures. In practice, these different means of containment will complement one another and documented substantial improvements in the ability to restrict the growth of bacterial hosts and vectors could permit modifications of the complementary physical containment requirements.

Stringent physical containment and rigorous laboratory procedures can reduce but not eliminate the possibility of spreading potentially hazardous agents. Therefore, investigators relying upon "disarmed" hosts and vectors for additional safety must rigorously test the effectiveness of these agents before accepting their validity as biological barriers.

III. Recommendations for Matching Types of Containment with Types of Experiments

No classification of experiments as to risk and no set of containment procedures can anticipate all situations. Given our present uncertainties about the hazards, the parameters proposed here are broadly conceived and meant to provide provisional guidelines for investigators and agencies concerned with research on recombinant DNAs. However, each investigator bears a responsibility for determining whether, in his particular case, special circumstances warrant a higher level of containment than is suggested here.

A. Types of containment

1. Minimal Risk. This type of containment is intended for experiments in which the biohazards may be accurately assessed and are expected to be minimal. Such containment can be achieved by following the operating procedures recommended for clinical microbiological laboratories. Essential features of such facilities are no drinking, eating, or smoking in the laboratory, wearing laboratory coats in the work area, the use of cotton-plugged pipettes or preferably mechanical pipetting devices, and prompt disinfection of contaminated materials.

2. Low Risk. This level of containment is appropriate for experiments which generate novel biotypes but where the available information indicates that the recombinant DNA cannot alter appreciably the ecological behavior of the recipient species, increase significantly its pathogenicity, or prevent effective treatment of any resulting infections. The key features of this containment (in addition to the minimal procedures mentioned above) are a prohibition on mouth pipetting, access limited to laboratory personnel, and the use of biological safety cabinets for procedures likely to produce aerosols (e.g., blending and sonication). Though existing vectors may be used in conjunction with low risk procedures, safer vectors and hosts should be adopted as they become available.

3. Moderate Risk. Such containment facilities are intended for experiments in which there is a probability of generating an agent with a significant potential for pathogenicity or ecological disruption. The principle features of this level of containment, in addition to those of the two preceding classes, are that transfer operations should be carried out in biological safety cabinets (e.g., laminar flow hoods), gloves should be worn during the handling of infectious materials, vacuum lines must be protected by filters, and negative pressure should be maintained in the limited access laboratories. Moreover, experiments posing a moderate risk must be done only with vectors and hosts that have an appreciably impaired capacity to multiply outside of the laboratory.

4. High Risk. This level of containment is intended for experiments in which the potential for ecological disruption or pathogenicity of the modified organism could be severe and thereby pose a serious biohazard to laboratory personnel or the public. The main features of this type of facility, which was designed to contain highly infectious microbiological agents, are its isolation from other areas by air locks, a negative pressure environment, a requirement for clothing changes and showers for entering personnel, and laboratories fitted with treatment systems to inactivate or remove biological agents that may be contaminants in exhaust air and liquid and solid wastes. All persons occupying these areas should wear protective laboratory clothing and shower at each exit from the containment facility. The handling of agents should be confined to biological safety cabinets in which the exhaust air is incinerated or passed through Hepa filters. High risk containment includes, in addition to the physical and procedural features described above, the use of rigorously tested vectors and hosts whose growth can be confined to the laboratory.

B. Types of experiments

Accurate estimates of the risks associated with different types of experiments are difficult to obtain because of our ignorance of the probability that the anticipated dangers will manifest themselves. Nevertheless, experiments involving the construction and propagation of recombinant DNA molecules using DNAs from (*i*) prokaryotes, bacteriophages, and other plasmids, (*ii*) animal viruses, and (*iii*) eukaryotes have been characterized as minimal, low, moderate, and high risks to guide investigators in their choice of the appropriate containment. These designations should be viewed as interim assignments which will need to be revised upward or downward in the light of future experience.

The recombinant DNA molecules themselves, as distinct from cells carrying them, may be infectious to bacteria or higher organisms. DNA preparations from these experiments, particularly in large quantities, should be chemically inactivated before disposal.

1. Prokaryotes, Bacteriophages, and Bacterial Plasmids. Where the construction of recombinant DNA molecules and their propagation involves prokaryotic agents that are known to exchange genetic information naturally, the experiments can be performed in minimal risk containment facilities. Where such experiments pose a potential hazard, more stringent containment may be warranted.

Experiments involving the creation and propagation of recombinant DNA molecules from DNAs of species that ordinarily do not exchange genetic information, generate novel biotypes. Because such experiments may pose biohazards greater than those associated with the original organisms, they should be performed, at least, in low risk containment facilities. If the experiments involve either pathogenic organisms or genetic determinants that may increase the pathogenicity of the recipient species, or if the transferred DNA can confer upon the recipient organisms new metabolic activities not native to these species and thereby modify its relationship with the environment, then moderate or high risk containment should be used.

Experiments extending the range of resistance of established human pathogens to therapeutically useful antibiotics or disinfectants should be undertaken only under moderate or high risk containment, depending upon the virulence of the organism involved.

2. Animal Viruses. Experiments involving linkage of viral genomes or genome segments to prokaryotic vectors and their propagation in prokaryotic cells should be performed only with vector-host systems having demonstrably restricted growth capabilities outside the laboratory and with moderate risk containment facilities. Rigorously purified and characterized segments of non-oncogenic viral genomes or of the demonstrably non-transforming regions of oncogenic viral DNAs can be attached to presently existing vectors and propagated in moderate risk containment facilities; as safer vector-host systems become available such experiments may be performed in low risk facilities.

Experiments designed to introduce or propagate DNA from non-viral or other low risk agents in animal cells should use only low risk animal DNAs as vectors (e.g., viral, mitochondrial) and manipulations should be confined to moderate risk containment facilities.

3. *Eukaryotes.* The risks associated with joining random fragments of eukaryote DNA to prokaryotic DNA vectors and the propagation of these recombinant DNAs in prokaryotic hosts are the most difficult to assess.

A priori, the DNA from warm-blooded vertebrates is more likely to contain cryptic viral genomes potentially pathogenic for man than is the DNA from other eukaryotes. Consequently, attempts to clone segments of DNA from such animal and particularly primate genomes should be performed only with vector-host systems having demonstrably restricted growth capabilities outside the laboratory and in a moderate risk containment facility. Until cloned segments of warm-blooded vertebrate DNA are completely characterized, they should continue to be maintained in the most restricted vector-host system in moderate risk containment laboratories; when such cloned segments are characterized, they may be propagated as suggested above for purified segments of virus genomes.

Unless the organism makes a product known to be dangerous (e.g., toxin, virus), recombinant DNAs from cold-blooded vertebrates and all other lower eukaryotes can be constructed and propagated with the safest vector-host system available in low risk containment facilities.

Purified DNA from any source that performs known functions and can be judged to be non-toxic, may be cloned with currently available vectors in low risk containment facilities. (Toxic here includes potentially oncogenic products or substances that might perturb normal metabolism if produced in an animal or plant by a resident microorganism.)

4. *Experiments to be Deferred.* There are feasible experiments which present such serious dangers that their performance should not be undertaken at this time with the currently available vector-host systems and the presently available containment capability. These include the cloning of recombinant DNAs derived from highly pathogenic organisms (i.e., Class III, IV, and V etiologic agents as classified by the United States Department of Health, Education and Welfare), DNA

containing toxin genes, and large scale experiments (more than 10 liters of culture) using recombinant DNAs that are able to make products potentially harmful to man, animals, or plants.

IV. Implementation

In many countries steps are already being taken by national bodies to formulate codes of practice for the conduct of experiments with known or potential biohazard.††,‡‡ Until these are established, we urge individual scientists to use the proposals in this document as a guide. In addition, there are some recommendations which could be immediately and directly implemented by the scientific community.

A. Development of safer vectors and hosts

An important and encouraging accomplishment of the meeting was the realization that special bacteria and vectors which have a restricted capacity to multiply outside the laboratory can be constructed genetically, and that the use of these organisms could enhance the safety of recombinant DNA experiments by many orders of magnitude. Experiments along these lines are presently in progress and in the near future, variants of λ bacteriophage, non-transmissible plasmids, and special strains of *Escherichia coli* will become available. All of these vectors could reduce the potential biohazards by very large factors and improve the methodology as well. Other vector-host systems, particularly modified strains of *Bacillus subtilis* and their relevant bacteriophages and plasmids, may also be useful for particular purposes. Quite possibly safe and suitable vectors may be found for eukaryotic hosts such as yeast and readily cultured plant and animal cells. There is likely to be a continuous development in this area and the participants at the meeting agreed that improved vector-host systems which reduce the biohazards of recombinant DNA research will be made freely available to all interested investigators.

††Advisory Board for the Research Councils, "Report of the Working Party on the Experimental Manipulation of the Genetic Composition of Micro-Organisms. Presented to Parliament by the Secretary of State for Education and Science by Command of Her Majesty, January 1975." London: Her Majesty's Stationery Office, 1975, 23pp.

‡‡National Institutes of Health Recombinant DNA Molecule Program Advisory Committee.

B. Laboratory procedures

It is the clear responsibility of the principal investigator to inform the staff of the laboratory of the potential hazards of such experiments before they are initiated. Free and open discussion is necessary so that each individual participating in the experiment fully understands the nature of the experiment and any risk that might be involved. All workers must be properly trained in the containment procedures that are designed to control the hazard, including emergency actions in the event of a hazard. It is also recommended that appropriate health surveillance of all personnel, including serological monitoring, be conducted periodically.

C. Education and reassessment

Research in this area will develop very quickly and the methods will be applied to many different biological problems. At any given time it is impossible to foresee the entire range of all potential experiments and make judgments on them. Therefore, it is essential to undertake a continuing reassessment of the problems in the light of new scientific knowledge. This could be achieved by a series of annual workshops and meetings, some of which should be at the international level. There should also be courses to train individuals in the relevant methods since it is likely that the work will be taken up by laboratories which may not have had extensive experience in this area. High priority should also be given to research that could improve and evaluate the containment effectiveness of new and existing vector-host systems.

V. New Knowledge

This document represents our first assessment of the potential biohazards in the light of current knowledge. However, little is known about the survival of laboratory strains of bacteria and bacteriophages in different ecological niches in the outside world. Even less is known about whether recombinant DNA molecules will enhance or depress the survival of their vectors and hosts in nature. These questions are fundamental to the testing of any new organism that may be constructed. Research in this area needs to be undertaken and should be given high priority. In general, however, molecular biologists who may construct DNA recombinant molecules do not undertake these experiments and it will be necessary to facilitate collaborative research between them and groups skilled in the study of bacterial infection or ecological microbiology. Work should also be undertaken which would enable us to monitor the escape or dissemination of cloning vehicles and their hosts.

Nothing is known about the potential infectivity in higher organisms of phages or bacteria containing segments of eukaryotic DNA and very little about the infectivity of the DNA molecules themselves. Genetic transformation of bacteria does occur in animals, suggesting that recombinant DNA molecules can retain their biological potency in this environment. There are many questions in this area, the answers to which are essential for our assessment of the biohazards of experiments with recombinant DNA molecules. It will be necessary to ensure that this work will be planned and carried out; and it will be particularly important to have this information before large scale applications of the use of recombinant DNA molecules is attempted.

The work of the committee was assisted by the National Academy of Sciences-National Research Council Staff: Artemis P. Simopoulos (Executive Secretary) and Elena O. Nightingale (Resident Fellow), Division of Medical Sciences, Assembly of Life Sciences, and supported by the National Institutes of Health (Contract NO1–OD–5–2103) and the National Science Foundation (Grant GBMS75–05293).

A five-page document had been handed out to the participants of Asilomar. It was then rewritten 26–27 February 1975 and submitted as a thirteen-page document to the Executive Committee, Assembly of Life Sciences of the National Academy of Sciences, under the title "Provisional statement of the conference proceedings" (PS). As a result of the discussions held and votes taken during the NAS session, substantial revisions were made in PS; the resulting document, written between March and May of 1975, was entitled "Summary statement of the Asilomar Conference on recombinant DNA molecules" (SS). The following remarks will compare these two documents.

The term "utmost caution" was used in PS to characterize how the rDNA work should be undertaken because of the considerable difficulty in evaluating the potential biohazards. The phrase was changed in section I of SS (SS.I) to "considerable caution," which has the effect of mitigating the urgency.

In section I of PS (PS.I) it is stated that "the work should proceed but with appropriate safeguards." At the final session of Asilomar, some participants were critical of that phrase. They believed, following the recommendation of the Plasmid Working Group, that some experiments should be postponed. The Organizing Committee modified its report to reflect the consensus view. In SS.I the following phrasing appeared: "[T]he participants at the Conference agreed that most of the work on construction of recombinant DNA molecules should proceed provided that appropriate safeguards . . . are employed." A statement endorsing the recommendation of the Plasmid Working Group stated, "Furthermore, it was agreed that there are certain experiments in which the potential risks are of such a serious nature that they ought not to be done with presently available containment facilities."

The sections in both documents referring to potential risks are substantially the same.[1] The key results and recommendations are (1) few agree the methodology is free of any risks; (2) containment should be built into the experimental strategy—combined use of physical and biological containment should be sought; (3) the effectiveness of the containment should be matched to the estimated risk; (4) estimating risks will be intuitive (in PS, "subjective") at first, but improved with new knowledge; (5) large-scale experiments are riskier than small-scale experiments.

In PS, three levels of containment were defined. Low containment consisted exclusively of "good medical microbiological techniques." Moderate containment was designed to provide both physical and biological barriers for the spread of DNA recombinants. For high containment, PS spoke of isolated facilities—the most advanced physically contained systems with filtered air, showers for persons exiting the facility, and negative air pressure. To each of the three levels of containment, specific classes of rDNA experiments were assigned. Table 10 illustrates how the levels of containment were matched with experiments in the provisional statement.

SS recommended four levels of containment: minimal, low, moderate, and high. It was acknowledged that the classification was "broadly conceived" and that each investigator had responsibility for determining whether higher containment was warranted. Essentially, the Organizing Committee divided category I in PS into two separate categories, low and minimal. Table 11 gives the containment levels matched with some classes of experiments as formulated in SS. Also, the category of deferred experiments was added to SS. Additional attention was given in SS to experiments in the moderate-risk category, which involved the use of biological safety cabinets, negative air pressure, and host-vector systems that are debilitated.

Table 10
Containment for selected classes of experiments from provisional statement of Asilomar

Level of containment	Type of rDNA experiment
I. Low	Organisms that normally exchange genetic information; involve no novel biotypes; most experiments involving the fusion of prokaryotic vectors with DNA from prokaryotes, lower eukaryotes, plants, invertebrates and cold-blooded vertebrates
II. Moderate	Placing bacterial genes into species in which such genes would not be found naturally; low-risk animal viruses used as vectors to place new genetic material into animal cells; joining of DNA of warm-blooded vertebrates to prokaryotic vectors; construction of recombinants between animal vectors and any DNA; viral genes linked to prokaryotic vectors and introduced into prokaryotic cells; rigorously purified fragments of nontransforming regions of oncogenic viral DNAs attached to plasmids and inserted into *E. coli*
III. High	High-risk viruses; fusion of eukaryotic or prokaryotic genes to prokaryotic vectors when the resultant organism is likely to express a toxic or pharmacologically active agent

The plenary session clearly played a role in shaping the final recommendations. However, for the vast majority of possible experiments listed in the moderate- or high-risk categories, there was very little change between the two documents. It is also clear from the two tables that the classification of risks by category did not arise out of a theoretical analysis of the problem, but reflected an inductive procedure incorporating disparate principles, conjectures, and special interests.

The criterion for low-risk experiments for prokaryotes as given by PS states, "Experiments involving organisms that *normally exchange genetic information, involve no novel biotypes* and pose no hazards that cannot be contained by the standard microbiological laboratory techniques appropriate for handling these organisms" (emphasis supplied). In SS, the criterion is simplified; minimal risk is identified with "prokaryotic agents that are known to exchange genetic information naturally."

For moderate- or high-risk experiments, the classifications in both PS and SS are substantially the same, with some changes in emphasis

Table 11
Containment for selected classes of experiments from summary statement of Asilomar

Level of containment	Type of rDNA experiment
I. Minimal	DNA recombinants from prokaryotes known to exchange genetic information naturally
II. Low	Creation and propagation of rDNA molecules from species that ordinarily do not exchange genetic information (novel biotypes)
III. Moderate	DNA recombinants involving pathogenic organisms, capable of increasing pathogenicity of recipient species, or that can confer new metabolic activities not native to the species; or that can extend the range of resistance of established human pathogensto therapeutically useful antibiotics or disinfects[a]
	Rigorously purified and characterized segments of nononcogenic viral genomes; nontransforming regions of oncogenic viral DNAs
	DNA from nonviral agents in animal cells
	DNA from primate genomes (with special host-vector systems)
	Uncharacterized DNA of warm-blooded vertebrates
IV. High	Some types of experiments with pathogenic organisms (unspecified) or experiments that could extend the host range of some virulent organisms

a. Refers to moderate or high containment, depending upon virulence of organism or potential for pathogenicity.

in SS: Moderate- and high-risk experiments are those for which the DNA transfers do not occur naturally. Both drafts cite a concern about experiments with antibiotic resistance genes, but SS has a clearer statement devoted exclusively to that issue: "Experiments extending the range of resistance of established human pathogens to therapeutically useful antibiotics or disinfectants should be undertaken only under moderate or high risk containment depending upon the virulence of the organism involved" (III.B.1).

Considerably more attention was placed on the risks of eukaryotic DNA experiments in SS than in PS. The following is a summary of what appeared in PS: (1) lower-risk experiments—lower eukaryotic, plants, invertebrates; (2) moderate-risk experiments—warm-blooded vertebrates joined to prokaryotic vectors, recombinants between animal vectors and any DNA; (3) high-risk experiments—eukaryotic genes that might express a toxic or pharmacologically active agent.

A special warning is offered in SS concerning experiments with eukaryotes that does not appear in PS: "[T]he DNA from warm-blooded vertebrates is more likely to carry "cryptic viral genomes" potentially pathogenic for man than DNA from other eukaryotes. Consequently, attempts to clone segments of DNA from such animal and particularly primate genomes should be performed only with the vector-host systems having demonstrably restricted growth capabilities outside the laboratory and in moderate risk facility."[2]

The term "pharmacologically active agent" was removed from PS and the term "toxic" was used by itself in SS. In the plenary session, some scientists wanted a deletion of the term "pharmacologically active" because, they argued, a substance can possess that property and still not be toxic. Insulin was the specific example used. It is pharmacologically active but, according to expert opinion, in the human gut it is not toxic. The term "toxic" was spelled out in SS, while it was left undefined in PS: "Toxic here includes potentially oncogenic products of substances that might perturb normal metabolism if produced in an animal or plant by a resident microorganism."

PS made a single reference to deferred experiments: "[A]ny large scale industrial, commercial, agricultural or other applications should be deferred pending the issuance of appropriate official guidelines by national scientific bodies." The debate at the plenary session whether the moratorium should be continued for some classes of experiments seemed to have an impact on those members of the Organizing Committee who redrafted the statement. They devoted an entire paragraph in SS to deferred experiments.[3] The statement is given as it appeared:

> There are feasible experiments which present such serious dangers that their performance should not be undertaken at this time with currently available vector-host systems and the present available containment capacity. These include the cloning of recombinant DNAs derived from highly pathogenic organisms (i.e., class III, IV, V etiologic agents as classified by the United States Department of Health, Education and Welfare), DNA containing toxic genes and large scale experiments (more than 10 liters of culture) using recombinant DNAs that are able to make products potentially harmful to man, animals or plants.

The change from PS to SS on deferred experiments was significant. There was a shift in emphasis on limiting rDNA activities from the industrial scale to include the scale of scientific experiments. It was a rare occasion in which scientists set some outer limits for their research, albeit a temporary boundary line for many who voted in favor of the provision.

The review process for overseeing laboratory compliance with the guidelines was spelled out more carefully in PS than it was in SS. The model proposed in PS called for institutional review committees empowered to grade the experiments according to containment needs and certifying the containment rating of the laboratory.[4] The certification rating would be appended to all funding requests to agencies that support rDNA work. Where there are doubts about a particular certification, the NIH Advisory Committee would be asked to render a decision.

In the NAS-approved report of Asilomar, no mention was made of institutional review committees.[5] Responsibility for implementation of the guidelines was not specified, although the principal investigator's role in promoting safe laboratory conditions was emphasized.

Both PS and SS have sections entitled "New knowledge," but that is where the similarity ends. In PS, this section consists of a list of six questions whose answers would presumably improve the risk assessment. These were listed in the form of research problems that could be translated into specific experiments, that is, "Can free DNA molecules infect animals or plants?" This entire section in PS was rewritten for SS. The outline of what needs to be known was given in more general terms. The influence of those trained in the ecology of microorganisms can be felt in this section far more than in PS. The key elements as they appeared in SS are given as follows: Little is known about the survival of laboratory strains of bacteria and bacteriophages in different ecological niches in the outside world. Even less is known about whether rDNA molecules will enhance or depress the survival of their vectors and hosts in nature. Nothing is known about the potential infectivity in higher organisms of phages or bacteria containing segments of eukaryotic DNA and very little about the infectivity of the DNA molecules themselves.

SS added a recommendation that special risk experiments be initiated to "monitor the escape or dissemination of cloning vehicles and their hosts." It ended with a special warning: "[I]t will be particularly important to undertake such tests before large scale applications of the use of recombinant DNA molecules is attempted."

Appendix C

Chronology of Potential Risks, 1971–1979

A *chronological list* gives the principal potential risks raised 1971–1979 at workshops, conferences, special reports, and congressional hearings, and in personal correspondence. They are numbered (1)–(19). Following this is a *categorical list* of potential risks; these categories are numbered (i)–(xvi). The categories are not meant to be logically exclusive. For example, the risks associated with the creation of new types of hybrid plasmids [category (iii)] and the risks associated with the spread of plasmids with antibiotic resistance [category (iv)] cover some of the same types of experiments. The particular categories were chosen because they represent the dominant mode of "risk discourse" for the period in question. The chronological list allows one to see how the identification of potential hazards developed through time. Nineteen episodes appear in the chronological list, but the appearance of an episode implies nothing about the consensus view concerning the likelihood of that hazard being realized.

Some scenarios were raised merely to examine worst-case possibilities. The chronology shows how new risks entered into the debate and which risks were the progenitors of others. Each of the sixteen categories of potential risks is followed by an indexing system that maps episodes in the chronological list to their respective categories. The number in, for example, episode 3B refers to the third episode in the chronological list, while the letter refers to the potential risk cited in that event by an alphabetical ordering (B is the second citation). Following the chronological and categorical lists are *summary discussions* of each of the sixteen categories and nineteen episodes.

Chronological list of potential hazards

(1) CSH Tumor Virus Workshop, June 1971

Animal tumor viruses cloned in *E. coli*

(2) Adeno-SV40 virus episode, 1971–1973

Genetic recombinant of a common human pathogen and a tumor virus

(3) Gordon Conference, June 1973

New types of hybrid plasmids or viruses

Large scale preparation of animal viruses

(4) Berg letter, July 1974

Plasmids with R factors

Cloning toxin genes

Cloning animal viruses

Shotgun cloning of animal DNA

(5) Curtiss memo, August, 1974

Creating plasmids with new traits

Cloning viral DNA from plants and animals in bacteria

Prokaryote-eukaryote hybrids

E. coli pathogenesis

(6) Pre-Asilomar letters, September 1974–January 1975

Bacteria with plant toxin genes

Genetic alteration of plants

Cellulose-degrading *E.coli*

Immunological hazards

Disturbance of biochemical pathways in ecological systems

Troublesome proteins in *E.coli*

(7) Asilomar Conference, February 1975

Eukaryotic toxins in bacteria

Altering the host range of bacteria

Animal viruses used as vectors

Military misapplications of rDNA technology

Genes in bacteria coding for pharmacologically active agents

Cloning animal virus DNA

Shotgunning eukaryote DNA in prokaryotes

Spread of antibiotic resistance plasmids

(8) Senate Health Subcommittee, April 1975

New combinations of antibiotic resistance genes

DNA recombinants inducing infections unsusceptible to known therapies

Animal tumor virus genes in bacteria

(9) DAC meeting, February 1976

E.coli pathogenesis

New routes of infection

Novel pathogens by fusing eukaryotic genes into prokaryotes

(10) Senate Subcommittee on Health and Administrative Practice and Procedure, September 1976

Spreading of "experimental cancer"

Increasing *E.coli* infections

Virulent hybrid of two benign viruses

Breaching species barriers: unknown impact on the biosphere

(11) House Subcommittee on Science, Research and Technology, March–September 1977

Pathogenic potential of *E.coli* K12

New environmental niches for bacteria

Colonizability of *E.coli* K12 in compromised hosts

Altering the course of evolution

Technological fix for complex problems

Transmission of *E.coli* K12 recombinants to wild strains

Recombinant organisms interfering with normal physiological functions and host defense mechanisms

Selective advantage of recombinant organisms

Expression of any foreign gene in *E.coli* could cause problems

(12) NAS Forum, March 1977

Active polypeptides in *E.coli*

Latent tumor viruses

Extraintestinal *E.coli* infections

Insulin genes in *E.coli*

Spread of antibiotic resistance

Breakdown products of recombinant DNA molecules

Technological fix for complex problems

Unknown evolutionary impacts of hybrid organisms

(13) Senate Subcommittee on Health, Science and Research, April 1977

Any gene can be harmful

Autoimmune response

E. coli K12 pathogenicity

Gene products not subject to host regulatory mechanisms

Unanticipated hazards

Transmission of genes from *E.coli* K12 to wild strains

Recombinant disturbing metabolic pathways

Recombinants conferring selective advantage to pathogenic strains

(14) Falmouth Workshop, June 1977

Transfer of DNA from *E.coli* K12 to wild strains

Creation of more pathogenic bacteria

Transforming *E.coli* K12 into a pathogen

Autoimmune response

Increased spread of R-factors

Breaching species barriers posing unforseen and unpredictable hazards

(15) EIS-NIH 1976 guidelines, October 1977

Irreversibility of a biohazard

Transmitting virulence factors in the gut

Increasing infectivity, virulence or host specificity

Biological warfare

Novel self-replicating pathogens

New ecological niches established

Any gene has the potential of being harmful

(16) Ascot Workshop, January 1978

Defective viral recombinants aided by helper virus

New routes of virus infection

Altering immune response

Penetration of recombinants into gut lining

Extra-intestinal infections

Cloning of DNA copies of viroids (naked RNA viruses) or complete genome of DNA viruses in *E.coli*

(17) Workshop on Agricultural Pathogens, March 1978

More virulent strains of plant pathogens

E.coli transformed into a plant pathogen

Hazards resulting from plant pathogens as host-vector systems

Extending the host range of plant pathogens that adversely affect plants and humans

(18) Beckwith letter, October 1978

Immunogenic effects of eukaryotic proteins in enteric organisms

(19) British Select Committee on Science and Technology, December 1978–March 1979

Increased virulence or range of pathogens

Changes in genetic constitution affecting cellular differentiation

Cryptic viral elements might be mobilized

Autoimmune reaction

Disturbances of endocrine function

Ecological consequences

Disturbances of cell metabolism

Wheat gluten immunogen in *E.coli*

Increased antibiotic resistance
Plasmid transfer

Categorical list of potential hazards

(i) Cloning viruses
(1);(3B);(4C);(5B);(7C);(7F);(8C);(10A);(12B);(15A);(15B);(16A);(16F);(19A);(19C)

(ii) Recombinant of a common human pathogen and tumor virus
(2);(9C);(10C)

(iii) New types of hybrid plasmids
(3A);(5A);(4C)

(iv) Spread of Plasmids with R factors
(4A);(7H);(8A);(12E);(14E);(19I)

(v) Creating *E. coli* pathogens
(4B);(4D);(6C);(7A);(8B);(9A);(9B);(9C);(10B);(11A);(12C);(13C);(14B);(14C);(15B);(15E);(16B);(16D);(17B)

(vi) Shotgun cloning mammalian DNA
(4D);(7G);(19C)

(vii) Breaching species barriers
(4C);(9C);(10D);(12H);(19B)

(viii) Autoimmune effects
(6D);(11G);(13B);(14D);(15C);(16C);(18);(19D);(19H)

(ix) *E. coli* with troublesome proteins
(6F);(7E);(12A);(12D);(13D);(13G);(16C);(18);(19B);(19E);(19G);(19H)

(x) Technological fix
(11E);(12G)

(xi) Transmission of DNA from *E. coli* K12 to wild strains
(11F);(12F);(13F);(14A);(19J)

(xii) Any gene can be harmful
(11I);(13A);(15G)

(xiii) Creation or enhancement of plant pathogens
(6A);(14B);(17A);(17C);(17D)

(xiv) Ecological consequences
(6B);(6E);(7B);(7E);(11B);(11C);(11H);(13H);(14F);(15A);(15C);(15F);(17C);(17D);(17A)

(xv) Altering the course of evolution
(11D);(14F);(19F)

(xvi) Military misapplications
(7D);(15D)

Summary discussions of categories (i)–(xvi)

(i) This prototype risk raised early in the debate was over experiments that would implant the DNA of animal tumor viruses into *E. coli*, whereupon the DNA would be reproduced with the bacterial progeny. Concern was over the establishment of animal tumor viruses in a bacterial host that is a close symbiont to man. Specific questions were, Would the virus become active in its new bacterial host? If the *E. coli* carrying the tumor virus genes colonized the human gut, could it result in new forms of pathology? The virus most cited was SV40, an organism known to transform human cells in tissue culture, cause tumors in mammals, and infect humans. Subsequent concerns were directed at other animal viruses not specifically associated with oncogenicity.

(ii) It was feared that by combining a known human pathogen such as a strain of the adenoviruses with an animal tumor virus, the resulting hybrid might have the pathogenic characteristics of the tumor virus and the host range of the human virus. Thus, a virus like SV40 or polyoma might be given a new route of infection into humans. An example of this was the natural laboratory recombination of adenovirus 2 (normally infecting infants and young children but not a serious pathogen) with SV40 virus.

(iii) Plasmids are circular extrachromosomal elements of DNA that exist in the cytoplasm of some cells. Natural genetic exchange takes place between the cells of some organisms through plasmid transfer by a process called conjugation. Plasmid engineering became the mainstay of the new rDNA technology. It was feared that by carefully engineering new plasmids or transposing plasmids, genetic material from species that were widely separated from each other phylogenetically could be brought together for reprogramming cells with potential hazards of (1) creating new types of pathogens or (2) enhancing or extending the host range of pathogens. An example is the proposed implanting of a cellulose-degrading plasmid into *E. coli* with the unknown effects of having this enteric organism with the cellulase gene occupying the human gut.

(iv) While similar to category (iii) the concerns associated by this issue were sufficiently intense to warrant special attention. A convenient way for molecular biologists to "mark" targeted genes so that one could follow fragments of genetic material before and after recombination is to "tag" the organism with a gene that confers on it resistance to a specific antibiotic. The organisms carrying such resistance could easily be selected out from a grouping of clones since they alone would survive treatment with the antibiotic. The creation of new plasmids with antibiotic resistance could potentially compromise human hosts if the plasmids are transferred from the laboratory to human cells. Particular attention was given to clinically used antibiotics. The principal question of interest was whether recombinant activities using antibiotic resistance factors (R factors) in plasmids would increase the human resistance to antibiotics by the transfer of the plasmids into bacteria that become resident in the intestinal flora.

(v) Since *Escherichia coli* was the most utilized host for rDNA experiments, there was intense controversy over having a symbiont to man as the recipient of foreign DNA. Critics speculated about scenarios in which the laboratory strain of *E. coli* K12 would become a pathogen. They also questioned whether genetically altered *E. coli* would take on novel properties and become competitive with more vigorous strains that persist in the environment.

When the criticism was answered by citing the poor ability of K12 to survive in nature, some scientists expressed skepticism about *E. coli* K12's lack of ability to colonize humans or pointed to their ability to transmit potentially hazardous genes to wild strains of *E. coli*.

(vi) Shotgun cloning refers to a practice of exposing the whole genome of a donor organism to restriction enzymes, thereby producing an undefined number of unspecified fragments. These fragments are then combined with an appropriate vector and inserted randomly into the host organism, allowing the researcher to examine the random fragments for special functional characteristics. Since the donor organism's genome is not completely characterized, the risk is associated with cloning and propagating a genetic sequence which could prove harmful in its new cell environment. Cryptic animal tumor viruses were among the things some scientists feared would be implanted into *E. coli* by shotgunning animal DNA.

(vii) Like many of the hypothetical risks associated with rDNA research, those associated with the breaching of species barriers rested on the question whether nature had been undertaking its own form of genetic intercourse between prokaryotes and eukaryotes. If that has in fact been happening, in what manner and at what frequency? The debate over risks was caught in the semantics of the nature of such a natural species barrier. Those who postulated such barriers were fearful that the promiscuous disregard of discrete species would result in unpredictable changes in the direction of evolution itself and that the pace of those changes would be too great for homeostatic conditions to prevail.

(viii) A risk scenario that was taken seriously only late in the debate, although it had been raised by scientists on earlier occasions, was directed to immunologists rather than infectious-disease epidemiologists. This is the concern that a recombinant molecule implanted in an enteric organism will initiate some autoimmune response. The scenario is as follows: The DNA coding for some animal protein is cloned in *E. coli*. The *E. coli* challenges a human host and produces a protein product that is similar but not exactly like a naturally occurring substance in the human body. When the foreign protein is produced in the *E. coli* that has colonized a person, the individual's immune system begins to operate. But the immune system fails to discriminate between the foreign product and the natural product. In trying to eliminate the invading substance, the individual's antibodies attack his own vital materials. It was conjectured that this could be fatal in the case of an autoimmune response to insulin or growth hormone; in different circumstances, it was conjectured, such a response could lead to chronic or nonspecific illness.

(ix) The concern here is that foreign proteins produced in *E. coli* could disrupt a normal human metabolic function or break an immunological threshold. Also considered was that a recombinant product would make a particular infection process more likely or more severe. Some suggested that if a recombinant product escaped from the recognition of the normal defense mechanisms, it could alter the nature of the organism's self-regulating process. This risk was two sided; immune system recognition of a unique foreign protein could alter the host regulatory process, while nonrecognition would mean new types of infection might result.

(x) The notion of the technological fix was applied to almost all applications of rDNA research and brought forth an archetypal question: "Just because we are *able* to do something, does that mean we *should* do it?" The concerns of critics using this argument were of two basic types: (1) Most of the applied problems for which the proponents wished to seek solutions with rDNA were not technological in origin—they were sociopolitical—and thus more resources were needed in other areas. (2) In a corollary to the first point, critics questioned the *true* need for the benefits of rDNA (e.g., the proposed production of human insulin), either because the benefits did not justify the risks or because the benefits could be achieved by other means.

In summary, the critics felt that a certain reductionist force was in play here that would amplify the move to "change the gene" instead of "changing society or cleaning up the environment." A potential risk was raised that the only answer to the problems brought about by technology running amok was more technology.

(xi) Even if *E. coli* K12 could not survive or colonize the human gut (which was debated), there was evidence that genetic information could be exchanged between strains during interaction in the gut. One process of exchange between species of bacteria through plasmid transfer was known as conjugation. The frequency and relevance of plasmid transfer was in question and a typical risk scenario went as follows: *E. coli* K12 carrying harmful genes is unwittingly ingested by the human. The organisms die out quickly in the gut, but some exchange plasmids with the natural bacterial inhabitants in the intestinal flora; thus the genetic information becomes colonized in the gut even though it was not colonized by the organism of origin.

(xii) While most of the risk arguments related to the nature of the genetic material provided by the donor, the vector, and the host organisms, this unique argument was based upon a contextualist view of genes according to which the particular properties of a transplanted gene depends upon *where* the gene is in relation to (1) other genes and (2) the entire cellular environment. Thus, it was considered an unknown but nontrivial risk to take *any* gene out of its "natural" context and transport it to a new cellular environment. The new gene sequence could affect the new host cell; conversely, the new cell could alter the functional state of the gene. The argument was an attack on the principal dogma of biology—one gene, one product—and a critique of reductionist thinking: A gene does what it does, or does nothing. The notion of emergent properties weighed heavily in the argument.

(xiii) Because most of the early debate was anthropocentric in character, there were few risks brought up concerning nonhuman species per se. However, at conferences that focused on plant pathogens, scenarios were considered through which rDNA-created or -enhanced pathogens threatened economically important plant species. There were basically three areas of concern: (1) Could recombinants of human pathogens (or subpathogens) be altered enough to extend their pathogenic properties to plants? (2) Since there were plant pathogens being used in host-vector systems, some people raised the risk of spreading plant pathogenesis. (3) Concern was also raised that some organisms are pathogenic for both plants and humans.

(xiv) This category of potential risks questioned whether the appearance of new recombinant organisms would upset the "balance of nature" by disturbing certain biochemical cycles, changing the microflora in some ecological system, or making severe evolutionary changes. Examples cited were a wild oil-eating bacteria or establishing the nif (nitrogen fixation) gene in undesirable vegetation.

(xv) This line of argument, associated with scientists like Robert Sinsheimer and George Wald, maintained that the new technology would place in the hands of science the power to direct evolutionary changes. They believed that natural evolutionary changes took place in complex interactions between all the species on earth over millions of years. They considered the manipulation of evolutionary change by man as potentially disruptive to the checks and balances that appear in natural evolutionary processes.

(xvi) This concern was raised prior to and during Asilomar but soon after drew little attention. Those who expressed fear of the military abuse of DNA technology were thinking of the secret chemical- and biological-warfare (CBW) research supported by the military at Fort Detrick. The CBW activity at Fort Detrick was discontinued when an international treaty was signed restricting chemical and biological warfare.

Summary discussions of episodes (1)–(19)

(1) This was a three-week-long workshop on tumor viruses held at the Cold Spring Harbor Laboratories in New York during the month of June 1971. A key event in the class taught by Robert Pollack on animal cell culture was his interaction with Janet Mertz, who was then a graduate student with Paul Berg. The episode represents one of the progenitors of the rDNA debate.

(2) Adenoviruses infect the upper respiratory tract, adenoids, and tonsils of humans especially children. SV40 is a virus that can cause tumor growth in mammals and transform human tissue culture cells. Andrew Lewis, a researcher with the National Institute of Allergy and Infectious Diseases, discovered that a hybrid was created that contained genetic sequences of both SV40 and adenoviruses. Lewis reported his findings at the Cold Spring Harbor Tumor Virus Meeting in August 1971. The report and the subsequent constraints placed on the dissemination of the hybrid helped to establish a climate for the rDNA debate. For the next couple of years a memorandum of understanding (which was originally composed by Lewis and fellow workers) was voluntarily read and signed by any researcher who wanted to use these hybrid viruses.

(3) The Gordon Conference, which has become an annual event, was officially called The Gordon Conference on Nucleic Acids when it was held at the New Hampton School, New Hampshire, in June 1973. The cochairpersons were Maxine Singer and Dieter Söll. At the conference Herbert Boyer discussed unpublished results from experiments that he and Stanley Cohen performed utilizing recombinant methods. In response to this notice, this Gordon Conference generated what is known as the Singer-Söll letter, which expressed some fears concerning rDNA research.

(4) Paul Berg of Stanford University, leader of an ad hoc committee of scientists, known as the Berg committee, wrote a letter to the scientific community expressing concerns about research with rDNA molecules. The letter also asked scientists to refrain voluntarily from two classes of rDNA work until risks could be properly ascertained; the risk assessment was to begin at an international conference suggested by the signers of the letter. The Berg committee officially was known as the Committee on Recombinant DNA Molecules of the Assembly of Life Sciences, National Research Council, National Academy of Sciences. The letter appeared in *Science, Nature* and the *Proceedings of the National Academy of Sciences, USA*.

(5) The Curtiss memo was a sixteen-page letter by Roy Curtiss, III, of the Microbiology Department of the University of Alabama Medical Center. Written 6 August 1974, it eventually reached more than one thousand scientists worldwide. Generally, this letter was a response to, and an affirmation of, the risks and proposed moratorium recommended by the Berg letter. There was also an elaborate description of

Curtiss's own work—the study of bacteria that are involved in tooth decay. Curtiss conjectured on how the changes of the ecology of bacteria by recombinant work might be harmful.

(6) These letters were generally responses to inquiries made by the organizers of the Asilomar Conference. They expressed concerns about risks, responded to particular scenarios, and discussed policy issues and the format of the conference. The letters included in this study are dated from after the Berg letter and Curtiss memo (September 1974) to the Asilomar Conference itself (February 1975).

(7) Sometimes referred to as Asilomar II (to distinguish it from a biohazards conference held in 1973 at the same location), but often simply called the Asilomar Conference or the International Conference on Recombinant DNA Molecules. It was held at the Asilomar Conference Center near Monterey, California, from 24 to 27 February 1975. The conference was organized in response to the Berg committee and Berg letter and other scientific commentaries that raised questions of possible risks, regulations, guidelines, and the scope and duration of a voluntary moratorium.

(8) After the framework for guidelines emerged from the Asilomar Conference, the long process of developing guidelines was set into motion. The first congressional hearing on rDNA research was before the Senate Subcommittee on Health of the Committee on Labor and Public Welfare. Testimony was drawn from four speakers primarily on the relation of a free society to its scientific community. The hearings were held 22 April 1975 and presided over by Senate Subcommittee Chairman Edward M. Kennedy (D-MA).

(9) DAC stands for the Director's Advisory Committee. It is a group that occasionally advised the director of the NIH (then Donald Fredrickson) on issues concerning public-policy-related scientific matters. Fredrickson reactivated this committee for the occasion to advise him on important policy decisions concerning rDNA research. The meetings were held 9–10 February 1976. For the transcripts of the two-day meeting see US DHEW-NIH, *Recombinant DNA Research*, vol. 1, August 1976.

(10) Known as the *Oversight Hearings on Implementation of the NIH Guidelines Governing Recombinant DNA Research* these hearings were held on September 22, 1976 before the Sub-Committee on Health of the Committee on Labor and Public Welfare and the Sub-Committee

on Administrative Practice and Procedure of the Committee on the Judiciary of the U.S. Senate. The chairman for both Subcommittees was Senator Edward M. Kennedy.

(11) Published under the title "Science policy implications of DNA recombinant molecule research," these hearings were before the Subcommittee on Science, Research and Technology of the Committee on Science and Technology of the US House of Representatives. The hearings were dated 29–31 March, 27–28 April, 3–5 and 25–26 May, and 7–8 September 1977. The subcommittee chairman was Representative Ray Thornton of Arkansas. These were the most extensive hearings carried out by a congressional subcommittee and included a broad range of scientific experts, policy analysts, and laypersons.

(12) The National Academy of Sciences Forum on Research with Recombinant DNA was held in Washington, DC, 7–9 March 1977. The forum was divided into five parts: (1) mapping the mammalian genome; (2) the dangers of planned or inadvertant laboratory infections and epidemics; (3) pharmaceutical applications; (4) food production and rDNA techniques; (5) genetic engineering: the future. The forum proceedings were published by the NAS in 1977.

(13) The hearings before the subcommittee on Health, Science and Research of the Committee on Human Resources of the US Senate were to review Senate bill S. 1217, The Recombinant DNA Regulation Act of 1977. The hearings took place 6 April 1977; the chairman of the subcommittees was Senator Edward M. Kennedy.

(14) The Workshop on Risk Assessment of Recombinant DNA Experimentation with *Escherichia coli* K12 was held at Falmouth, Massachusetts, 20 and 21 June 1977. The workshop was significant as it addressed a specific area of risk in the debate, the potential pathogenicity of *E. coli* K12, which was the host organism of choice for most researchers; also, the workshop brought together experts from a wider range of scientific disciplines than had been convened on previous occasions, such as infectious-disease epidemiology and bacterial ecology, to address the risks of genetic engineering. The proceedings of the workshop were published in a special volume of *The Journal of Infectious Diseases* 137 (May 1978), Sherwood L. Gorbach, ed.

(15) The final environmental impact statement (EIS) appeared in October 1977 and was a culmination of the draft EIS [*Federal Register* (9 September 1976)], the NIH *Guidelines*, and numerous written com-

ments concerning research with rDNA molecules. The official title is "The National Institute of Health environmental impact statement on NIH guidelines for research involving recombinant DNA molecules." The EIS appeared in two parts (DHEW publication numbers 1489, 1490).

(16) On 26–28 January 1978 a joint US-EMBO (European Molecular Biologists Organization) Workshop to Assess Risks for Recombinant DNA Experiments Involving the Genomes of Animal, Plant and Insect Viruses was held in Ascot, England. The workshop was sponsored by the NIH in response to discussions concerning viruses at the Director's Advisory Committee (DAC) meeting of December 1977. The chairmen for the Ascot workshop were Malcolm Martin and Wallace Rowe of the National Institute of Allergy and Infectious Diseases (United States) and John Tooze of EMBO. A report of the workshop can be found in the *Federal Register* 43:33159–33167 (28 July 1978).

(17) The Workshop on Risk Assessment of Agricultural Pathogens was conducted by the Recombinant DNA Molecule Advisory Committee and sponsored by the Department of Agriculture, the National Science Foundation, and the National Institutes of Health. It was held in Washington, DC, 20–21 March 1978. The chairmen were Milton Zaitlin and Peter Day. A report on the workshop can be found in the *Federal Register* 43:33174–33178 (28 July 1978).

(18) A correspondence by Jonathan Beckwith of the Harvard Medical School to members of the Recombinant DNA Molecule Advisory Committee (RAC) expressed concern about potential adverse immunological reactions due to DNA recombinant molecules. The letter and various responses highlighted immunological effects as an area of potential concern.

(19) The potential risks are taken from the publication entitled *The Second Report from the Select Committee on Science and Technology of the United Kingdom*, the published proceedings of hearings from December 1978 to March 1979 of this British Committee of Parliament. It also included minutes of evidence taken before the Genetic Engineering Sub-Committee, a report from visits to the United States, an update from work done by the Williams Working Party in 1974, and a memorandum submitted by the Association of the British Pharmaceutical Industry.

Notes

Introduction

1. Plato, *The Republic*, translated by F. M. Cornford (Oxford: Oxford University Press, 1965).

2. Mannheim's views on rationality and public decision making appear in two works: *Ideology and Utopia* (New York: Harcourt, 1936); *Man and Society in an Age of Reconstruction* (New York: Harcourt, 1949).

3. John Friedman, *Retracking America: A Theory of Transactive Planning* (Garden City, NY: Doubleday, 1973).

4. Thomas S. Kuhn, *The Structure of Scientific Revolutions* (Chicago: University of Chicago Press, 1970).

5. In an article by Harold P. Green entitled "The recombinant DNA controversy: A model of public influence," he mistakenly develops the analogy between gene splicing and nuclear engineering in his statement: "Scientific experiments and resultant technology have the potential in both cases for enormous benefit and catastrophic disaster." The risks associated with recombinant DNA techniques are generally regarded as hypothetical. See *The Bulletin of the Atomic Scientists* 34:12 (November 1978).

Chapter 1

1. The Gordon Conferences are informal occasions for the exchange of new and unpublished scientific information. They are held in the summer. The Nucleic Acid Conference took place in June 1973.

2. A letter was published in *Science* 181:1114 (21 September 1973).

3. A standard definition of "recombinant DNA technique" is the formation of molecules that consist of different segments of DNA that have been joined together in cell-free systems and that have the capacity to infect and replicate in some host cell, either autonomously or as an integrated part of the host's genome. (The genome is the totality of an organism's genetic information.)

4. Besides the electronic detection of infiltrators, the plan called for the seeding of the zone with minefields to blow off their feet.

5. Joseph Weizenbaum, *Computer Power and Human Reason: From Judgment to Calculation* (San Francisco: W. H. Freeman, 1976), pp. 274–275.

6. For an account of chemical and biological activities during the Viet Nam

war, see Seymour M. Hersh, *Chemical and Biological Warfare: America's Hidden Arsenal* (Indianapolis: Bobbs-Merrill, 1968).

7. For an informative review, see David Nichols, "The associational interest groups of American science," in Albert H. Teich, ed., *Scientists and Public Affairs*, (Cambridge, MA: MIT Press, 1974), pp. 123–170.

8. Aaron Novick, "A plea for atomic freedom," *New Republic* 114:400 (25 March 1946). Quoted *ibid.*, p. 165, note 16.

9. Jeremy J. Stone, "'Greening of America' raises questions for FAS," *F.A.S. Newsletter* 23:2 (December 1970). Quoted in Teich, *op. cit.*, p. 137. A bracketed word in a quotation indicates it does not appear in the original text but was introduced for stylistic or grammatical purposes. A bracketed upper-case letter in a quotation indicates that it was changed in the original text from a lower-case letter.

10. Teich, *op. cit.*, p. 145, note 7.

11. *Ibid.*, pp. 144–145.

12. *Ibid.*, pp. 146–147.

13. *Ibid.*, pp. 149 *et passim*.

14. *Ibid.*, pp. 150ff.

15. *Ibid.*, pp. 151–152.

16. For example, MCHR provided medical care to the August 1968 demonstrators at the Democratic National Convention in Chicago and to numerous antiwar demonstrations locally and nationally. In several cities, local MCHR chapters worked closely with the Black Panther party to set up free clinics or conduct community sickle-cell or lead-screening programs.

17. In 1970, for instance, MCHR presented the following demands to the National AMA: (1) End racism in the health-care delivery system. (2) End oppression of women by and in the system. (3) End war collaboration by the health industry. (4) Socialize the health-care system. Demands given in Quentin D. Young, "Welcome to Chicago," *The Body Politic* 1:4 (July–August 1970). Quoted in Nichols, *op. cit.*, p. 153, note 7.

18. Nichols, *op. cit.*, pp. 155ff, note 7.

19. *Science for the People* 3:6–7 (February 1971). Quoted in Nichols, *op. cit.*, p. 156.

20. Nichols, *op. cit.*, p. 157, note 7.

21. *Ibid.*, p. 159.

Chapter 2

1. Paul Berg, transcript of an interview by Charles Weiner, 17 May 1975, Recombinant DNA Controversy, Oral History Collection, Institute Archives

and Special Collections, MIT Libraries, Cambridge, MA [hereafter IASC, MIT], p. 17 *et passim.*

2. Janet Mertz, transcript of an interview by Mary Terrall, 9 March 1977, Recombinant DNA Controversy, Oral History Collection, IASC, MIT, p. 11.

3. David Baltimore, transcript of an interview by Charles Weiner and Rae Goodell, 13 May and 22 July 1975, Recombinant DNA Controversy, Oral History Collection, IASC, MIT, p. 37.

4. Norton Zinder, transcript of an interview by Charles Weiner, 2 September 1975, Recombinant DNA Controversy, Oral History Collection, IASC, MIT, p. 54.

5. Baltimore, *op. cit.* pp. 37–38, note 3.

6. Zinder, *op. cit.*, pp. 58–59, note 4.

7. *Ibid.*, p. 56.

8. Mertz, *op. cit.*, p. 10, note 2.

9. Berg, *op. cit.*, pp. 18–19, note 1. In accordance with his wishes, we have agreed to accept Dr. Berg's edited version of the citation in the transcript of his interview. The editing does not alter the substance of that interview, but was done to improve clarity and style. All other direct citations from his interview have been similarly edited.

10. S. E. Luria, J. E. Darnell, D. Baltimore, Allan Campbell, *General Virology*, 3rd ed. (New York: John Wiley and Sons, 1978), pp. 234–235.

11. Mertz, *op. cit.*, p. 5, note 2.

12. J. Beckwith, "Gene expression in bacteria and some concerns about the misuse of science," *Bacteriological Review* 34:222–27 (September 1970); also, Jim Shapiro et al., [letter], *Nature* 224:1337 (27 December 1969).

13. Mertz, *op. cit.*, p. 6, note 2.

14. *Ibid.*, p. 7.

15. Robert Pollack, transcript of an interview by Mary Terrall, 27 March 1976, Recombinant DNA Controversy, Oral History Collection, IASC, MIT, p. 22.

16. Mertz, *op. cit.*, p. 17, note 2.

17. *Ibid.*, p. 18.

18. Pollack, *op. cit.*, p. 27, note 15.

19. Mertz, *op. cit.*, pp. 19–20, note 2.

20. Berg, *op. cit.*, p. 25, note 1.

21. Baltimore, *op. cit.*, pp. 38–39, note 3.

22. Daniel Singer, transcript of an interview by Rae Goodell, 28 July 1975, Recombinant DNA Controversy, Oral History Collection, IASC, MIT, p. 5.

Singer interprets these comments in his interview as meaningful only if laboratory personnel face nontrivial risks. Personal correspondence, 8 May 1979.

23. Berg, *op. cit.*, p. 26, note 1.

24. Leon Kass to Paul Berg [letter], 30 October 1970, Recombinant DNA Controversy, Oral History Collection, IASC, MIT.

25. Berg, *op. cit.*, p. 26, note 1.

26. *Ibid.*, p. 27.

27. Mertz, *op. cit.*, p. 22, note 2.

28. *Idem.*

29. *Ibid.*, pp. 23–24.

30. Berg, *op. cit.*, p. 27, note 1.

Chapter 3

1. Vernon Knight, "Common viral respiratory illnesses," in G. W. Thorn, R. D. Adams, E. Braunwald, K. J. Isselbacher, and R. G. Petersdorf, eds., *Harrison's Principles of Internal Medicine*, 8th ed. (New York: McGraw-Hill, 1977), chapter 195, pp. 991–992.

2. A. M. Lewis, Jr., "Experience with SV40 and Adenovirus-SV40 hybrids," in A. Hellman, M. N. Oxman, and R. Pollack, eds., *Biohazards in Biological Research: Proceedings of a Conference held at the Asilomar Conference Center, Pacific Grove, California, 22–24 January 1973*, (Cold Spring Harbor, Long Island, NY: Cold Spring Harbor Laboratory, 1973), pp. 96–113. For a good account of the adenovirus-SV40 hybrid episode, see also John Lear, *Recombinant DNA: The Untold Story* (New York: Crown, 1978), chapter 2.

3. Andrew Lewis, transcript of an interview by Rae Goodell, 30 July 1975, Recombinant DNA Controversy, Oral History Collection, IASC, MIT, p. 7.

4. Cf. W. P. Rowe, "Lymphochoriomeningitis virus infection," in Hellman et al., eds., *op. cit.*, pp. 41–46, note 2.

5. Lewis, *op. cit.*, p. 5, note 3.

6. *Idem.*

7. Lewis, *op. cit.*, p. 103, note 2.

8. Lewis, *op. cit.*, p. 10, note 3.

9. *Ibid.*, p. 11.

10. *Ibid.*, p. 12.

11. A. M. Lewis, Jr., M. H. Levin, W. H. Wiese, C. S. Crumpacker, and P. Henry, "A non-defective (competent) Adenovirus-SV40 hybrid isolated from the Ad 2-SV40 hybrid population," *Proceedings of the National Academy of Sciences* 63:1128 (1969).

12. Lewis, *op. cit.*, p. 13, note 3.

13. *Ibid.*, pp. 13–14.

14. *Ibid.*, pp. 15–16.

15. *Ibid.*, pp. 18–19.

16. *Ibid.*, pp. 16–17.

17. *Ibid.*, pp. 16–18.

18. *Ibid.*, p. 19.

19. *Ibid.*, p. 20.

20. *Ibid.*, p. 21.

21. *Ibid.*, pp. 21–22.

22. Andrew Lewis, Jr., letter to Carel Mulder, 1 September 1971, Recombinant DNA Controversy, Oral History Collection, IASC, MIT.

23. *Idem.*

24. Andrew Lewis, Jr., letter to Carel Mulder, 15 September 1971, Recombinant DNA Controversy, Oral History Collection, IASC, MIT.

25. *Idem.*

26. Lewis, *op. cit.*, p. 23, note 3.

27. *Ibid.*, p. 25.

28. *Idem.*

29. Andrew Lewis, Jr., memo to Dorland J. Davis, 27 November 1972, Recombinant DNA Controversy, Oral History Collection, IASC, MIT.

30. *Idem.*

31. *Idem.*

32. *Idem.*

33. *Idem.*

34. *Idem.*

35. The NIAID scientists were Sheldon Wolff, Robert Chanock, and James Rose. The members of the Board of Scientific Counselors were Jules Younger and John Wallace.

36. Scientific director, NIAID (John R. Seal), memo to Sheldon Wolff et al. (cf. note 35), 4 December 1972, Recombinant DNA Controversy, Oral History Collection, IASC, MIT.

37. *Idem.*

38. Clinical director, NIAID (Sheldon M. Wolff), memo to scientific director, NIAID, 8 December 1972, Recombinant DNA Controversy, Oral History Collection, IASC, MIT.

39. J. S. Younger, memo to John R. Seal, 8 December 1972, Recombinant DNA Controversy, Oral History Collection, IASC, MIT.

40. Head, Molecular Structure Section, LBV (James A. Rose), memo to scientific director, NIAID, 13 December 1972, Recombinant DNA Controversy, Oral History Collection, IASC, MIT.

41. *Idem.*

42. Director, NIAID (Dorland J. Davis), memo to associate director for collaborative research, NIH, 20 December 1972, Recombinant DNA Controversy, Oral History Collection, IASC, MIT.

43. *Idem.*

44. *Idem.*

45. Deputy director for science, NIH (Robert W. Berliner), memo to scientific directors (NIH), 20 December 1972, Recombinant DNA Controversy, Oral History Collection, IASC, MIT.

46. Director, Center for Disease Control (David J. Sencer), memo to director, National Institute of Allergy and Infectious Diseases, NIH (Dorland J. Davis), 4 January 1973, Recombinant DNA Controversy, Oral History Collection, IASC, MIT.

47. Special assistant, Office of the Director, NIAID (Earl C. Chamberlayne), memo to file, 26 March 1973, Recombinant DNA Controversy, Oral History Collection, IASC, MIT.

48. *Idem.*

49. *Idem.* A copy of the memorandum can be found appended to Memo, director, NIAID, to scientific director, NIAID, 16 October 1973, Recombinant DNA Controversy, Oral History Collection, IASC, MIT.

50. Chief, Laboratory of Biology of Viruses (Norman P. Salzman), memo to scientific director, NIAID, 26 March 1973, Recombinant DNA Controversy, Oral History Collection, IASC, MIT.

51. *Ibid.*

52. Biohazards control officer, Office of Biosafety, CDC (John H. Richardson), memo to Andrew Lewis, 27 March 1973, Recombinant DNA Controversy, Oral History Collection, IASC, MIT.

53. Head, Environmental Control Section, Office of Biohazard and Environmental Control, VO, DCCP, NCI (W. Emmett Barkley), memo to Earl C. Chamberlayne, 22 May 1973, Recombinant DNA Controversy, Oral History Collection, IASC, MIT.

54. Chairman, Biohazard Committee, memo to deputy director for science, NIH, 23 July 1973, Recombinant DNA Controversy, Oral History Collection, IASC, MIT.

55. *Ibid.*, note 49.

56. See, for example, *ibid.*, note 54, and Lewis, *op. cit.*, pp. 32–35, note 3.

57. Lewis, *op. cit.*, p. 31, note 3.

58. *Op. cit.*, note 54.

Chapter 4

1. Samuel S. Epstein, *The Politics of Cancer* (San Francisco: Sierra Club Books, 1979), p. 20.

2. *Annual Report*, Viral Oncology Program, Division of Cancer Cause and Prevention, National Cancer Institute, 1 July 1976–30 September 1977, p. 1243.

3. *Ibid.*, p. 1233.

4. Paul Berg, transcript of an interview by Charles Weiner, 17 May 1975, Recombinant DNA Controversy, Oral History Collection, IASC, MIT.

5. *Ibid.*, p. 27.

6. *Ibid.*, p. 28.

7. The full citation of the published proceedings is A. Hellman, M. Oxman, and R. Pollack, eds., *Biohazards in Biological Research: Proceedings of a Conference on Biohazards in Cancer Research, at the Asilomar Conference Center, Pacific Grove, California, 22–24 January 1973* (Cold Spring Harbor Laboratory, 1973).

8. R. N. Hull, "Biohazards associated with simian viruses" and Andrew Lewis, Jr., "Experience with SV40 and Adenovirus-SV40 hybrids," both in *Biohazards in Biological Research, op. cit.*, note 7.

9. *Ibid.*, p. 106.

10. *Ibid.*, p. 347.

11. Report of the Cambridge Experimentation Review Board on Recombinant DNA Experimentation, *The Bulletin of the Atomic Scientists* 33:23–27 (May 1977).

12. *Biohazards, op. cit.*, p. 237, note 7.

13. *Ibid.*, p. 347.

14. *Ibid.*, p. 348.

15. *Ibid.*, p. 350.

16. *Ibid.*, p. 351.

17. Robert Pollack, transcript of an interview by Mary Terrall, 27 March 1976, Recombinant DNA Controversy, Oral History Collection, IASC, MIT, p. 21.

Chapter 5

1. U.S. Department of Health, Education and Welfare (DHEW-NIH), *Recombinant DNA Research*, vol. 3: *Documents Relating to NIH Guidelines for Research Involving Recombinant DNA Molecules, November 1977–September 1978* (Washington, DC: US Government Printing Office, 1978), p. 599.

2. Alexander M. Cruickshank, "Gordon research conferences: Winter program, 1979," *Science* 202:337 (20 October 1978).

3. See, for example, Diana Crane, *Invisible Colleges* (Chicago: The University of Chicago Press, 1972).

4. *Science* 202:337 (20 October 1978).

5. Maxine Singer, transcript of an interview by Rae Goodell, 31 July 1975, Recombinant DNA Controversy, Oral History Collection, IASC, MIT, pp. 9, 11.

6. *Ibid.*, p. 21.

7. *Congressional Record—Senate*, 23 July 1970, p. 25603.

8. *Idem.*

9. Singer, *op. cit.*, p. 25.

10. *Ibid.*, p. 21.

11. Letter from Maxine Singer to participants in the 1973 Gordon Conference on Nucleic Acids, dated 21 June 1973, Recombinant DNA Controversy, Oral History Collection, IASC, MIT.

12. *Idem.*

13. In a letter to Maxine Singer from Robert W. Chambers, of the Department of Biochemistry, New York University Medical Center, Chambers requested that the letter be sent to the National Academy of Sciences, stating that the hazards were small but finite. See a copy of the Chambers letter, dated 2 July 1973, IASC, MIT.

14. Edward Ziff, "Benefits and hazards of manipulating DNA," *New Scientist* 60:274–275 (25 October 1973).

15. *Ibid.*, 274.

16. *Ibid.*, p. 236, note 14.

Chapter 6

1. Paul Berg, transcript of an interview by Charles Weiner, 17 May 1975, Recombinant DNA Controversy, Oral History Collection, IASC, MIT, p. 40.

2. *Idem.*

3. Theodore Friedman and Richard Roblin, "Gene therapy for human genetic disease," *Science* 175:949–955 (3 March 1972).

4. Richard Roblin, transcript of an interview by Charles Weiner and Rae Goodell, 21 April and 2 May 1975, Recombinant DNA Controversy, Oral History Collection, IASC, MIT, p. 28. Roblin quotes from a letter he received from Paul Berg, dated 8 April 1974, which gives the agenda for the 17 April meeting.

5. David Baltimore, lecture entitled "Where does molecular biology become

more of a hazard than a promise?" delivered 6 November 1974 to the Technology Studies Workshop at MIT. The transcript of the lecture is held in the Recombinant DNA Controversy Collection, IASC, MIT.

6. Paul Berg et al. [letter], "Potential biohazards of recombinant DNA molecules," *Science* 185:303 (26 July 1974). A modified version of the letter with the last paragraph deleted appeared in *Nature* 250:175 (19 July 1974). The complete version of the letter was also published in the *Proceedings of the National Academy of Science* 71:2593–2594 (July 1974).

7. Theodore Friedman, transcript of an interview by Charles Weiner, 21 October 1977, Recombinant DNA Controversy, Oral History Collection, IASC, MIT, p. 6.

8. Baltimore, *op. cit.*, note 5.

9. *Idem.*

10. Roblin, *op. cit.*, note 4.

11. Roy Curtiss III, memo to Paul Berg et al., 6 August 1974, Recombinant DNA Controversy Collection, IASC, MIT.

12. *Ibid.*, p. 11.

13. *Ibid.*, p. 7.

14. *Ibid.*, p. 8.

Chapter 7

1. Michael Rogers represented the weekly music journal, *Rolling Stone*, at Asilomar, and published *Biohazard* (New York: Alfred A. Knopf, 1977); Nicholas Wade of *Science* wrote *The Ultimate Experiment* (New York: Walker & Co., 1977). Both books have accounts of the Asilomar meeting.

2. Ephraim Anderson of England wrote to Stanley Cohen that he (Anderson) could not get his travel expenses reimbursed unless he was invited to speak. Anderson asked Cohen whether Cohen could arrange the invitation. Letter from E. S. Anderson to Stanley Cohen, 17 December 1974, Recombinant DNA Controversy Collection, Institute Archives and Special Collections (IASC), MIT.

3. David Baltimore, transcript of a tape-recorded interview by Charles Weiner and Rae Goodell, 13 May and 22 July 1975, Recombinant DNA Controversy Collection, IASC, MIT.

4. A copy of the original tapes of the Asilomar proceedings has been deposited in the MIT Institute Archives and forms a part of the extensive holdings comprising the Recombinant DNA Controversy Collection.

5. Tapes of the International Conference on Recombinant DNA Molecules (Asilomar Tapes) 24–27 February 1975, Asilomar Conference Center, Pacific Grove, California, Recombinant DNA Controversy Collection, IASC, MIT.

6. Baltimore was neither disinterested nor uninformed about the military

uses of biological weapons. Prior to Asilomar, he worked with other scientists in support of Congressional ratification of a treaty banning chemical and biological warfare (CBW). Over a period of three years, he helped organize informal sessions at the official meetings of the American Society for Microbiology (ASM). In part, through his efforts, ASM took a stand in opposition to CBW.

7. Baltimore interview, *op. cit.*, p. 132, note 3.

8. *Ibid.*, p. 49.

9. The initial subject areas can be found in a draft of a report of the Asilomar Organizing Committee, circa April 1975, Recombinant DNA Controversy Collection, IASC, MIT.

10. Paul Berg, transcript of an interview by Charles Weiner, 17 May 1975, Recombinant DNA Controversy Collection, IASC, MIT, pp. 75–76.

11. *Ibid.*, p. 74.

12. The report of the nomenclature working group was published in *Bacteriological Reviews* 40:168–189 (March 1976).

13. Paul Berg, letter to E. S. Anderson, 18 October 1974, Recombinant DNA Controversy Collection, IASC, MIT. After Asilomar, Roy Curtiss III maintained that his plasmid working group never received a specific charge as a group from Paul Berg. See transcript of an interview with Roy Curtiss III by Charles Weiner, 30 April 1976, Recombinant DNA Controversy Collection, IASC, MIT, p. 182. Another working group participant, Andrew Lewis, who served on the virus panel, claimed he was not apprised until several days before Asilomar that his panel was required to submit a written position statement. Andrew Lewis, transcript of an interview by Rae Goodell, 30 July 1975, Recombinant DNA Controversy Collection, IASC, MIT, p. 47.

14. Nicholas Wade, *The Ultimate Experiment* (New York: Walker & Co., 1977), p. 41.

15. Richard Novick, transcript of an interview by Rae Goodell, May 1, 1975, Recombinant DNA Controversy Collection, IASC, MIT, p. 40.

16. Berg interview, *op. cit.*, p. 76, note 10.

17. Baltimore interview, *op. cit.*, p. 84, note 3.

18. Donald Brown, letter to Ethan Signer, 23 October 1974, Recombinant DNA Controversy Collection, IASC, MIT.

19. The essay Signer authored is entitled "Gene manipulation: Progress and prospects," in E. D. Hay, T. J. King, and J. Papaconstantinou, eds., *Macromolecules Regulating Growth and Development, The 30th Symposium for the Society for Developmental Biology*, (New York: Academic Press, 1974), pp. 217–241.

20. Baltimore interview, *op. cit.*, p. 85, note 3; Berg interview, *op. cit.*, p. 55, note 10.

Chapter 8

1. The essay was written prior to the Asilomar Conference, but was not published until April 1975.

2. The lecture was delivered on 6 November 1974.

3. Paul Berg, letter to Irving P. Crawford, 9 December 1974, Recombinant DNA Controversy Collection, IASC, MIT.

4. Irving Crawford [letter], "Recombinant DNA molecules," *Genetics* 78:573–574 (October 1974).

5. *Idem.*

6. Ray Chaleff, letter to Donald Brown, 9 October 1974, Recombinant DNA Controversy Collection, IASC, MIT.

7. Arthur W. Galston, letter to Donald Brown, 15 November 1974, Recombinant DNA Controversy Collection, IASC, MIT.

8. *Idem.*

9. R. H. Burris, letter to Donald Brown, 16 October 1974, Recombinant DNA Controversy Collection, IASC, MIT.

10. Roy Curtiss III, letter to A. M. Chakrabarty, 16 September 1974, Recombinant DNA Controversy Collection, IASC, MIT.

11. Roy Curtiss III, letter to A. M. Chakrabarty, 20 September 1974, Recombinant DNA Controversy Collection, IASC, MIT.

12. A. M. Chakrabarty, letter to Roy Curtiss III, 2 October 1974, Recombinant DNA Controversy Collection, IASC, MIT.

13. As an example, see Donald D. Brown, letter to Harvey Lodish, 17 September 1974, Recombinant DNA Controvsery Collection, IASC, MIT. At the time, Brown was in the Department of Embryology, Carnegie Institution of Washington, DC, and Lodish was a member of the Department of Biology, MIT.

14. Donald Brown, transcript of an interview by Mary Terrall, 21 May 1976, Recombinant DNA Controversy Collection, IASC, MIT, p. 21.

15. Frederick C. Robbins, letter to Aaron Shatkin, 27 September 1974, Recombinant DNA Controversy Collection, IASC, MIT.

16. Robbins cited a study in *Science* entitled "Simian virus 40 in polio vaccine: Follow-up of newborn recipients," by J. F. Fraumeni, Jr., et al. 167:59–60 (2 January 1970).

17. Renato Dulbecco, letter to Aaron Shatkin, 16 October 1974, Recombinant DNA Controversy Collection, IASC, MIT.

18. Howard Temin, letter to William Reznikoff, member, Subcómmittee on Safety Factors Involved in Chimeric DNA Molecule Research, University of Wisconsin, 4 November 1974, Recombinant DNA Controversy Collection, IASC, MIT.

19. Paul Berg, letter to E. S. Anderson, 18 October 1974, Recombinant DNA Controversy Collection, IASC, MIT.

20. Allan Campbell, letter to Donald Brown, 14 November 1974, Recombinant DNA Controversy Collection, IASC, MIT.

21. Naomi Datta, letter to Paul Berg, 17 October 1974, Recombinant DNA Controversy Collection, IASC, MIT. Datta was originally chosen to serve on the Plasmid Working Group, but she did not participate in writing the report.

22. Martin Gellert, letter to Carolyn Ruby, 18 November 1974, Recombinant DNA Controversy Collection, IASC, MIT.

23. Sumner C. Kraft, letter to Donald B. Brown, 21 October 1974, Recombinant DNA Controversy Collection, IASC, MIT.

24. Bernard D. Davis, letter to Paul Berg, 5 September 1974, Recombinant DNA Controversy Collection, IASC, MIT.

Chapter 9

1. The other members on the panel were Royston C. Clowes, Institute for Molecular Biology, University of Texas at Dallas; Stanley N. Cohen, Department of Medicine, Stanford University Medical School; Roy Curtiss III, Department of Microbiology, University of Alabama Medical Center; and Stanley Falkow, Department of Microbiology, University of Washington School of Medicine. Naomi Datta of the Department of Bacteriology, University of London Royal Post Graduate School, was a member of the plasmid nomenclature group. Datta was invited to join the working group, but declined the invitation because she did not believe there were any biohazards inherent in her work.

2. Plasmid Working Group, *Proposed Guidelines on Potential Biohazards Associated with Experiments Involving Genetically Altered Microorganisms*, Asilomar Conference, 24 February 1975, Recombinant DNA Controversy Collection, IASC, MIT.

3. *Ibid.*, p. 2.

4. *Ibid.*, p. 18.

5. *Ibid.*, p. 12.

6. *Ibid.*, p. 19.

7. The other members on the viral panel were J. Michael Bishop, Department of Microbiology, University of California Medical Center at San Francisco; David A. Jackson, Department of Microbiology, University of Michigan Medical School; Daniel Nathans, Department of Microbiology, Johns Hopkins University School of Medicine; Bernard Roizman, Department of Microbiology and Biophysics, University of Chicago; Joe Sambrook, Cold Spring Harbor Laboratory; Duard Walker, Department of Medical Microbiology, University of Wisconsin; and Andrew M. Lewis, Cellular Biology Section, Division of Biological and Medical Sciences, National Science Foundation.

8. The other participants on the eukaryote panel were Sydney Brenner, Medical Research Council Laboratory for Molecular Biology, Cambridge, England; Robert H. Burris, Department of Biochemistry, University of Wisconsin; Dana Carroll, Department of Embryology, Carnegie Institution of Washington, Baltimore, Maryland; Ronald W. Davis, Department of Biochemistry, Stanford University Medical School; David S. Hogness, Department of Biochemistry, Stanford University; Kenneth Murray, Department of Molecular Biology, University of Edinburgh; and Raymond C. Valentine, Department of Chemistry, University of California.

Chapter 10

1. The complete list of signers of the GEG/SESPA statement and their affiliations at the time is Jonathan King, MIT; Jon Beckwith, Harvard Medical School; Luigi Gorini, Harvard Medical School; Kostia Bergman, MIT; Kaaren Janssen, MIT; Ethan Signer, MIT; Annamaria Torriani, MIT; Fred Ausubel, Harvard University; and Paoli Strigini, Boston University.

2. Harold Green, transcript of a talk entitled "A public policy perspective," 24 February 1975, at Pacific Grove, California, Recombinant DNA Controversy Collection, IASC, MIT, p. 8.

3. *Ibid.*, p. 9.

4. Alex Capron, transcript of a talk to the Asilomar Conference, 26 February 1975, Recombinant DNA Controversy Collection, IASC, MIT, p. 7.

5. Roger Dworkin, "Science, society, and the expert town meeting: Some comments on Asilomar," *Southern California Law Review* 51(6):1474 (September 1978).

6. Robert Sinsheimer, transcript of an interview by Charles Weiner, 26 December 1975, Recombinant DNA Controversy Collection, IASC, MIT, p. 60.

7. John Lear, *Recombinant DNA: The Untold Story*. (New York: Crown, 1978), p. 141.

8. Richard Roblin, transcript of an interview by Charles Weiner and Rae Goodell, 21 April 1975, Recombinant DNA Controversy Collection, IASC, MIT, p. 77.

9. Paul Berg, transcript of an interview by Rae Goodell, 17 May 1975, Recombinant DNA Controversy Collection, IASC, MIT, p. 83.

10. David Baltimore, speech at Asilomar, tapes of the Asilomar Conference, Recombinant DNA Controversy Collection, IASC, MIT.

11. Berg interview, *op. cit.*, p. 93, note 9.

12. Maxine Singer, transcript of an interview by Rae Goodell, 31 July 1975, Recombinant DNA Controversy Collection, IASC, MIT, p. 71.

13. An early draft of the summary statement written by the Organizing Committee was entitled "Provisional statement of the conference proceedings." In a later draft, there were penciled revisions in which the words "organizing

committee" were substituted for "conference proceedings," reflecting a shift in thinking about whose views were to be represented in the final statement.

14. Berg interview, *op. cit.*, p. 95, note 9.

15. Asilomar tapes, note 10.

16. Berg interview, *op. cit.*, p. 95, note 9.

17. Singer interview, *op. cit.*, p. 73, note 12.

18. Horace Freeland Judson attended the Asilomar meeting as a member of the press. He wrote an essay on the events titled "Fearful of science," *Harper's Magazine* (June 1975). His observations on the final vote appear on p. 15 of his essay.

19. Transcript of an interview with Roy Curtiss III conducted by Charles Weiner, 30 April 1976, Recombinant DNA Controversy Collection, IASC, MIT, p. 192.

20. Stanley Falkow, transcript of an interview by Charles Weiner, 20 May 1976 and 26 February 1977, Recombinant DNA Controversy Collection, IASC, MIT.

21. *Ibid.*, p. 141.

22. In a letter to Stanley Cohen, dated 4 March 1975, Falkow expressed his disgruntlement with the outcome of the final sessions, stating that the plasmid panel's report was either ignored or misinterpreted.

23. Dworkin, *op. cit.*, note 5.

Chapter 11

1. Paul Berg et al., "Potential biohazards of recombinant DNA molecules," *Science* 185:303 (26 July 1974).

2. The announcement of the formation of the advisory committee appeared in the *Federal Register* 39 (6 November 1974).

3. *Idem.*

4. Paul Berg, letter to Robert S. Stone, 10 December 1974, Recombinant DNA Controversy Collection, IASC, MIT.

5. DeWitt Stetten, Jr., transcript of an interview by Rae Goodell, 3 August 1976 and 2 September 1976, Recombinant DNA Controversy Collection, IASC, MIT, p. 5.

6. *Idem.*

7. *Idem.*

8. Other members of the committee were James E. Darnell, Rockefeller University; and John W. Littlefield, Johns Hopkins Hospital.

9. Stetten interview, p. 50.

10. *Ibid.*, p. 51.

11. *Ibid.*, pp. 50–51.

12. DeWitt Stetten, Jr., letter to Richard Goldstein and Harrison Echols, 9 September 1975, Recombinant DNA Controversy Collection, IASC, MIT. Goldstein was situated at Harvard Medical School's Department of Microbiology and Molecular Genetics: Echols was at the Molecular Biology Department of the University of California, Berkeley.

13. *Idem.*

14. LeRoy Walters, transcript of an interview by Rae Goodell, 15 January 1977, Recombinant DNA Controversy Collection, IASC, MIT.

15. *Ibid.*, p. 15.

16. Waclaw Szybalski, letter to DeWitt Stetten, Jr., 12 March 1975, Recombinant DNA Controversy Collection, IASC, MIT.

17. Elizabeth Kutter, transcript of an interview by Charles Weiner, 19 May 1976, Recombinant DNA Controversy Collection, IASC, MIT., p. 109.

18. Stetten interview, p. 60.

19. *Nature* 257:637 (23 October 1975).

20. DeWitt Stetten, letter to Richard Goldstein and Harrison Echols, *op. cit.*, note 12.

21. Jane K. Setlow, letter to DeWitt Stetten, Jr., 4 September 1975. Recombinant DNA Controversy Collection, IASC, MIT.

22. Falkow wrote a letter to Stetten on 13 October 1975, offering his resignation from RAC. The stated reason: to continue his work with FDA on the issue of antibiotics in animal feed. His resignation was to be effective 1 January 1976. Also, Stanley Falkow, letter to Richard Goldstein, 5 September 1979, Recombinant DNA Controversy Collection, IASC, MIT.

23. Paul Berg, letter to DeWitt Stetten, 2 September 1975, Recombinant DNA Controversy Collection, IASC, MIT.

24. Paul Primakoff, letter to DeWitt Stetten, 25 September 1975, Recombinant DNA Controversy Collection, IASC, MIT.

25. Wallace P. Rowe, letter to Paul Primakoff, 8 October 1975, Recombinant DNA Controversy Collection, IASC, MIT.

26. Leon Jacobs, memorandum to members of RAC, 16 October 1975, Recombinant DNA Controversy Collection, IASC, MIT.

27. Jane K. Setlow, letter to Leon Jacobs with copies sent to RAC members, 31 October 1975, Recombinant DNA Controversy Collection, IASC, MIT.

Chapter 12

1. *Hearings before the Subcommittee on Health of the Committee on Labor and Public Welfare*, 94th Congress, 22 April 1975 (Washington, DC: US Government Printing Office, 1975).

2. *Ibid.*, p. 3.

3. Stanley N. Cohen, "The manipulation of genes," *Scientific American* 233:24–33 (July 1975): "[T]here are natural barriers that normally prevent the exchange of genetic information between unrelated organisms" (p. 25). Also, "Basic research and the public interest," *Stanford M.D.* 14 (2):9–12 (April 1975): "Ordinarily, natural barriers prevent the union of hereditary characteristics from widely dissimilar organisms . . . " (p. 9).

4. *Hearings*, p. 26.

5. *Ibid.*, p. 27.

6. Maxine Singer, letter to Willard Gaylin, 30 April 1975, Recombinant DNA Controversy Collection, IASC, MIT.

7. Daniel M. Singer, letter to Richard O. Roblin, III, 24 April 1975, Recombinant DNA Controversy Collection, IASC, MIT.

8. US Department of HEW, *NIH Public Advisory Groups*, DHEW Pub. No. (NIH) 78–10, 1 July 1978.

9. DeWitt Stetten, Jr., transcript of an interview by Rae Goodell, 3 August and 2 September 1976, Recombinant DNA Controversy Collection, IASC, MIT, p. 69.

10. Elizabeth Kutter, transcript of an interview by Charles Weiner, 19 May 1976, Recombinant DNA Controversy Collection, IASC, MIT, p. 212.

11. US DHEW-NIH, *Recombinant DNA Research*, vol. I, August 1976. *Proceedings of a Conference on NIH Guidelines for Research on Recombinant DNA Molecules, 9–10 February 1975* (hereafter *Proc. DAC Meeting*), pp. 257–258.

12. *Proc. DAC Meeting*, p. 308.

13. *Ibid.*, p. 258.

14. *Ibid.*, p. 249.

15. *Idem.*

16. *Ibid.*, p. 252.

17. *Ibid.*, p. 283.

18. *Ibid.*, p. 300.

19. *Ibid.*, p. 289.

20. *Ibid.*, p. 299.

21. *Idem.*

22. *Ibid.*, p. 296.

23. *Ibid.*, p. 297.

24. Nicholas Wade, *The Ultimate Experiment* (New York: Walker, 1977), p. 57.

Chapter 13

1. *Nature* 276:2–3 (2 November 1978); Sydney Brenner, unpublished paper, "A framework for assessing the potential risks of recombinant DNA research."

2. *Proc. DAC Meeting*, p. 36, see note 11, chapter 12.

3. National Institutes of Health, *Guidelines for Research Involving Recombinant DNA Molecules* [throughout, NIH *Guidelines*], June 23, 1976, p. 1.

Chapter 14

1. Stanley N. Cohen, "Recombinant DNA: Fact and fiction," a statement prepared for a meeting of the Committee on Environmental Health, California Medical Association, 18 November 1976.

2. Barbara J. Culliton, "Scientists dispute book's claim that human clone has been born," *Science* 199:1314–1316 (24 March 1978).

3. Susan Wright, "DNA: Let the public choose," *Nature* 275:468 (12 October 1978).

4. Annette Oestreicher, "Somatostatin produced by 'programmed' *E. coli*," *Internal Medicine News* (1 March 1978).

5. Robert Cooke, "Insulin via bacteria is reported," *Boston Globe* (18 June 1978).

6. Robert Cooke, "Human insulin made in lab," *Boston Globe* (7 September 1978).

7. *Science* 202:610 (10 November 1978).

8. Eliot Marshall, "Environmental groups lose friends in effort to control DNA research," *Science* 202:1265–1269 (22 December 1978).

9. Leslie Dach [letter], *Science* 203:312 (26 January 1979).

10. *Science* 199:274 (20 January 1978).

11. NIH-HEW, "Recombinant DNA research, revised guidelines," *Federal Register* 43:60080–60105 (22 December 1978).

12. Science 203:9 (5 January 1979).

13. Philip H. Abelson [editorial], "Recombinant DNA" *Science* 197:721 (August 1977).

14. Spyros Andreoupoulos, "Gene cloning by press conference," *New England Journal of Medicine* 302:743–746 (28 March 1980).

15. William Stockton, "On the brink of altering life," *New York Times Magazine* (17 February 1980).

16. Susan Wright, "Recombinant DNA policy: From prevention to crisis intervention," *Environment* 21:34–37, 42 (November 1979).

Chapter 15

1. Editor's note, *Recombinant DNA Technical Bulletin* 2(4):ii (April 1980).

2. C. E. Dolman, "Escherich, Theodor," in Charles C. Gillispie, ed., *Dictionary of Scientific Biography*, vol. 4 (New York: Scribner, 1971), pp. 403–406.

3. J. Lederberg, "Genetic studies with bacteria," in L. C. Dunn, ed., *Genetics in the 20th Century* (New York: Macmillan, 1951), pp. 263–289.

4. B. J. Bachman, "Pedigrees of some mutant strains of *Escherichia coli* K12," *Bacteriological Reviews* 36:528–557 (1972).

5. *Idem.*

6. Stanley Falkow, memorandum to Recombinant DNA Advisory Committee (RAC) members, "The ecology of *Escherichia coli*," appended to the RAC minutes of the 12–13 May 1975 meeting.

7. Falkow memorandum, *op. cit.*, p. 8, note 6.

8. *Ibid.*, p. 7.

Chapter 16

1. S. L. Gorbach, letter to Donald Fredrickson, 14 July 1977, reprinted in US DHEW-NIH, *Environmental Impact Statement on NIH Guidelines for Research Involving Recombinant DNA Molecules*, Part II, October 1977, appendix M.

2. Summary minutes of Director's Advisory Committee Meeting, NIH, 15–16 December 1977. Reprinted in US DHEW-NIH, *Recombinant DNA Research*, vol. 3: *Documents Relating to "NIH Guidelines for Research Involving Recombinant DNA Molecules,"* November 1977–September 1978, p. 247.

3. *Ibid.*, p. 247.

4. The committee was composed of David Botstein, MIT; Harry Feldman, State University of New York-Upstate Medical Center; S. B. Formal, Walter Reed Army Institute of Research; Richard Goldstein, Harvard Medical School; Malcolm Martin, NIH; and Wallace Rowe, NIH.

5. Summary minutes, *op. cit.*, p. 239, note 2.

6. Gorbach, letter to Fredrickson, *op. cit.*, note 1.

7. NIH, "Environmental impact assessment of a proposal to release revised NIH guidelines fo research involving recombinant DNA molecules," *Federal Register* 43:33096. Reprinted in *Recombinant DNA Research*, vol. 3, *op. cit.* p. 57, note 2.

8. S. L. Gorbach, "Recombinant DNA: An infectious disease perspective," *Journal of Infectious Diseases* 137:615–623 (1978).

9. R. L. Guerrant, "Yet another pathogenic mechanism for *Escherichia coli* diarrhea?," *New England Journal of Medicine* 302:113–115 (1980).

10. C. C. J. Carpenter, "Diarrheal disease caused by *Escherichia coli*," P. D.

Hoeprich, ed., *Infectious Diseases*, 2nd ed. (Hagerstown, MD: Harper and Row, 1977), chapter 58, p. 547.

11. Gorbach, *op. cit.*, note 8.

12. J. Z. Montgomerie, "Factors affecting virulence in *Escherichia coli* urinary tract infections," *Journal of Infectious Diseases* 137:645–647 (1978).

13. T. McKeown, *The Modern Rise of Population* (London: Edward Arnold, 1976).

14. R. B. Sack, "The epidemiology of diarrhea due to enterotoxigenic *Escherichia coli*," *Journal of Infectious Diseases* 137:639–640 (1978).

15. B. H. Minshew, J. Jorgensen, M. Swanstrum, G. A. Grootes-Reuvecamp, and S. Falkow, "Some characteristics of *Escherichia coli* strains from extraintestinal infections of humans," *Journal of Infectious Diseases* 137:648–654 (1978).

16. E. Gangarosa, "Epidemiology of *Escherichia coli* in the United States," *Journal of Infectious Diseases* 137:634–638 (1978).

17. Gorbach, *op. cit.*, note 8.

18. Gangarosa, *op. cit.*, note 16.

19. H. W. Smith, "Is it safe to use *Escherichia coli* K12 in recombinant DNA experiments?," *Journal of Infectious Diseases* 137:655–660 (1978).

20. E. S. Anderson, "Plasmid transfer in *Escherichia coli*," *Journal of Infectious Diseases* 137:686–687 (1978).

21. S. L. Gorbach, discussion in "Risk assessment of recombinant DNA experimentation with Escherichia coli K12," *Journal of Infectious Diseases* 137:700 (1978).

22. Gorbach, *op. cit.*, note 8.

23. R. Freter, "Possible effects of foreign DNA on pathogenic potential and intestinal proliferation of *Escherichia coli*," *Journal of Infectious Diseases* 137:624–629 (1978).

24. *Ibid.*, pp. 625–626.

25. *Ibid.*, p. 626.

26. *Idem.*

27. *Ibid.*, p. 628.

28. Gorbach, letter to Fredrickson, *op. cit.*, note 1.

29. Gorbach, *op. cit.*, note 8 (emphasis added).

30. Minshew et al., *op. cit.*, note 15.

31. Smith, *op. cit.*, note 19.

32. S. B. Formal and R. B. Hornick, "Invasive *Escherichia coli*," *Journal of Infectious Diseases* 137:641–644 (1978).

33. *Ibid.*, p. 644.

34. Smith, *op. cit.*, note 19.

35. J. King, "Recombinant DNA and autoimmune disease," *Journal of Infectious Diseases* 137:663–666 (1978).

36. Anderson, *op. cit.*, note 20.

37. Gorbach, *op. cit.*, pp. 621–622, note 8.

38. "Risk assessment protocols for recombinant DNA experimentation," *Journal of Infectious Diseases* 137:704–708 (1978).

39. *Ibid.*, p. 704.

40. *Ibid.*, p. 705.

41. *Ibid.*, p. 706.

42. *Ibid.*, p. 707.

Chapter 17

1. Transcript of proceedings of the 15–16 December 1977 meeting of the Advisory Committee to the director, NIH, on the proposed revision of the NIH *Guidelines on Recombinant DNA Research* at Bethesda, MD. Reprinted in US DHEW-NIH, *Recombinant DNA Research*, vol. 3, September 1978, pp. 215–216.

2. NIH, "Recombinant DNA research: Proposed revised guidelines," *Federal Register* 42:49596–49609, 27 September 1977. Reprinted in *Recombinant DNA Research*, vol. 3, pp. 164ff.

3. Nancy Pfund, transcript, p. 24, *op. cit.*, p. 228, note 1.

4. *Ibid.*, p. 228.

5. NIH, "Recombinant DNA research: Revised guidelines," *Federal Register* 43:60108–60131, 22 December 1978. Reprinted in US DHEW-NIH, *Recombinant DNA Research*, vol. 4, December 1978, pp. 29–53.

6. W. P. Rowe, "Statement to the NIH Recombinant DNA Advisory Committee," 21–23 May 1979, RAC Meeting, item #671.

7. *Idem.*

8. Roy Curtiss, letters to D. Fredrickson, 4 October 1979 and 11 May 1979. Both reprinted in US DHEW-NIH, *Recombinant DNA Research*, vol. 5, March 1980, pp. 339–340, 260–264.

9. Curtiss, letter of 4 October 1979, *ibid.*

10. B. P. Sagik and C. A. Sorber, "The survival of host-vector systems in domestic sewage treatment plants," *Recombinant DNA Technical Bulletin* 2:55–61 (July 1979).

11. Proceedings, *Ad Hoc NIAID Working Group on Risk Assessment*, Bethesda, MD, 30 August 1979.

12. *Ibid.*, pp. 34–35.

13. M. A. Chatigny, M. T. Hatch, H. Wolochow, T. Adler, J. Hresko, J. Macher, and D. Besemer, "Studies on release and survival of biological substances used in recombinant DNA laboratory procedures," *Recombinant DNA Technical Bulletin* 2:62–67 (July 1979).

14. R. Freter, H. Brickner, J. Fekete, P. C. M. O'Brien, and M. M. Vickerman. "Testing of host-vector systems in mice," *Recombinant DNA Technical Bulletin* 2:68–76 (July 1979).

15. S. B. Levy and B. Marshall, "Survival of *E. coli* host-vector systems in the human intestinal tract," *Recombinant DNA Technical Bulletin* 2:77–80 (July 1979).

16. P. S. Cohen, R. W. Pilsucki, M. L. Myhal, C. A. Rosen, D. C. Laux, and V. J. Cabelli, "Fecal *E. coli* strains in the mouse GI tract," *Recombinant DNA Technical Bulletin* 2:106–113 (November 1979).

17. NIH, "Recombinant DNA research: Proposed actions under guidelines," *Federal Register* 44:69234–69251. Reprinted in US DHEW-NIH, *Recombinant DNA Research*, vol. 5, March 1980, pp. 220–237.

18. Susan Gottesman, letter to Donald Fredrickson, 20 December 1979. Reprinted in *Recombinant DNA Research*, vol. 5, p. 563.

Chapter 18

1. Minutes of meeting, Recombinant DNA Molecule Program Advisory Committee, 4–5 December 1975. The committee's name was changed to the Recombinant DNA Advisory Committee (RAC) around 1977.

2. Report of US EMBO Workshop to Assess Risks for Recombinant DNA Experiments Involving the Genomes of Animal, Plant and Insect Viruses in *Federal Register* 43:33159–33169 (28 July 1978). The conference, held in Ascot, England, on 26–28 January 1978, was a joint venture between the United States and the European Molecular Biology Organization (EMBO). The workshop was attended by twenty-seven scientists from the United States, the United Kingdom, West Germany, Finland, France, Sweden, and Switzerland.

3. Minutes, meeting of the Recombinant DNA Advisory Committee, 1–2 April 1976.

4. Minutes, Recombinant DNA Molecule Program Advisory Committee, 13–14 September 1976.

5. Minutes, meeting of the Recombinant DNA Advisory Committee, 22–23 June 1977. Also see DHEW-NIH, *Recombinant DNA Research*, vol. 3. *Documents Relating to NIH Guidelines for Research Involving Recombinant DNA Molecules, November 1977–September 1978*, appendix C.

6. Malcolm Martin and Wallace P. Rowe, memorandum to NIH Recombinant DNA Molecule Program Advisory Committee, October 14, 1977. "As we

have stressed repeatedly, the goal of the polyoma experiment is best phrased as being to assess the factors (bacterial, vector, and animal host) that determine if a DNA molecule can be transferred intact from an *E. coli* host into mammalian cells. The goal is not, and cannot be for any single type of experiment, to determine if recombinant DNA research with *E. coli* K12 is safe or not" (IASC, MIT).

7. M. A. Israel, H. W. Chan, W. P. Rowe, and M. A. Martin, "Molecular cloning of polyoma virus DNA in *Escherichia coli*: Plasmid vector system," *Science* 203:883–887 (2 March 1979); H. W. Chan, M. A. Israel, C. F. Garon, W. P. Rowe, and M. A. Martin, "Molecular cloning of polyoma virus DNA in *Escherichia coli*: Lambda phage vector system," *Science* 203:887–892 (2 March 1979).

8. Israel et al., *op. cit.*, p. 883.

9. The sentence that introduces the results states: "In contrast to recombinant phage containing a single copy of a viral genome, those containing a head-to-tail dimer insert of PY DNA would, in theory at least, be more likely to be infectious in mouse cells since infectious PY genomes could be generated by intramolecular recombination" (Chan et al., *op. cit.*, p. 890).

10. The authors summed up their results as follows (Chan et al., *op. cit.*, p. 892):

In many discussions with virologists during the planning of these studies there was general consensus that the experiments would show *E. coli* carrying PY-lambda or PY-plasmids to induce no PY infections when given by mouth, to give some infections when given parenterally, and to be quite infectious by parenteral injection if the insert were an oligomer. The most striking feature of our results, then, is the extremely low or absent infectivity of the recombinant molecules. In no instance was a recombinant molecule with a monomeric insert infectious, and in no instance did oral or parenteral administration of massive doses of live recombinant-containing *E. coli* induce PY infection. We thus view these results as being highly reassuring with respect to the safety of cloning viral genomes in *E. coli*.

11. *Nature* 276:4–7 (2 November 1978).

12. Malcolm A. Martin and Wallace P. Rowe, protocol for risk-assessment experiments entitled "Proposed experiments on the biological properties of polyoma recombinant DNA cloned in bacterial cells," in US DHEW-NIH, *Recombinant DNA Research*, vol. 3, *Documents Relating to NIH Guidelines for Research Involving Recombinant DNA Molecules, November 1977–September 1978*, appendix C, p. 79.

13. W. A. Thomasson, "Recombinant DNA and regulating uncertainty," *The Bulletin of the Atomic Scientists* 35:26–32 (December 1979).

14. M. A. Israel, H. W. Chan, M. A. Martin, and W. P. Rowe, "Molecular cloning of polyoma virus DNA in *Escherichia coli*: Oncogenicity testing in hamsters," *Science* 205:1140–1142 (14 September 1979).

15. *Ibid.*, p. 1141.

16. Barbara Hatch Rosenberg letter to Jane Setlow, 25 May 1979. Rosenberg

ends her letter with the statement, "Experiments performed by inoculation with bacteria are not a valid test of the risks entailed in accidental exposure of an animal to bacteria containing recombinant DNA."

17. Barbara Rosenberg and Lee Simon, "Recombinant DNA: Have recent experiments assessed all the risks?," *Nature* 282:773–774 (20–27 December 1979).

18. The results of the polyoma experiments were reported in the *New York Times*, 2 May 1979, p. A13; David Dixon, *Nature* 227:505 (1979); E. Marshall, *Science* 203:1223 (1979). Letters also appeared: D. Stetten, *Science* 203:1292 (1979); S. A. Newman, *Nature* 281:176 (20 September 1979). Jonathan King and Ethen Signer, both professors of biology at MIT, published a letter in the *New York Times* (3 May 1979) and referred to a *Times* news story that stated that the mouse polyoma experiments showed there were virtually no hazards. The King-Signer letter countered that interpretation: "Thus the Rowe-Martin results are far from showing 'virtually no hazard.' Rather, they show that at least one application of the recombinant DNA technology results in the creation of a laboratory hybrid not found in nature which does represent a new source infection, even if the majority of applications do not. . . . [N]either the public nor the scientific community is served when a positive result indicating a danger is buried in a mass of negative data and ignored. Such a situation is truly a hazard to us all."

19. Malcolm Martin and Wallace Rowe, memorandum to NIH Recombinant DNA Molecule Program Advisory Committee, 14 October 1977, IASC, MIT.

20. Malcolm Martin's personal notes on the Enteric Bacteria Meeting, 31 August 1976, p. 52.

Chapter 19

1. Robert Sinsheimer, lecture at the University of California at Los Angeles, November 1976, published in David A. Jackson and Stephen P. Stich, eds., *The Recombinant DNA Debate* (Englewood Cliffs, NJ: Prentice Hall, 1979), p. 86.

2. Robert Sinsheimer, "Troubled dawn for Genetic engineering," *New Scientist* 68:151 (16 October 1975).

3. Robert L. Sinsheimer, "The Galilaean imperative," a paper presented at a conference held at the University of Georgia, 15–16 April 1977, on the Ethical and Methodological Dimensions of Scientific Research, focusing on rDNA as a case study. See John Richards, ed., *Recombinant DNA: Science, Ethics, and Politics* (New York: Academic Press, 1978), pp. 17–32.

4. *Ibid.*, p. 26.

5. Robert Sinsheimer and Gerard Piel, "Inquiry into inquiry: Two opposing views," *Hastings Center Report*, 6:18–19 (August 1976).

6. Robert Sinsheimer, "The presumptions of science," *Daedalus* 107:23–45 (spring 1978).

7. Erwin Chargaff [letter], *Science* 192:938–939 (4 June 1976).

8. *Idem.*

9. Erwin Chargaff, "Recombinant DNA research: A debate on the benefits and risks," *Chemical and Engineering News* 55:35 (30 May 1977). He places rDNA research in the context of nucleic acid research in his essay "Profitable wonders," *The Sciences* 15:21–26 (August-September 1975).

10. George Wald, "The case against genetic engineering," *The Sciences* 16:6–11 (September-October 1976).

11. Robert L. Sinsheimer [editorial], "Recombinant DNA—on our own," *BioScience* 26:599 (October 1976); also "An evolutionary perspective for Genetic engineering," *New Scientist* 73:150–152 (20 January 1977).

12. Ernst Mayr, *Populations Species and Evolution* (Cambridge, MA: Harvard University Press, 1971), p. 13.

13. *Ibid.*, p. 20.

14. J. D. Watson, *The Molecular Biology of the Gene*, 3rd ed. (Menlo Park, CA: W. A. Benjamin, 1976).

15. Allan M. Campbell, "How viruses insert their DNA into the DNA of the host cell," *Scientific American* 235(6):102–113 (December 1976). Reprinted in David Freifelder, ed., *Recombinant DNA* (a collection of *Scientific American* articles; San Francisco: W. H. Freeman, 1978), pp. 70–80. Also, Allan Campbell, "Tests for gene flow between eukaryotes and prokaryotes," *Journal of Infectious Diseases* 137:681–685 (May 1978).

16. Richard Novick, "The dangers of unrestricted research: The case of recombinant DNA," *Recombinant DNA: Science, Ethics, and Politics*, pp. 71–102, note 3. Note: Novick here is speaking of eukaryotic viruses as carriers of genes into eukaryotic cells.

17. Shing Chang and Stanley N. Cohen, "*In vivo* site-specific genetic recombination promoted by the Eco RI restriction endonuclease," *Proceedings of the National Academy of Science* 74:4811–4815 (November 1977).

18. *Ibid.*, p. 4815.

19. Novick, *op. cit.*, p. 88, note 16.

20. Paul Berg and Maxine Singer, "Seeking wisdom in recombinant DNA research," testimony before the Senate Subcommittee on Health and the Subcommittee on Labor and Public Welfare, 22 September 1976. *Oversight Hearings on Implementation of the NIH Guidelines Governing Recombinant DNA Research* (Washington, DC: US Government Printing Office, 1976), p. 151.

21. Bernard D. Davis, "Recombinant DNA scenarios: Andromeda strain, chimera, and golem," *American Scientist* 65:547–555 (September-October 1977).

22. Novick, *op. cit.*, p. 80, note 16.

23. *Ibid.*, p. 81.

24. Davis, *op. cit.*, p. 551, note 21. "[O]nly if the entire class is novel is there likely to be a novel danger for which evolution has not prepared us."

25. Novick, *op. cit.*, p. 82, note 16.

Chapter 20

1. Jonathan King, "Recombinant DNA and autoimmune disease," *Journal of Infectious Diseases* 137:663–666 (May 1978).

2. Burke K. Zimmerman, "Comments on the draft environmental impact statement for the NIH guidelines for research involving recombinant DNA molecules," 18 October 1976. Zimmerman's comments were reprinted in US DHEW-NIH, *Environmental Impact Statement on NIH Guidelines Involving Recombinant DNA Molecules*, Part II, October 1977, appendix K, pp. 55–76. The relevant passage is "There is the risk that *any* foreign protein, even if it has no enzymatic function, could, if expressed in or on the surface of mammalian cells in sufficient quantity, elicit an immune response from the host, leading to the destruction of the cells, as in known autoimmune diseases following certain bacterial infections."

3. Jonathan Beckwith, letter to William Gartland, 26 October 1978.

4. *Idem.*

5. Wallace Rowe, letter to Frank Dixon, 22 March 1979.

6. Baruj Benacerraf, letter to Wallace Rowe, 29 March 1979.

7. NIAID Recombinant DNA Risk Assessment Workshop, 11–12 April 1980, notes of meeting.

8. *Ibid.*, p. 5.

9. *Ibid.*, p. 11.

10. *Ibid.*, p. 42

Chapter 21

1. Freeman Dyson [letter], "Costs and benefits of recombinant DNA research," *Science* 193:6 (2 July 1976).

2. Nicholas Wade, "Cloning gold rush turns basic biology into big business," *Science* 208:688–693 (16 May 1980).

3. *Nature* 283:130–131 (10 January 1980).

4. Quoted from *Newsweek*, 17 March 1980, p. 62.

5. A. M. Chakrabarty, "Recombinant DNA: Areas of potential applications," in David A. Jackson and Stephen P. Stich, eds., *The Recombinant DNA Debate* (Englewood Cliffs: Prentice Hall, 1979), pp. 56–66.

6. National Academy of Sciences, *Research with Recombinant DNA* (Washington, DC: National Academy of Sciences, 1977), p. 169.

7. A description of the molecular biology of interferon production is given in Walter Gilbert and Lydia Villa-Komaroff, "Useful proteins from recombinant bacteria," *Scientific American* 242:74–94 (April 1980).

8. The worldwide market potential for interferon was placed at $3 billion in 1980. *European Chemical News* 34:18 (14 April 1980).

9. Harold M. Schmeck, Jr., "Interferon: Studies put cancer use in doubt," *New York Times* (27 May 1980).

10. Clifford Grobstein, *A Double Image of the Double Helix* (San Francisco: W. H. Freeman, 1979), p. 64. For a summary of the scientific basis of nitrogen fixation see K. T. Shanmugam and Raymond C. Valentine, "Molecular biology of nitrogen fixation," *Science* 187:919–924 (14 March 1975); P. R. Day, "Plant genetics: Increasing crop yield," *Science* 197:1334–1339 (1977).

11. John M. Pemberton, "A biological answer to environmental pollution by phenoxyherbicides," *Ambio* 8(5):202–205 (1979).

12. Chakrabarty, *op. cit.*, p. 61, note 5.

13. Nicholas Wade, "Recombinant DNA: Warming up for the big payoff," *Science* 206:663, 665 (9 November 1979); "Where genetic engineering Will change industry," *Business Week* (22 October 1979), pp. 160–172.

14. Harold M. Schmeck, Jr., "US to process 100 applications for patents on living organisms," *New York Times* (18 June 1980).

15. A discussion of the Supreme Court ruling in *Chakrabarty vs. Diamond* is given in Sheldon Krimsky, "Patents for life forms *sui generis*: Some new questions for science, law, and society," *Recombinant DNA Technical Bulletin* 4(1):11–15 (April 1981).

16. Jonathan King, "Arguments against patenting modified life forms," Robert F. Acker and Moselio Schaechter, eds., *Patentability of Microorganisms: Issues and Questions* (Washington, DC: American Society for Microbiology, 1981), pp. 36–41.

17. Office of Technology Assessment, *Impacts of Applied Genetics* (Washington, DC: US Government Printing Office, April 1981).

Chapter 22

1. A summary of the events at the University of Michigan is reported by Alvin Zander, "The discussion of recombinant DNA at the University of Michigan," in David K. Jackson and Stephen P. Stich, eds., *The Recombinant DNA Debate* (Englewood Cliffs, NJ: Prentice-Hall, 1979).

2. June Goodfield, *Playing God* (New York: Random House, 1977), p. 164.

3. Robert L. Sinsheimer, letter to Robben Fleming, 12 May 1976, IASC, MIT.

4. Frederick C. Neidhardt, "Statement to the Board of Regents of the University of Michigan on recombinant DNA research, May 20, 1976." The forum referred to was held in the afternoons and evenings of 3 and 4 March 1976.

5. Among the modifications were that the Board of Regents must approve any experiments requiring containment levels above P3 and that experiments done at P3 must use an EK2 host-vector system.

6. Zander, *op. cit.*, pp. 34–35, note 1.

7. *Science Policy Implications of DNA Recombinant Molecule Research*, hearings before the Subcommittee on Science, Research and Technology of the Committee on Science and Technology, 95th Congress, 3 May 1977.

8. Everett Mendelsohn, "Frankenstein at Harvard: The public policies of recombinant DNA research," E. Mendelsohn, D. Nelkin, and P. Weingart, eds., *The Social Assessment of Science* (Bielefeld: Univ. of Bielefeld, 1978), pp. 57–78; Sheldon Krimsky, "A citizen court in the recombinant DNA debate," *Bulletin of the Atomic Scientists* 34:37–43 (October 1978); John Lear, *Recombinant DNA: The Untold Story* (New York: Crown, 1978); William Bennett and Joel Gurin, "Science that frightens scientists," *Atlantic Monthly* (February 1977); Rae Goodell, "Public involvement in the DNA controversy: The case of Cambridge, Massachusetts," *Science, Technology, and Human Values* 4:36–43 (spring 1979).

9. The critique was distributed to all members of the Recombinant DNA Advisory Committee on 24 November 1975. The group consisted of Margaret Duncan, Richard Goldstein, and Paul Primakoff of the Harvard Medical School and Christien Orrego of Brandeis University.

10. Ruth Hubbard, letter to Daniel Branton, 8 April 1976, IASC, MIT.

11. Mendelsohn, *op. cit.*, p. 65, note 8.

12. The story was written by Charles Gottlieb and Ross Jerome and appeared 8 June 1976.

13. Lear, *op. cit.*, p. 155, note 8.

14. The Cambridge hearings on rDNA research were transcribed and are available from the city clerk in a two-volume set.

15. Lear, *op. cit.*, p. 163, note 8.

16. Donald Fredrickson, "The public governance of science," *Man and Medicine* 3(2):77–88 (1978).

17. A shortened version of the CERB report was published in *The Bulletin of the Atomic Scientists* 33:23–27 (May 1977).

18. For an analysis and comparison of the six enacted laws, see Sheldon Krimsky, "A comparative view of state and municipal laws regulating the use of recombinant DNA molecules technology," *Recombinant DNA Bulletin* 2:121–125 (November 1979).

19. Subcommittee on Biohazardous Research, *Recommendations for the Conduct of Research with Biohazardous Materials at Princeton University* (6 December 1976), p. 17.

Chapter 23

1. Erwin Chargaff, "On the dangers of genetic meddling," *Science* 192:938 (4 June 1976).

2. *Idem.*

3.Maxine F. Singer and Paul Berg [letter], *Science* 193:186–188 (16 July 1976).

4. "Recombinant DNA research guidelines," *Federal Register* 41:27902–27943 (7 July 1976).

5. "Recombinant DNA research guidelines: Draft environmental impact statement," *Federal Register* 41:38426–38483 (9 September 1976).

6. DeWitt Stetten, Jr., letter to Charles A. Thomas, 5 April 1976, IASC, MIT.

7. Charles A. Thomas, letter to DeWitt Stetten, 8 April 1976, IASC, MIT. Also, letter of 9 April 1976.

8. Lorna Salzman, Friends of the Earth, letter to Donald Fredrickson, 15 October 1976, IASC, MIT.

9. "Genetic manipulation: Enter the environmentalists," *Nature* 264:206 (18 November 1976).

10. George Jacobson, transcript of an interview by Aaron Seidman, MIT Oral History Program, 2 February 1978, IASC, MIT.

11. Richard Ottinger subsequently introduced S. 621 into the House 7 February 1977 as H.R. 3191.

12. Jacobson interview, p. 37, note 10.

13. The committee's charter appears in US DHEW-NIH, *Recombinant DNA Research*, vol. 2 (March 1978), p. 181.

14. *Interim Report of the Federal Interagency Committee on Recombinant DNA Research: Suggested Elements for Legislation*, 15 March 1977, reprinted in US DHEW-NIH *RDNA Research*, vol. 2, pp. 279–319.

15. Donald S. Fredrickson, letter to Larry G. Horowitz, 20 July 1977.

16. Barbara J. Culliton, "Recombinant DNA bills derailed: Congress still trying to pass a law," *Science* 199:274–277 (20 January 1978).

17. Nicholas Wade, "Congress set to grapple again with gene splicing," *Science* 199:1319–1322 (24 March 1978).

18. Harlyn Halvorson, transcript of an interview by Aaron Seidman, 10 and 22 May 1978, Recombinant DNA Controversy Collection, IASC, MIT.

19. Harlyn O. Halvorson [letter], *Science* 196:1154 (10 June 1977).

20. The other organizations represented are American Institutes of Biological Sciences, American Society of Allied Health Professions, Association of American Medical Colleges, Federation of American Societies for Experi-

mental Biology, National Society for Medical Research, and American Society for Medical Technology.

21. Halvorson interview, p. 63.

22. *Ibid.*, pp. 117–119.

23. David Dickson, "Friends of DNA fight back," *Nature* 272:664 (20 April 1978).

24. Walter Gilbert [letter], *Science* 197:208 (15 July 1977).

25. Eliot Marshall, "Environmental groups lose friends in effort to control DNA research," *Science* 202:1265–1269 (22 December 1978). A response to Marshall's article by environmentalists appeared in *Science* 203:312–313 (26 January 1979). See also Lewis Thomas, "Hubris in science," *Science* 200:1459–1462 (30 June 1978).

26. Roy Curtiss III, letter to Donald Fredrickson, 12 April 1977, IASC, MIT.

27. Curtiss letter, 1977.

28. National Academy of Sciences, Academy Forum, *Research with Recombinant DNA* (Washington, DC: National Academy of Sciences, 1977), p. 185.

29. Sherwood L. Gorbach, ed., "Risk assessment of recombinant DNA experimentation with *Escherichia coli* K12," *Journal of Infectious Diseases* 137 (May 1978).

30. Sherwood L. Gorbach, letter to Donald S. Fredrickson, 14 July 1977. The letter is contained in US DHEW-NIH, *Environmental Impact Statement on N.I.H. Guidelines for Research Involving Recombinant DNA Molecules*, Part II, October, 1977.

31. Bruce R. Levin, letter to Donald Fredrickson, 29 July 1977.

32. Jonathan King and Richard N. Goldstein, letter to participants of the Cold Spring Harbor Phage Meeting, 22 August 1977.

33. Stanley N. Cohen, letter to Donald S. Fredrickson, 6 September 1977.

34. *Idem.*

35. Burke K. Zimmerman, "Beyond recombinant DNA: Two views of the future," in John Richards, ed., *Recombinant DNA: Science, Ethics and Politics* (New York: Academic Press, 1978), pp. 273–301.

Chapter 24

1. James D. Watson, "An imaginary monster," *Bulletin of the Atomic Scientists* 33(5):12–13 (May 1977).

2. James D. Watson, "Why the Berg letter was written," *Recombinant DNA and Genetic Experimentation*, J. Morgan and W. J. Whelan [eds.] (Oxford: Pergamon Press, 1979), pp. 187–194.

3. Congressional Research Service, *Genetic Engineering; Human Genetics and Cell Biology, Evolution of Technological Issues; Biotechnology.* Report for the Sub-

committe on Science, Research and Technology. (Washington, DC: US Government Printing Office, 1980), p. 48.

4. Salvador Luria, "Modern biology: A terrifying power," *The Nation* (20 October 1969).

5. Rae Goodell, "The gene craze," *Columbia Journalism Rev.* 19(4):41–45 (November/December 1980).

6. Nicholas Wade, *The Ultimate Experiment* (New York: Walker, 1977).

7.John Lear, *Recombinant DNA: The Untold Story* (New York: Crown, 1978).

8. Michael Rogers, *Biohazard* (New York: Alfred A. Knopf, 1977).

9. Richard Hutton, *Bio-Revolution: DNA & The Ethics of Man-Made Life*, (New York: Mentor, 1978).

10. Ted Howard and Jeremy Rifkin, *Who Should Play God?* (New York: Dell, 1977).

11. Roger L. Shinn [ed.], *Faith and Science in an Unjust World*, World Council of Churches (WCC) Conference, vol. 1, Plenary Presentations (Geneva: WCC, 1980), p. 282.

12. *Ibid.*, vol. 2, reports and recommendations, p. 53.

13. In its proposed first annual update of a risk-assessment program for rDNA research, NIH published the following comment in the *Federal Register* 45:61874 (17 September 1980): "Despite intensive study by the RAC Subcommittee on Risk Assessment and NIH staff, several conferences and workshops to consider specific issues and several experiments, no risks of recombinant DNA research have been identified that are not inherent in the microbiological and biochemical methodology used in such research."

14. David Baltimore, "The Berg letter: Certainly necessary, possibly good," *The Hastings Center Report* 10(5):15 (October 1980).

15. *New York Times* Editorial, 13 November 1980. See also Nicholas Wade, "Harvard marches up hill and down again," *Science* 210:1104 (December 1980).

16. *The Chronicle of Higher Education* 21(20): (26 January 1980).

Appendix B

1. Provisional statement of the conference proceedings, section II (PS.II); summary statement of the Asilomar Conference on Recombinant DNA Molecules, section II (SS.II); Recombinant DNA Controversy Collection, IASC, MIT.

2. SS.III.B.3.

3. SS.III.B.4.

4. PS.V.

5. SS.IV.B.

Bibliography

Part I

Ausubel, Frederick, et al. "The politics of genetic engineering: Who decides who's defective?" *Psychology Today* 8(1):30–43 (June 1974).

Baltimore, David. "Where does molecular biology become more of a hazard than a promise?" Unpublished transcript of a lecture at the Technology Studies Seminar, MIT, IASC, MIT, 6 November 1974.

Barker, G. R. "Plasmid engineering" [letter]. *Nature* 250:530 (16 August 1974).

Beckwith, J., T. Shapiro, and L. Eron. "More alarms and excursions" [letter]. *Nature* 224:1337 (27 December 1969).

Berg, Paul, et al. "Potential biohazards of recombinant DNA molecules" [letter]. *Science* 185:303 (26 July 1974); also *Proc. Nat'l. Acad. Sci.* 71:2593–2594 (July 1974).

Brownlee, G. G. "Genetic engineering with viruses." *Nature* 251:463 (11 October 1974).

Bylinsky, Gene "Industry is finding more jobs for microbes." *Fortune* 89(2):96–102 (February 1974).

Cohen, Stanley, Letter. *Chem. & Eng. News* 52:3 (7 October 1974).

Crawford, Irving. "Recombinant DNA molecules" [letter]. *Genetics* 78:573–574 (October 1974).

Davis, Bernard. "Genetic engineering: how great is the danger?" [editorial]. *Science* 186:309 (25 October 1974).

Edwards, J. H. "Bacterial engineering" [letter]. *Nature* 252:341 (29 November 1974).

Frances, Saul. "Recombinant DNA molecules" [letter]. *Science* 185:1001 (20 September 1974).

"Genetic manipulation." *Nature* 247:336–337 (8 February 1974).

"Hazards of genetic experiments" [editorial]. *Brit. Med. J.* (5929):483–484 (24 August 1974).

Lewin, Roger. "Ethics and genetic engineering." *New Scientist* 64:163 (17 October 1974).

Lewin, Roger. "The future of genetic engineering." *New Scientist* 64:166 (17 October 1974).

McDonald, H. "Implanting human values into genetic control." *Bulletin of the Atomic Scientists* 30:21–22 (February 1974).

Moment, Gairdner B. "Andromeda strain?" [editorial]. *BioScience* 24:487 (September 1974).

Murray, Ken. "Alternative experiments?" *Nature* 250:279 (26 July 1974).

"Nobelist warns of genetic research hazard," *Chem. & Eng. News* 52(37):15 (16 September 1974).

Shinn, Roger L. "Perilous progress in genetics." *Social Research* 41(1):83–103 (spring 1974).

"Should we publicize those experiments?" [editorial]. *Nature* 251:1 (6 September 1974).

Sieghart, Paul. "Biohazards and the law" [letter]. *Nature* 251:182 (20 September 1974).

Singer, Maxine, and Dieter Söll. "Guidelines for DNA hybrid molecules" [letter]. *Science* 181:1114 (21 September 1973).

Sinsheimer, Robert L. "The prospect for designed genetic change." *American Scientist* 57:134–142 (1969).

Skalka, A., et al. "Genetic recombination: Genetic, physical and biochemical aspects." *Science* 183:1218–1219 (22 March 1974).

Smithies, Oliver. "Recombinant DNA molecules" [letter]. *Genetics* 78:575–576 (October 1974).

Szilard, Gertrud W. "Plasmid moratorium" [letter]. *Nature* 252:93 (8 November 1974).

Wade, Nicholas. "Microbiology: Hazardous profession faces new uncertainties." *Science* 182:566–567 (9 November 1973).

Wade, Nicholas. "An embargo on genetic research?" *New Scientist* 63:170–171 (25 July 1974).

Watson, James D. "Moving toward the clonal man: Is this what we want?" *The Atlantic* 50–53 (May 1971).

Weinberg, Janet. "A safer road to engineering genes?" *Science News* 106:293–294 (9 November 1974).

Ziff, Edward. "Benefits and hazards of manipulating DNA." *New Scientist* 60:274–275 (25 October 1973).

Part II

Allen, Leland C. "Biomedical research: "Ethics and rights" [letter]. *Science* 189:502 (August 1975).

"Amber light for genetic manipulation." *Nature* 253:295 (31 January 1975).

Anderson, E. S. "Viability of, and transfer of a plasmid from *E. coli* K12 in the human intestine." *Nature* 255:502–504 (5 June 1975).

Arents, J., P. Siekevitz, and T. Shannon. "Controversial areas of research" [letters]; and one reply, D. Stetten. *Science* 190:324, 326, 328, 330 (24 October 1975). See also *Science* (19 September and 28 October 1975).

"Asilomar Conference on recombinant DNA molecules." *Nature* 255:442 (5 June 1975).

"Asilomar and the Pasteur Institute." *Nature* 256:5 (3 July 1975).

Berg, Paul. "Genetic engineering: Challenge and responsibility." *Stanford M.D.* 14(2):4–8 (spring 1975).

Berg, Paul, et al. "Asilomar Conference on recombinant DNA molecules." *Science* 188:991–994 (6 June 1975).

Berg, Paul, et al. "Summary statement of the Asilomar Conference on recombinant DNA molecules." *Proc. Nat'l. Acad. Sci.* 72(6):1981–1984 (June 1975).

Berg, Paul. "Genetic engineering: Challenge and responsibility." *ASM News* 42:273–277 (1976).

Bhattacharjee, J. K. "Recombinant DNA molecules." *ASM News* 41:445 (1975).

Brown, Donald. "Quality and relevance—scientists must decide." *Hastings Center Report* 5:7–8 (June 1975).

Chakrabarty, Anando M. "Which way genetic engineering?" *Industrial Research* 18:45–50 (January 1976).

Chargaff, Erwin. "Profitable wonders—a few thoughts on nucleic acid research." *The Sciences* 15:21–26 (August–September 1975).

Chedd, Graham. "Cautious agreement at Pacific Grove." *New Scientist* 65:546 (6 March 1975).

Chedd, Graham. "Genetic engineers discuss our futures." *New Scientist* 65:547 (6 March 1975).

Cohen, Stanley, "Basic research and the public interest." *Stanford M.D.* 14(2):9–12 (spring 1975).

Cohen, Stanley. "The manipulation of genes." *Scientific American* 233:24–33 (July 1975).

Culliton, Barbara. "Kennedy: Pushing for more public input in research." *Science* 188:1187–1189 (20 June 1975).

Dixon, Bernard. "Not good enough." *New Scientist* 65:186 (23 January 1975).

Dixon, Bernard. "Genetic engineering goes commercial." *New Scientist* 66:594 (12 June 1975).

Dixon, Bernard. "Safeguards for microbial manipulation." *New Scientist* 68:618 (11 December 1975).

Dixon, Bernard. "R-DNA rules without enforcement?" *New Scientist* 69:218 (29 January 1976).

Dyson, Freeman J. "The hidden cost of saying no." *Bulletin of the Atomic Scientists* 31(6):23–27 (June 1975).

Edwards, J. H. "Bacterial engineering" [letter]. *Nature* 252:341 (29 November 1974).

Ehrlich, Paul R. "The benefit of saying yes." *Bulletin of the Atomic Scientists* 31(7):49–51 (September 1975).

Eisinger, J. "The ethics of human gene manipulation." *Fed. Proc.* 34(6):1418–1420 (May 1975).

Falkow, Stanley. "The ecology of *Escherichia coli*" [unpublished paper]. Submitted to the RAC 12–13 May 1975.

Fraenkel-Conrat, H. "Taboo research?" [letter]. *Nature* 254:12 (6 March 1975).

Frazier, Kendrick. "Rise to responsibility at Asilomar." *Science News* 107:187 (22 March 1975).

Fried, Charles. "Public input in research" [letter]. *Science* 189:248 (25 July 1975).

Galston, A. W. "Here come the clones." *Natural History* 84:72–75 (February 1975).

Garb, Solomen. "Research and public funds" [letter]. *Science* 190:834 (28 November 1975).

"Genetic engineers look for safety." *New Scientist* 64(920):240 (24 October 1974).

Hecht, Frederick. "Biomedical research: ethics and right" [letter]. *Science* 189:502 (15 August 1975).

Holman, Halsted. "Experimental safety: Should the public decide?" *Stanford M.D.* 14(2):12–14 (spring 1975).

Holman, Halsted. "Scientists and citizens–the inexpertness of experts." *Hastings Center Report* 5:8 (June 1975).

Judson, Horace Freeland. "Fearful of science." *Harpers* 250:32, 36–41 (March 1975), 70–76 (June 1975).

Lederberg, Joshua. "DNA splicing—will fear rob us of its benefit?" *Prism* 3(10):33–37 (November 1975).

McBride, Gail. "Gene grafting experiments produce both high hopes and grave worries." *J. Amer. Med. Assoc.* 232(4):337–342 (28 April 1975).

McBride, Gail. "Scientists deal with risks posed by gene grafting research." *J. Amer. Med. Assoc.* 232(5):473–476, 478–479, 484, 563 (5 May 1975).

Norman, Colin. "Berg Conference favors use of weak strain." *Nature* 254:6–7 (6 March 1975).

Norman, Colin. "Genetic politics: The treaty of Pacific Grove." *Science and Government Report* 5:1–2 (15 March 1975).

Norman, Colin. "Genetic manipulation: Recommendations drafted." *Nature* 258:561–564 (18 December 1975).

Norman, Colin. "The public case is put." *Nature* 259:521 (19 February 1976).

O'Sullivan, Dermot, "Group gives go-ahead to genetic engineering." *Chemical and Engineering News* 53(5):17–18 (3 February 1975).

O'Sullivan, Dermot. "Genetic engineering to resume—cautiously." *Chemical and Engineering News* 53(10):19 (10 March 1975).

Parry, R. M. C. "Promethean situation: The applications and limitations of genetic engineering—the ethical implications, Davos." *Futures* 7:169–173 (April 1975).

Roblin, R. "Ethical and social aspects of experimental gene manipulation." *Fed. Proc.* 34(6):1421–1424 (May 1975).

Rogers, Michael. "The Pandora's box congress." *Rolling Stone* (19 June 1975).

Rogers, Michael. "And G. E. created life and it was hungry." *Rolling Stone* (1 January 1976).

Russell, Cristine. "Weighing the hazards of genetic research: A pioneering case study." *BioScience* 24:691 (December 1974).

Russell, Cristine. "Recombinant DNA molecules, biologists draft genetic research guidelines." *BioScience* 25:237–240, 277–278 (April 1975).

Russell, Cristine. "Disarming the doomsday bug: Scientists rally to right the real dangers of man-made germs." *Science Digest* 78(1):70–77 (July 1975).

"Safer road to engineering genes?" *Science News* 196:293–294 (9 November 1974).

Sherratt, David. "Biological safeguards in genetic engineering." *Nature* 259:526 (19 February 1976).

Singer, Daniel, and Maxine Singer. "Ethical problems at the frontiers of biological research." *Swarthmore College Bulletin* 11–13 (April 1975).

Singer, M. F. "Summary of the proposed guidelines." *ASM News* 42:277–287 (1976).

Sinsheimer, Robert. "Troubled dawn for genetic engineering." *New Scientist* 68:148–151 (16 October 1975).

Steinfels, Peter. "A note on Asilomar." *Hastings Center Report* 5(2):17 (April 1975).

Stetten, DeWitt. "Freedom of inquiry" [editorial]. *Science* 189:953 (14 September 1975).

Wade, Nicholas. "Genetic manipulation: Temporary embargo proposed on research." *Science* 185:332–334 (26 July 1974).

Wade, Nicholas. "Genetics: Conference sets strict controls to replace moratorium." *Science* 187:931–935 (14 March 1975).

Wade, Nicholas. "Biological warfare: Unexpectedly good." *Science* 189:772–773 (5 September 1975).

Wade, Nicholas. "Recombinant DNA: NIH group stirs storm by drafting laxer rules." *Science* 190:767–769 (21 November 1975).

Wade, Nicholas. "Go-ahead for r-DNA." *New Scientist* 68:682–684 (18 December 1975).

Wade, Nicholas. "R-DNA: NIH sets strict rules." *Science* 190:1175–1179 (19 December 1975).

Wade, Nicholas. "R-DNA: Guidelines debated at public hearing." *Science* 191:834–836 (27 February 1976).

Weinberg, Janet. "Asilomar decision: Unprecedented guidelines for gene-transplant research." *Science News* 107:148–149, 156 (8 March 1975).

Weinberg, Janet. "Decision at Asilomar." *Science News* 107:194–196 (22 March 1975).

Wilson, Richard. "The cost of safety." *New Scientist* 68:274–275 (30 October 1975).

Young, F. R. "Report from Asilomar." *ASM News* 41:260–261 (1975).

Ziff, Edward B. "Genetic engineering." *McGraw-Hill Year Book of Science and Technology, 1975 Review/1976 Preview.* New York: McGraw-Hill, 1976, pp. 66–78.

Part III

Abelson, J. "Recombinant DNA: Examples of present day research." *Science* 196:159–160 (April 1977).

Abelson, Philip H. "Recombinant DNA" [editorial]. *Science* 197:721 (19 August 1977).

Adelberg, E. A., D. L. Longo, and W. Szybalski. "The Frankenstein monster and recombinant DNA." *Hospital Practice* 14(3):21 (March 1979) [letters].

Andreopoulos, S. "Sounding board: Gene cloning by press conference." *New Eng. J. Med.* 302(13):743–746 (27 March 1980).

Ascot Conference. US-EMBO workshop to assess risks for rDNA experiments involving the genomes of animal, plant and insect viruses, 28 January 1978, Ascot, England. Report published in *Federal Register* 43:33159–33167 (28 July 1978).

Australian Academy of Science. *Recombinant DNA: An Australian Perspective.* Canberra: Australian Academy of Science, 1980.

Baker, Robert, and Wendy Clough. "The technological uses and methodology of recombinant DNA." *S. Calif. Law Rev.* 51:1009–1016 (September 1978).

Baltimore, David. "Genetic scaremongering." *Nature* 272:767 (27 April 1978). A review of *Improving on Nature: The Brave New World of Genetic Engineering*, by Robert Cooke.

Bareikis, Robert P. (ed.). *Science and the Public Interest: Recombinant DNA Research.* Bloomington: The Poynter Center, 1978. Proceedings of a forum held 10–12 November 1977 at Indiana University.

Becker, Frank. "Law vs. science: Legal control of genetic research." *Kentucky Law J.* 65:880–894 (1977).

Beckwith, Jonathan. "Recombinant DNA: Does the fault lie within our genes?" *Science for the People* 9(3):14–17 (May–June 1977).

Beckwith, Jonathan. "Recombinant DNA—the end of the beginning of a controversy." *Trends in Biochem. Sci.* 3:94–95 (April 1978). Book review.

Beers, Roland F., and Edward G. Bassett (eds.). *Recombinant Molecules: Impact on Science and Society.* Miles International Symposium Series No. 10. New York: Raven Press, 1977.

Bennett, William, and Joel Gurin. "Science that frightens scientists—the great debate over DNA." *Atlantic Monthly* 239:43–62 (February 1977).

Bent, Stephen A. "Living matter found to be patentable: In re Chakrabarty." *Conn. Law Rev.* 11(2):311–330 (winter 1979).

Bereano, Philip L. "Recombinant DNA: Issues on the regulation of basic scientific research." *Idea* 20(4):315–334 (1979).

Berg, Paul. "Genetic engineering: Challenge and responsibility." *ASM News* 42:273–277 (1976).

Berg, Paul, and Maxine Singer. "Seeking wisdom in recombinant DNA research." *Fed. Proc.* 35(14):2542–2543 (December 1976).

Berg, Paul. "Recombinant DNA research can be safe." *Trends in Biochem. Sci.* 2:N25–N27 (February 1977).

Biotechnology and the law: Recombinant DNA and the control of scientific research, Symposium. *S. Calif. Law Rev.* 51(6) (September 1978).

Boone, C. Keith. "Recombinant DNA and Nuremberg: Toward the new application of old principles." *Perspectives in Biology and Medicine* 23(2, Part 1):240–254 (winter 1980).

Boyer, Herbert W. "The age of molecular biology." *Amer. Patent Law Assoc. Quart. J.* 7(3&4):185–189 (1979).

Boyer, Herbert W., and S. Nicosia (eds.). *Genetic Engineering.* Amsterdam: Elsevier-North Holland, 1978.

Budrys, Algis. "The politics of deoxyribonucleic acid." *New Republic* 176:18–21 (16 April 1977).

Callaham, Michael B., and Kosta M. Tsipis. "Biological warfare and recombinant DNA." *Bulletin of the Atomic Scientists* 34:11,50 (November 1978).

Callahan, Daniel. "Recombinant DNA: Science and the public." *Hastings Center Report* 7:20–23 (April 1977).

Cambridge Experimentation Review Board. "Report of the committee to Cambridge City Council." *Bulletin of the Atomic Scientists* 33(5):23–27 (May 1977).

Capron, Alexander. "Reflections on issues posed by r-DNA molecule technology, III." *Annals New York Academy of Sciences* 265:71–81 (1976).

Caskey, C. Thomas. "Recombinant DNA—an approach to gene isolation and study." *Southern Med. J.* 71(1):66–68 (January 1978).

Cavalieri, Liebe F. "New strains of life—or death." *New York Times Magazine* (22 August 1976).

Cavalieri, Liebe F. " Science as technology." *S. Calif. Law Rev.* 51(6):1153–1165 (September 1978).

Cavalieri, Liebe F. *The Double-edged Helix.* New York: Columbia University Press, 1981.

Chalfant, J. C., M. E. Hartmann, and A. Balkeboro. "Recombinant DNA: A case study in regulation of scientific research. *Ecology Law Quart.* 8(1):55–131 (1979).

Christensen, Ken. "Local control of recombinant DNA research—only for accidents?" *Medicolegal News* 6(2):4–6, 13 (summer 1978).

Cohen, C. "When may research be stopped?" *New Eng. J. Med.* 296:1203–1210 (1977).

Cohen, Stanley. "Recombinant DNA: Fact and fiction." *Science* 195:654–657 (18 February 1977).

Cohen, S., et al. "Construction of biologically functional bacterial plasmids *in vitro.*" *Proc. Nat'l. Acad. Sci.* 70:3240–3244 (1973).

Cohen, S. N., and J. S. Shapiro. "Transposable genetic elements." *Scientific American* 242(2):40–49 (1980).

Cooke, Robert. *Improving on Nature: The Brave New World of Genetic Engineering.* New York: New York Times Books, 1977.

Crossland, Janice. "Hands on the code." *Environment* 18:6–16 (September 1976).

Curtiss, Roy. "Genetic manipulations of microorganisms: Potential benefits and biohazards." *Ann. Rev. Micro.* 30:507–533 (1976).

Curtiss, Roy, et al. "Biohazard assessment of recombinant DNA molecule research." In *Plasmids: Medical and Theoretical Aspects,* S. Mitsuhashi et al. (eds.) Prague: Avicenum, Czechoslovak Medical Press, 1977, pp. 375–387.

Davis, Bernard. "Evolution, epidemiology, and recombinant DNA." *Science* 193:442 (1976).

Davis, Bernard. "The recombinant DNA scenarios: Andromeda strain, chimera, and golem." *American Scientist* 65:547–555 (September–October 1977).

Davis, Bernard, E. Chargaff, and S. Krimsky. "Recombinant DNA research: A forum on the benefits and risks." *Chem. & Eng. News* 55:26–42 (30 May 1977).

Davis, Bernard, and Robert L. Sinsheimer. "Discussion forum: the hazards of recombinant DNA." *Trends in Biochem. Sci.* 1:N178–N180 (August 1976).

Dismukes, Key. "Recombinant DNA: A proposal for regulation." *Hastings Center Report* 7(2):25–30 (April 1977).

Dorman, Janis. "History as she is made: MIT's archives unfurl the recombinant DNA banner." *New Scientist* 85:86–88 (10 January 1980).

Dunner, Donald R., and C. E. Lipsey. "The patentability of life forms, new technologies and other flooks of nature." *Amer. Patent Law Assoc. Quart. J.* 7(3&4):190–219 (1979).

Dworkin, Roger B. "Science, society, and the expert town meeting: Some comments on Asilomar." *S. Calif. Law Rev.* 51(6):1471–1482 (September 1978).

Dyson, Freeman. "Costs and benefits of recombinant DNA research" [letter]. *Science* 193:6 (2 July 1976).

Edlin, Gordon. "The hazards of recombinant DNA: Neglected environmental effects" [letter]. *Trends in Biochem. Sci.* 1(12):N272 (December 1976).

Engelhardt, H. Tristram. "Taking risks: Some background issues in the debate concerning recombinant DNA research." *S. Calif. Law Rev.* 51(6):1141–1151 (September 1978).

Federow, Harold. "Recombinant DNA and nuclear energy." *Bulletin of the Atomic Scientists* 34(2):6–7 (February 1978).

Fletcher, Joseph. "Ethics and recombinant DNA research." *S. Calif. Law Rev.* 51(6):1131–1140 (September 1978).

Fredrickson, Donald S. "Recombinant DNA guidelines: Environmental impact statement" [letter]. *Science* 193:1192 (24 September 1976).

Fredrickson, Donald S. "The government role in biomedical research." *Fed. Proc.* 35(14):2538–2540 (December 1976).

Freifelder, D. (ed.). *Recombinant DNA: Readings from Scientific American.* San Francisco: W. H. Freeman, 1978.

Friedman, Jane M. "Health hazards associated with recombinant DNA technology: Should Congress impose liability without fault?" *S. Calif. Law Rev.* 51(6):1355–1379 (September 1978).

Gardner, Barbara J. "The potential for genetic engineering: A proposal for international legal control." *Vir. J. Int. Law* 16:403–429 (1976).

Gaylin, Willard. "Sounding board: The Frankenstein factor." *New Eng. J. Med.* 297:665–667 (22 September 1977).

Gies, Joseph C. "Regents make history: The DNA decision." *Assoc. of Governing Boards of Univ. & Colleges, Reports* 19:3–11 (July–August 1977).

Goldstein, Richard. "Public health policy and recombinant DNA." *New Eng. J. Med.* 296:1226–1228 (26 May 1977).

Goodell, Rae. "Literature guide: Review of recent books on the rDNA controversy." *News. on Sci., Tech. & Hum. Val.* 22:25–29 (January 1978).

Goodell, Rae. "Public involvement in the DNA controversy: The case of Cambridge, Massachusetts." *Sci., Tech. & Hum. Val.* 4(27) (spring 1979).

Goodfield, June. *Playing God: Genetic Engineering and the Manipulation of Life.* New York: Random House, 1977.

Green, Harold P. "The recombinant DNA controversy: A model of public influence." *Bulletin of the Atomic Scientists* 34(9):12–16 (November 1978).

Green, Harold P. "Law and genetic control: Public policy questions." *Annals New York Academy of Sciences* 265:170–175 (1976).

Greenberg, Daniel S. "Lessons of the DNA controversy." *New Eng. J. Med.* 297:1187–1188 (24 November 1977).

Grobstein, Clifford. "The recombinant DNA debate." *Scientific American* 237:22–33 (July 1977).

Grobstein, Clifford. "Recombinant DNA research: Beyond the NIH guidelines." *Science* 194:1133–1135 (1976).

Grobstein, Clifford. "Regulation and basic research: Implications of recombinant DNA." *S. Calif. Law Rev.* 51:1181–1200 (September 1978).

Grobstein, Clifford. *A Double Image of the Double Helix.* San Francisco: W. H. Freeman, 1979.

Gross, Paul R. "Xeroxing life" [letter]. *Science* 200:126 (14 April 1978).

Halvorson, Harlyn O. "Recombinant DNA legislation—what next?" [editorial]. *Science* 198:357 (28 October 1977).

Halvorson, Harlyn O. "ASM nine principles on legislation" [letter]. *Science* 196:1154 (10 June 1977).

Halvorson, Harlyn O. "DNA and the law." *S. Calif. Law Rev.* 51(6):1167–1180 (September 1978).

Halvorson, Harlyn O. Letter. *Chem. & Eng. News* 55(24):4 (13 June 1977).

Handler, Philip. "Recombinant DNA research" [editorial]. *Chem. & Eng. News* 55(19):3 (9 May, 1977).

Hardin, Garrett J. "The fateful quandary of genetic research." *Prism* 3:20–23 (March 1975).

Harris, R. J. "Common sense in genetic engineering." *Ethics in Science & Medicine* 5(2–4):57–64 (1978).

Helling, R. B., and S. L. Allen. "Freedom of inquiry and scientific responsibility: Recombinant DNA research." *BioScience* 26:609–610 (October 1976).

Holliday, R. "Should genetic engineers be contained?" *New Scientist* 73:399–401 (17 February 1977).

Holman, Halsted, and Diana Dutton. "A case for public participation in science policy formation and practice." *S. Calif. Law Rev.* 51(6):1505–1534 (September 1978).

Hopson, J. L. "Recombinant lab for DNA and my 95 days in it." *Smithsonian* 8:54–63 (June 1977).

Hoskins, B. B., et al. "Applications of genetic and cellular manipulations to agricultural and industrial problems." *BioScience* 27:188–191 (March 1977).

Howard, Ted, and Jeremy Rifkin. *Who Should Play God?* New York: Dell, 1977.

Hubbard, Ruth. "Gazing into the crystal ball: Recombinant DNA research." *BioScience* 26:608,611 (October 1976).

Hubbard, Ruth. "The hazards of recombinant DNA: Unfettered intuition" [letter]. *Trends in Biochem. Sci.* 1(12):N272–N273 (December 1976).

Hubbard, Ruth. "Unnatural selection" [letter]. *The Sciences* 16(6) (November-December 1976).

Hubbard, Ruth. "Recombinant DNA: Unknown risks" [letter]. *Science* 193:834, 836 (3 September 1976).

Hutt, Peter B. "Research on recombinant DNA molecules: The regulatory issues." *S. Calif. Law Rev.* 51:1435–1450 (September 1978).

Hutton, Richard. *Bio-Revolution: DNA and the Ethics of Man-Made Life.* New York: Mentor, 1978.

Jackson, David, and Stephen P. Stich (eds.). *The Recombinant DNA Debate.* Englewood Cliffs, NJ: Prentice Hall, 1979.

Jonas, Hans. "Freedom of scientific enquiry and the public interest." *Hastings Center Report* 6:15–17 (August 1976).

Kassinger, Ted, and Benna Solomon. "Recombinant DNA and technology assessment." *Georgia Law Rev.* 11:785–878 (summer 1977).

King, Jonathan. "New diseases and new niches." *Nature* 276:4–7 (1978).

King, Jonathan. "A science for the people." *New Scientist* 74:634–636 (16 June 1977).

King, Jonathan. "Recombinant DNA and autoimmune disease." *J. Infectious Diseases* 137(5):663–666 (May 1978).

King, Jonathan. "Arguments against patenting modified life forms." R. F.

Acker and M. Schaechter, eds. *Patentability of Microorganisms.* Washington, DC: ASM, 1981.

Kourilsky, Philippe. "The hazards of recombinant DNA" [letter]. *Trends in Biochem. Sci.* 2:N12 (January 1977).

Krimsky, Sheldon. "Paradigms and politics: the roots of conflict over recombinant DNA research." In *Science and the Public Interest: Recombinant DNA Research.* Robert P. Bareikis (ed.). Bloomington: The Poynter Center, 1978. Proceedings of a forum held 10–12 November 1977 at Indiana University.

Krimsky, Sheldon. "Regulating recombinant DNA research." In *Controversy: Politics of Technical Decisions.* Dorothy Nelkin (ed.). Beverly Hills, CA: Sage, 1978.

Krimsky, Sheldon. "A citizen court in the recombinant DNA debate." *Bulletin of the Atomic Scientists* 34:37–43 (October 1978).

Krimsky, Sheldon. "A comparative view of state and municipal laws regulating the use of recombinant DNA molecule technology." *Recombinant DNA Technical Bulletin* 2:121–125 (November 1979).

Krimsky, Sheldon. "Patenting life: Social and ethical issues." *Science for the People* 12:14–18 (September–October 1980).

Krimsky, Sheldon and David Ozonoff. "Recombinant DNA research: The scope and limits of regulation." *Amer. J. Pub. Health* 69:1252–1259 (December 1979).

Krimsky, Sheldon. "Patents for life forms *sui generis*: Some new questions for science, law, and society." *Recombinant DNA Technical Bulletin* 4(1):11–15 (April 1981).

Kukin, Marrick. "Research with recombinant DNA—potential risks, potential benefits, and ethical considerations. *New York State Jour. Med.* 78(2):226–235 (February 1978).

Lappe, Marc. *Genetic Politics: The Limits of Biological Control.* New York: Simon & Schuster, 1979.

Lappe, Marc, and Patricia A. Martin. "The place of the public in the conduct of science." *S. Calif. Law Rev.* 51(6):1535–1554 (September 1978).

Lappe, Marc, and Robert S. Morrison (eds.). *Ethical and Scientific Issues Posed by Human Uses of Molecular Genetics.* Annals New York Academy of Sciences, vol. 265. New York: New York Academy of Sciences, 1976.

Lear, John. *Recombinant DNA: The Untold Story.* New York: Crown Pub. Co., 1978.

Leeper, E. "Nepa and basic research: DNA debate prompts review of environmental impacts." *BioScience* 27(8):515–517 (1977).

Leeper, E. "Recombinant DNA forum—stellar cast; gripping plot; but no new message." *BioScience* 27:317–319 (May 1977).

Lefkowitz, Louis J. "A legal officer's dilemma." *Bulletin of the Atomic Scientists* 33(5):11 (May 1977).

Lewin, Roger. "Science and politics in genetic engineering." *New Scientist* 82:114–115 (April 1979).

Lewis, R. V. "Recombinant DNA and the future of medical research" [editorial]. *R. I. Med. J.* 62(7):284 (July 1979).

Lincoln, D. R., L. R. Landis, and H. A. Gray. "Containing recombinant DNA: How to reduce the risk of escape." *Nature* 281:421–423 (11 October 1979).

Luria, Salvador E. "The goals of science." *Bulletin of the Atomic Scientists* 33:28–33 (May 1977).

Macklin, Ruth. "On the ethics of not doing scientific research." *Hastings Center Report* 7:11–13 (December 1977).

Marx, J. L. "Molecular cloning: Powerful tool for studying genes." *Science* 191:1160–1162 (19 March 1976).

Marx, J. L. "Nitrogen fixation: Prospects for genetic manipulation." *Science* 196:638–641 (6 May 1977).

May, William F. "The right to know and the right to create." *News. on Sci., Tech. & Hum. Val.* 23:34–41 (April 1978).

McCaull, Julian. "Research in a box: Escape of new Life forms." *Environment* 19:31–37 (April 1977).

Medawar, P. B. "Fear and DNA." *New York Review of Books* 24:15–20 (27 October 1977).

Mendel, Agata. *Les manipulations genetiques.* Paris: Seuil, 1980.

Miller, J. A. "Gene-splicing research: Some safety advice from virus scientists." *Science News* 111:141 (26 February 1977).

Milton, Joyce. "The hazards of altering nature." *Nation* 225:361–365 (15 October 1977).

Monaghan, W. P., et al. "DNA recombinant research and you." *Amer. J. Med. Tech.* 44(1):62–65 (January 1978).

Morgan, J., and W. J. Whelan (eds.). *Recombinant DNA and Genetic Experimentation.* Oxford: Pergamon, 1979.

Nelkin, Dorothy. "Threats and promises: Negotiating the control of research." *Daedalus* 107:191–209 (spring 1978).

Newburger, David J. "Effects of legal regulation on scientific research using recombinant DNA technology." In *Regulation of Scientific Inquiry: Societal Concerns with Research.* Keith M. Wulff (ed.). Boulder: Westview Press, 1979, pp. 145–157.

Newman, S. A. "Tumour virus DNA: hazards no longer speculative" [letter]. *Nature* 281:176 (20 September 1979).

Novick, Richard P. "Present controls are just a start." *Bulletin of the Atomic Scientists* 33(5):16–22 (May 1977).

Philipson, Lennart, and Pierre Tiollais. "Rational containment on recombinant DNA." *Nature* 268:90, 91 (14 July 1977).

Pollack, Allan. "Engineering within the guidelines." *Trends in Biochem. Sci.* 2:N137–N138 (June 1977).

Pollack, Richard. "Recombinant DNA: The promise of genetic engineering." *Fusion* 1(2):12–19 (October–November 1977).

Powledge, Tabitha. "Recombinant DNA: Backing off on legislation." *Hastings Center Report* 7:8–10 (December 1977).

Powledge, Tabitha. "Recombinant DNA: The argument shifts." *Hastings Center Report* 7:18–19 (April 1977).

Powledge, Tabitha. "Recent social and ethical developments in genetic engineering." In *Genetics Now: Ethical Issues in Genetic Research.* John J. Buckley (ed.). Washington: Univ. Press Amer., 1978, pp. 7–23.

Ptashne, Mark. "The defense doesn't rest." *The Sciences* 16(5):11–12 (September–October 1976).

Randall, Judith. "Life from the labs: Who will control the new technology?" *Progressive* 41:16–20 (March 1977).

Randall, Judith. "DNA debate." *Progressive* 41:11–12 (May 1977).

Randall, Judith. "Who will oversee the gene jugglers?" *Change* 9:54–55 (May 1977).

Ravetz, J. R. "DNA research as 'high-intensity' science." *Trends in Biochem. Sci.* 4:N97–N98 (May 1979).

"Recombinant DNA hearing at Harvard." *Boston Univ. Jour.* 24:5–23 (fall 1976).

"Recombinant DNA." *Scientific American.* San Francisco: W. H. Freeman, 1978.

Rifkin, Jeremy. "One small step beyond mankind." *Progressive* 41:21 (March 1977).

Rifkin, Jeremy, and Ted Howard. "Patenting life." *Progressive* 34–37 (September 1979).

Robertson, John A. "The scientist's right to research: A constitutional analysis." *S. Calif. Law Rev.* 51:1203–1279 (September 1978).

Roblin, Richard. "Reflections on issues posed by recombinant DNA molecule technology." *Annals New York Academy of Sciences* 265:59–65 (1976).

Rogers, Michael. *Biohazard.* New York: Alfred A. Knopf, 1977.

Rosenberg, Barbara, and Lee Simon. "Recombinant DNA: Have the experiments assessed all the risks?" *Nature* 282:773–774 (December 1979).

Rosner, F. "Recombinant DNA, cloning, genetic engineering and Judaism." *New York State J. Med.* 79(9):1439–1444 (August 1979).

Rowe, Wallace. "Recombinant DNA: What happened?" [editorial]. *New Eng. J. Med.* 297:1176–1177 (24 November 1977).

Rowe, Wallace. "Guidelines that do the job." *Bulletin of the Atomic Scientists* 33(5):14–15 (May 1977).

Rowe, Wallace. "Recombinant DNA guidelines: Scientific and political questions." *Science* 198:563 (11 November 1977).

Seidman, Aaron. "The U.S. Senate and recombinant DNA research." *News. on Sci., Tech. & Hum. Val.* 22:30–32 (January 1978).

Seidman, Aaron. "Legislature report: The U.S. House of Representatives and DNA." *News. on Sci., Tech. & Hum. Val.* 23:23–24 (1978).

Siekevitz, P. "Recombinant DNA research: A Faustian bargain?" [letter]. *Science* 194:256–257 (15 October 1976).

Simring, Francine. "The double helix of self-interest." *The Sciences* 17:10–13, 27 (May-June 1977).

Simring, Francine. "Folio for folly: NIH guidelines for rDNA research." *Man & Medicine* 2(2) (1977).

Simring, Francine. "On the dangers of genetic meddling" [letter]. *Science* 192:940 (June 1976).

Simring, Francine. "Monkeying with genes." *Humanist* 38(5):52–53 (September 1978).

Singer, Maxine. "Scientists and the control of science." *New Scientist* 74:631–634 (June 1977).

Singer, Maxine. "The recombinant DNA debate" [editorial]. *Science* 196:127 (8 April 1977).

Singer, Maxine. "Spectacular science and ponderous process" [editorial]. *Science* 203:9 (5 January 1979).

Singer, Maxine, and Paul Berg. "Recombinant DNA: NIH guidelines" [letter]. *Science* 193:186–188 (16 July 1976).

Sinsheimer, Robert L. "On our own" [editorial]. *BioScience* 26:599 (October 1976).

Sinsheimer, Robert L. "Recombinant DNA." *Ann. Rev. Biochem.* 46:415–438 (1977).

Sinsheimer, Robert L. "An evolutionary perspective for genetic engineering." *New Scientist* 73:150–152 (20 January 1977).

Sinsheimer, Robert L. "On coupling inquiry and wisdom." *Fed. Proc.* 35:2540 (1976).

Sinsheimer, Robert L. "Genetic engineering and gene therapy: Some implications." In *Genetic Issues in Public Health & Medicine.* B. H. Cohen et al. (eds.). Springfield, IL: Charles C. Thomas, 1978, pp. 439–461.

Sinsheimer, Robert L., and Gerard Piel. "Inquiry into inquiry: two opposing views." *Hastings Center Report* 6:18–19 (August 1976).

Skalka, A. M. "Recombinant DNA research." *Research Management* 21:9–13 (January 1978).

Spece, Roy G. "A purposive analysis of constitutional standards of judicial review and a practical assessment of the constitutionality of regulating recombinant DNA research." *S. Calif. Law Rev.* 51(6):1281–1351 (September 1978).

Steinfels, Peter. "Biomedical research and the public: a report from the Airlie House conference." *Hastings Center Report* 6:21–25 (June 1976).

Stetten, DeWitt, Jr. "A parable on recombination." *Nature* 266:488 (7 April 1979).

Stetten, DeWitt, Jr. "Freedom on enquiry." *Genetics* 81:415–425 (1975).

Stich, Stephen P. "The recombinant DNA debate." *Phil. & Pub. Affairs* 7(3):187–205 (spring 1978).

Straton, David. "The genetic engineering debate." *Ecologist* 7:381–388 (December 1977).

Swazey, J. P., J. Sorenson, and C. Wong. "Risks and benefits, rights and responsibilities: A history of the recombinant DNA research controversy." *S. Calif. Law Rev.* 51(6):1019–1078 (September 1978).

Szabo, G. S. A. "Patents and recombinant DNA." *Trends in Biochem. Sci.* 2:N246–N249 (November 1977).

Szybalski, Waclaw. "Dangers of regulating the recombinant DNA technique." *Trends in Biochem. Sci.* 3:N243–N247 (November 1978).

Szybalski, Waclaw. "Risks of recombinant DNA regulations." *Nature* 278:10 (1 March 1979).

Szybalski, Waclaw. "Much ado about recombinant DNA regulations." In *Biomedical Scientists and Public Policy.* H. H. Fudenberg and V. L. Melnick (eds.). New York: Plenum Press, 1978.

Terall, Mary. "Recombinant DNA: Other cities look to Cambridge." *Harvard Magazine* (March 1977).

Thomas, Lewis. "Notes of a biology-watcher: The hazards of science." *New Eng. J. Med.* 296:324–328 (10 February 1977).

Thomas, Lewis. "Hubris in science." *Science* 200:1459–1462 (30 June 1978).

Thomasson, W. A. "Recombinant DNA and regulating uncertainty." *Bulletin of the Atomic Scientists* 35:26–32 (December 1979).

Tooze, John. "Practical guidelines." *Trends in Biochem. Sci.* 1:N246–N247 (November 1976).

Tooze, John. "Genetic engineering in Europe." *New Scientist* 73:592–594 (10 February 1977).

US Congress. House Committee on Science and Technology. Subcommittee on science, research and technology. *Science Policy Implications of DNA Recombinant Molecule Research.* Washington, DC: US Government Printing Office, 1978.

US Library of Congress. Science Policy Research Division. "Genetic Engineering, Human Genetics and Cell Biology." Washington, DC: US Government Printing Office, 1976.

Valentine, Raymond C. "Genetic blueprints for new plants." *The Sciences* 18:10–13 (February 1978).

Vickers, T. "Flexible DNA regulation: The British model." *Bulletin of the Atomic Scientists* 34:4–5 (January 1978).

Wade, Nicholas. *The Ultimate Experiment: Man-Made Evolution.* New York: Walker & Co., 1977.

Wade, Nicholas. "Recombinant DNA: Chimeras set free under guard." *Science* 193:215–217 (16 July 1976).

Wade, Nicholas. "Recombinant DNA: A critic questions the right to free inquiry." *Science* 194:303–306 (15 October 1976).

Wade, Nicholas. "Gene-splicing: Critics of research get more brickbats than bouquets." *Science* 195:466–467, 469 (4 February 1977).

Wade, Nicholas. "Gene-splicing: At grass-roots level a hundred flowers bloom." *Science* 195:558–560 (11 February 1977).

Wade, Nicholas. "Gene-splicing: Congress starts framing law for research." *Science* 196:39–40 (1 April 1977).

Wade, Nicholas. "Gene-splicing: Senate bill draws charges of Lysenkoism." *Science* 197:348–350 (22 July 1977).

Wade, Nicholas. "Recombinant DNA: NIH rules broken in insulin gene project." *Science* 197:1342–1345 (30 September 1977).

Wade, Nicholas. "The recombinant DNA debate in retrospect." *Trends in Biochem. Sci.* 3:N251–N252 (November 1978).

Wade, Nicholas. "Major relaxation in DNA rules." *Science* 205:1238 (21 September 1979).

Wade, Nicholas. "Cloning gold rush turns basic biology into big business." *Science* 208:688–692 (16 May 1980).

Wald, George. "The case against genetic engineering." *The Sciences* 16:6–11 (September–October 1976).

Watson, James D. "An imaginary monster." *Bulletin of the Atomic Scientists* 33(5):12–13 (May 1977).

Watson, James D. "Recombinant DNA" [letter]. *New Times* 8(2):4 (21 January 1977).

Watson, James D. "In defense of DNA." *New Republic* 176:11–14 (25 June 1977).

Watson, James D. "Remarks on recombinant DNA." *The CoEvolution Quarterly* 14:40 (summer 1977).

Watson, James D. "DNA folly continues." *New Republic* 12, 14–15 (13 January 1979).

Watson, James D. "Why the Berg letter was written." In *Recombinant DNA and Genetic Experimentation*. J. Morgan and W. J. Whelan (eds.), Oxford: Pergamon Press, 1979, p. 187–194.

Watson, James D., and John Tooze. *The DNA Story*, San Francisco: W. H. Freeman, 1981.

Weissmann, Gerald. "Is there an alternative to recombinant DNA?" *The Sciences* 17(4):6–9, 30, 31 (1977).

Weizenbaum, Joseph. "Costs & benefits of recombinant DNA research" [letter]. *Science* 193:6, 8 (2 July 1976).

Wolff, A. "Danger: Biologists at work: Recombinant DNA rules." *Saturday Review* 3:55 (18 September 1976).

Wright, Susan. "Doubts over genetic engineering controls." *New Scientist* 72:520–521 (2 December 1976).

Wright, Susan. "Recombinant DNA technology: Who should regulate?" *Bulletin of the Atomic Scientists* 33(8):4–5 (October 1977).

Wright, Susan. "Recombinant DNA research." *Science* 195:131–132 (14 January 1977).

Wright, Susan. "Molecular politics in Great Britain and the United States: The development of policy for recombinant DNA technology." *S. Calif. Law Rev.* 51(6):1383–1434 (September 1978).

Wright, Susan. "Quis custodiet custodes?" *Times Higher Educational Supplement* 281 (11 March 1977).

Wright, Susan. "Setting science policy: The case of recombinant DNA." *Environment* 20(4):7–15, 39–41 (May 1978).

Wright, Susan. "DNA: Let the public choose." *Nature* 275:468 (12 October 1978).

Wright, Susan. "Recombinant DNA policy: From prevention to crisis intervention." *Environment* 21:34–37, 42 (November 1979).

Wright, Susan. "Recombinant DNA policy: Controlling large-scale processing." *Environment* 22(7):29–33 (September 1980).

Wright, Susan. "Recombinant DNA: Dismantling controls." *Environment* 22(10):4–5 (December 1980).

Yoxen, Richard. "Regulating the exploitation of recombinant genetics." In *Directing Technology: Policies for Promotion and Control.* Ron Johnston and Phillip Gummett (eds.). London: St. Martin's Press, 1979.

Zimmerman, Burke K. "Self-discipline or self-deception." *Man & Medicine* 2:120–132 (1977).

Zimmerman, Burke K. "The right of free inquiry: Should the government impose limits?" In *Science and the Public Interest: Recombinant DNA Research.* Robert P. Bareikis (ed.). Bloomington: The Poynter Center, 1978. Proceedings of a forum held 10–12 November 1977 at Indiana University.

Zimmerman, Burke K. "Beyond recombinant DNA—two views of the future." In *Recombinant DNA: Science, Ethics, and Politics.* John Richards (ed.). New York: Academic Press, 1978.

Index